U0197839

计算机技术开发与应用丛书

深度探索Flutter
企业应用开发实战

赵　龙◎编著

清華大学出版社

北京

内 容 简 介

本书采用由浅入深的方式讲解 Flutter 技术原理，内容翔实、面面俱到。通过阅读本书，读者能够掌握如何使用 Flutter 编写跨平台移动客户端应用，并且对应用到的组件有透彻的理解。

本书分为三篇，共 12 章。基础知识篇（第 1～5 章）概述 Dart 核心语言基础，通过视频讲解来概述 Flutter 开发环境搭建，纵向概述 Flutter 开发中使用的基础组件（如 Text、Image）、UI 布局组件（如 Column）、滑动组件（如 NestScrollView）、功能性组件（如手势识别）等；核心功能篇（第 6～10 章）涵盖动画、弹框、绘图、文件操作、Flutter 与 Android 和 iOS 原生的消息通信、数据持久化、网络请求等；实战应用篇（第 11、12 章），将前两篇讲解内容加以应用，并补充开发细节，如应用图标配置、打包发布、权限请求、各种工具类封装等，此外还提供一个 App 的基础架构。

本书面向 Flutter 初学者，可供 Web 前端、iOS 开发、Android 的开发人员，以及想更进一步了解 Flutter 并进阶实战的技术人员阅读。

图书在版编目（CIP）数据

深度探索 Flutter：企业应用开发实战/赵龙编著. —北京：清华大学出版社，2022.10
（计算机技术开发与应用丛书）
ISBN 978-7-302-61217-9

Ⅰ. ①深…　Ⅱ. ①赵…　Ⅲ. ①移动终端－应用程序－程序设计　Ⅳ. ①TN929.53

中国版本图书馆 CIP 数据核字（2022）第 111289 号

责任编辑：赵佳霓
封面设计：吴　刚
责任校对：焦丽丽
责任印制：杨　艳

出版发行：清华大学出版社
　网　　址：http://www.tup.com.cn，http://www.wqbook.com
　地　　址：北京清华大学学研大厦 A 座　　**邮　　编**：100084
　社 总 机：010-83470000　　**邮　　购**：010-62786544
　投稿与读者服务：010-62776969，c-service@tup.tsinghua.edu.cn
　质量反馈：010-62772015，zhiliang@tup.tsinghua.edu.cn
　课件下载：http://www.tup.com.cn，010-83470236
印 装 者：三河市龙大印装有限公司
经　　销：全国新华书店
开　　本：186mm×240mm　　**印　张**：24.75　　**字　　数**：558 千字
版　　次：2022 年 10 月第 1 版　　**印　　次**：2022 年 10 月第1 次印刷
印　　数：1～2000
定　　价：99.00 元

产品编号：096503-01

前言

PREFACE

Flutter 是谷歌的移动 UI 框架，可以快速在 iOS 和 Android 上构建高质量的原生用户界面，至本书发稿前，Flutter 稳定版本发布到了 2.10 版本，现在已经支持移动端（Android、iOS）、Web 端和 PC 端，本书内容是基于 2.10.2 版本构建的。本书的特色是由浅至深，从 Dart 语言基础至功能基础，再至核心架构搭建，并配有大量的视频讲解。

在本书的每章、每节落笔前的构思中，笔者都在考虑如何才能把各个知识点由简到详更有条理地论述，也在考虑如何才能使读者以简单易懂的方式快速理解每个知识点以致在实际项目中的开发使用，也在担心自己的理解有偏差而误导了读者。在本书的发行期间，Flutter SDK 也在不断地更新迭代中，本书中的内容可能与最新 Flutter 版本略有不同，请读者以最新版本为主。

本书主要内容

本书由浅入深讲解 Flutter 技术的原理，基础知识篇（第 1～5 章）主要介绍 Dart 语言基础，以及 Flutter UI 构建；核心功能篇（第 6～10 章）详细讲解动画、弹框、绘图、Flutter 与 Android 和 iOS 原生的消息通信等，其中第 7 章详细讲述状态管理 GetX 的使用，并搭配了 GetX 与 Provider 状态管理系列视频教程；实战应用篇（第 11、12 章）是实战演示，第 11 章主要讲解如何搭建一个基础 App 架构，包括启动页面、倒计时页面、首页面的应用逻辑搭建，第 12 章通过 GetX 搭建 App 架构，在此基础上实现了列表视频播放案例。

本书开发依赖包括以下几点。

（1）开发工具：MacBook Pro（Retina，15-inch，Mid 2015）版本 10.15.6（19G73）。

（2）开发软件工具：Android Studio 2021.1.1 Patch 2、Xcode Version 13.2.1。

（3）测试手机：

- Android 小米 Max MIUI 11.0.3 稳定版本，尺寸 6.9 英寸、分辨率为 2160×1080 像素，Android 版本 9。
- iPhone 11，系统版本 13.5.1，6.1 英寸，分辨率为 1792×828 像素。

本书开发的语言环境如下：

```
Flutter 2.10.2 · channel stable · https://github.com/flutter/flutter.git
Framework · revision 097d3313d8
Engine · revision a83ed0e5e3
Tools · Dart 2.16.1 · DevTools 2.9.2
```

本书源代码

扫描下方二维码,即可下载本书源代码。

致谢

感谢广大读者以往对作者的支持,感谢给笔者前期的书籍提交过问题反馈与建议的读者;感谢 Flutter 开发社区诸多开发者的提议及实践经验分享;感谢清华大学出版社的赵佳霓编辑,本书能够顺利出版离不开她细心、负责任的工作态度。

由于写作水平与时间所限,书中难免存在不妥之处,请读者见谅,并提宝贵意见。

赵　龙

2022 年 4 月

目录
CONTENTS

基础知识篇

核心功能篇

实战应用篇

基础知识篇

第1章 Flutter 开发起步

CHAPTER 1

Flutter 是谷歌的移动 UI 框架，可以快速在 iOS 和 Android 上构建高质量的原生用户界面。现在，主流的移动开发平台是 Android 和 iOS，每个平台上的开发技术不一样，如在 Android 中支持 Java 与 Kotlin，在 iOS 中支持 Object-C 与 Swift，针对每个开发平台都需要特定的人员来开发。

Flutter 是最新的跨平台开发技术，可以一套代码同时适配 Android、iOS、macOS、Windows、Linux 等多个系统，本书是零基础开发人员的学习路线指引书，也是有基础开发者完善知识体系的说明书，读者可观看视频讲解【1.1 Flutter 学习路线分享】。

4min

1.1 Flutter 开发入门基础

本节讲述 Flutter 环境配置相关内容，如果读者的计算机已安装了 Flutter 环境，则可以通过本节了解如何升级 Flutter，如果读者是新入门开发者，则可以通过本节来了解配置系统开发环境。

1.1.1 Flutter 环境搭建概述

Flutter 支持 3 种环境：Windows、macOS 和 Linux。这里我们主要讲解 Windows 及 macOS 的环境搭建，无论在哪个平台，第一步都是下载 Flutter SDK，可到 Flutter 官网下载对应系统的 SDK，官网下载网址如下：

```
https://docs.flutter.dev/development/tools/sdk/releases
```

第一步下载好 Flutter SDK 后，第二步就是在不同计算机系统上配置环境变量，Windows 计算机系统配置步骤如下：

(1) 右击"此计算机"→"属性"→"高级系统设置"→"环境变量"，打开环境变量配置。

(2) 在用户变量中找到 path（如果没有，则新建）：C:\src\flutter\bin（计算机中安装

Flutter 的地址，注意：一直到 bin 目录）。

（3）在用户变量中新建 PUB_HOSTED_URL：https://pub.flutter-io.cn（临时镜像配置，不是必需的）。

（4）在用户变量中新建 FLUTTER_STORAGE_BASE_URL：https://storage.flutter-io.cn（临时镜像配置，不是必需的，建议配置）。

（5）配置成功之后，按快捷键 Win+R，输入 cmd 命令，输入 flutter doctor 执行命令，查看结果（第一次运行时间可能较长）。

4min

对于 macOS 系统中的 Flutter 开发环境，本章中不做详细概述，读者可观看视频讲解【1.1.1 Flutter 开发环境搭建】。

配置好 Flutter SDK 环境变量后，接下来需要配置开发 Flutter 的编辑器工具，可开发 Flutter 项目的编辑器包括 Android Studio、IntelliJ IDEA、Visual Studio Code(VS Code)，也可以选择其中的一个来使用，也可自行搜索安装软件，本书不做概述，本书使用的是 Android Studio，Flutter 项目在打包 iOS 时，需要使用 Xcode，所以对应的 macOS 系统计算机也需要安装 Xcode。安装好开发工具后，最后一步就是在开发工具中配置 Flutter 开发插件。

启动 Android Studio，在 macOS 系统中，选择 Preferences→Plugins on macOS；在 Windows 系统中，选择 File→Settings→Plugins on Windows，选择 Browse repositories→Flutter 插件并单击 Install 按钮，重启 Android Studio 后插件生效。

启动 VS Code，选择 View→Command Palette，然后输入 install，选择 Extensions：Install Extension action，在搜索框输入 flutter，在搜索结果列表中选择 Flutter，单击 Install 按钮，然后单击 OK 按钮，安装完成后，需重新启动 VS Code。

已有 Flutter 环境升级，打开计算机终端控制台或者开发工具的终端，输入命令如下：

```
Flutter --version
```

如果有新版本可用，则终端会提示有新版本，可以输入命令 flutter--version 进行升级，如图 1-1 所示。

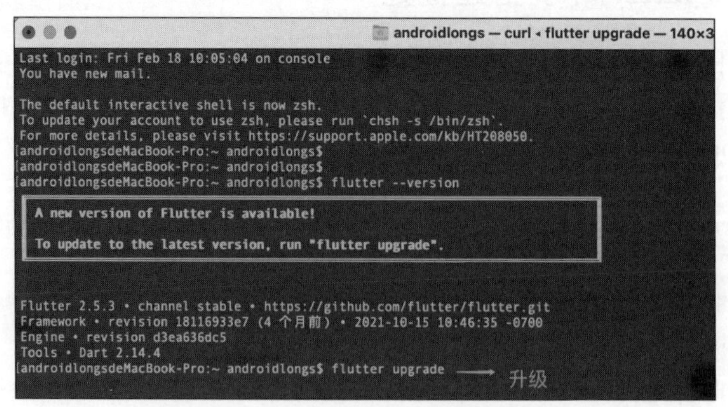

图 1-1 Flutter 当前版本查看效果图

可以运行命令 flutter channel 查看当前 Flutter 渠道，如图 1-2 所示，包括 stable（稳定版）、beta（预览版）、dev（开发版）、master（最新版，但未经充分测试），可通过执行命令 flutter channel beta 来切换到对应的分支。

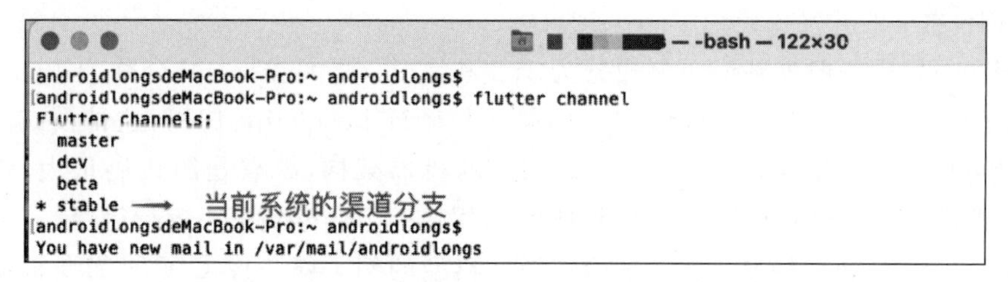

图 1-2 当前系统 Flutter 渠道说明图

运行命令 flutter upgrade 更新 Flutter SDK 版本，可以在 Flutter 项目根目录下执行 flutter packages get 命令获取 pubspec.yaml 文件中所有的依赖关系，通过 flutter packages upgrade 命令获取 pubspec.yaml 文件中所有列表的依赖项的最新版本。

如果需要桌面支持添加到现有的 Flutter 项目，则需要从根项目目录在终端中运行的命令如下：

```
flutter create --platforms=Windows,macOS,Linux .
```

1.1.2 Dart 语言与 Flutter 概述

Dart 语言的最初定位是一种运行在浏览器中的脚本语言，使用 JavaScript 开发的程序混乱不堪，没有严谨的程序范式与语言数据类型限定，所以 Dart 语言的出现最初的定位也是一种运行在浏览器中的脚本语言，为解决 JavaScript 存在的、在语言本质上无法改进的缺陷问题。

在最开始，谷歌公司在自家的 Chrome 浏览器中内置了 Dart VM，可以直接高效地运行 Dart 代码。2015 年前后，由于少有项目使用 Dart 语言，所以谷歌公司将 Dart VM 引擎从 Chrome 中移除，之后，在谷歌公司内部孵化了开发移动框架 Flutter，并且在谷歌公司的操作系统 Fuchsia 中，Dart 被指定为官方的开发语言，前端开发框架 Angular 也在持续迭代对应的 Dart 版本的 Angular Dart。

Dart 属于应用层编程语言，它有自己的 Dart VM，通常情况下运行在自己的 VM 上，但是在特定情况下它也可以被编译成 Native Code 运行在硬件上，例如在移动开发框架中，Flutter 会将代码编译成指定平台的 Native Code 以提高性能。

Flutter 是谷歌公司推出并开源的移动应用开发框架，Flutter 开发框架采用的开发语言是 Dart，开发者可以通过 Dart 语言开发 App，一套代码同时运行在 iOS 和 Android 平台，目前 Flutter 默认支持 Android、iOS、Fuchsia 操作系统三个移动平台，也支持 Web 开发（Flutter for Web）和 PC 开发，本书的示例和介绍主要基于 iOS 和 Android 平台。

Flutter 的第 1 个版本支持 Android 操作系统，开发代号称作 Sky，于 2015 年 4 月的

Flutter 开发者会议上公布，2021 年 3 月 4 日，Flutter 团队公布开放了 Flutter 2.0.0 States 版本，对桌面版添加了许多小组件，2021 年 9 月 9 日，发布了 Flutter 2.5，根据官方的介绍，这是一个大版本更新，一共解决了 4600 个问题，从 252 个贡献者和 216 个审阅者合并了 3932 个 PR，本书的开发依赖为 Flutter 2.10.2。

在程序开发中一般来讲编译模式分为 JIT 和 AOT 两大类。JIT 全称为 Just In Time（即时编译），如 V8 JS 引擎，它能够即时地编译和运行 JavaScript 代码，这种模式的优点是可以直接将代码分发给用户，而不用考虑用户的机器架构，缺点是源代码量大，将会花费 JIT 编译器大量的时间和内存来编译和执行，应用体验上会感到程序运行得慢。

AOT 全称为 Ahead Of Time（事前编译），典型的例子就是像 C/C++需要被编译成特殊的二进制才可以通过进程加载和运行，这种模式的优势是速度快，通过事先编译好的二进制代码，加载和执行的速度都会非常快，在密集计算或者图形渲染的场景下能够获得比较好的用户体验。

在对 Flutter App 进行代码开发时（DeBug 模式），使用热更新（Hot Reload）可以方便快速地刷新 UI，同时也需要比较高的性能进行视图渲染，所以 Flutter 在 DeBug 模式下使用了 Kernel Snapshot（可以归类为 AOT 编译）编译模式，将 Dart 源码进行编译。

1.2 Dart 语言核心基础

本节将详细描述 Dart 语言的变量、基础语法及基础的数据结构，是 Flutter 项目开发的一个基础，使用的调试工具为在线工具 dartpad，在线调试的网址如下：

```
https://dartpad.dev/
```

也可以使用 Android Studio 开发工具创建 Dart 项目工程来测试，如图 1-3 所示，在 Android Studio 中创建工程，选择 Dart，依次单击“下一步”完成项目创建，本节中的测试代码在本书配套源码中的 dart_base 项目中。

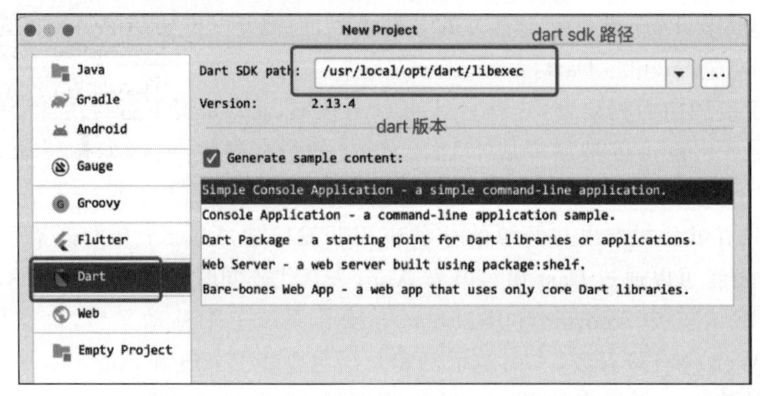

图 1-3　Android Studio 创建 Dart 项目工程图

如图 1-4 所示，在 dart 项目工程的 bin 目录下创建 dart 文件，对应会生成程序入口函数 main 函数，单击 main 函数旁边的小三角可以运行调试程序。

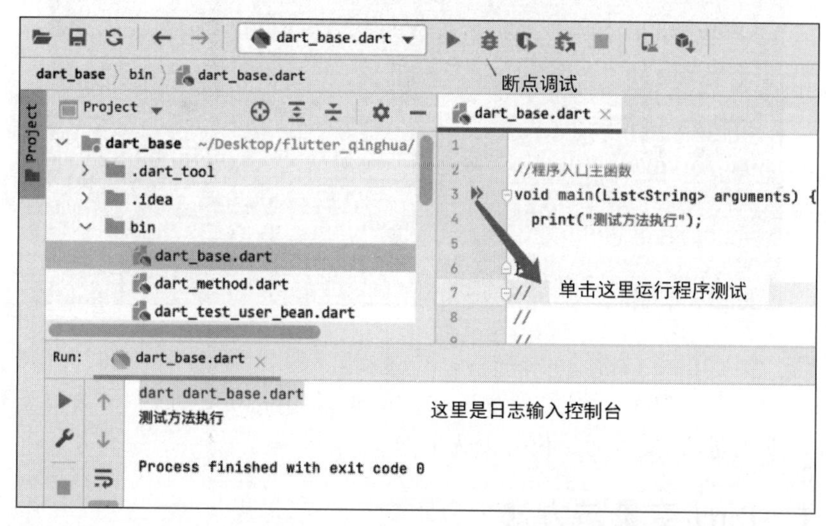

图 1-4　Android Studio Dart 项目运行程序测试效果图

Dart 2.12 和 Flutter 2 中提供了空安全特性(Sound null safety)，对应到 Flutter 项目中，则需要在 pubspec.yaml 文件中添加如下配置：

```
environment:
  sdk: ">=2.12.0 <3.0.0"
```

本书的开发依赖为 Flutter 2.10.2，Dart 2.16.1，所以在本书内所有的代码中，是基于空安全特性来讲解的。

定义一个 int 类型的变量，代码如下：

```
int age = null;
```

在没有空安全前，上面的代码是没有问题的，但当使用空安全后，在编译阶段会出现异常，如图 1-5 所示。

这是空安全与以前最大的不同，默认情况下，变量不能为 null(空安全以前任何类型都可以设置为 null)，空安全在编译阶段会出现异常，无法通过编译。正确的写法如下：

```
//设置默认值
int age = 22;

//声明为空类型
int? age;
```

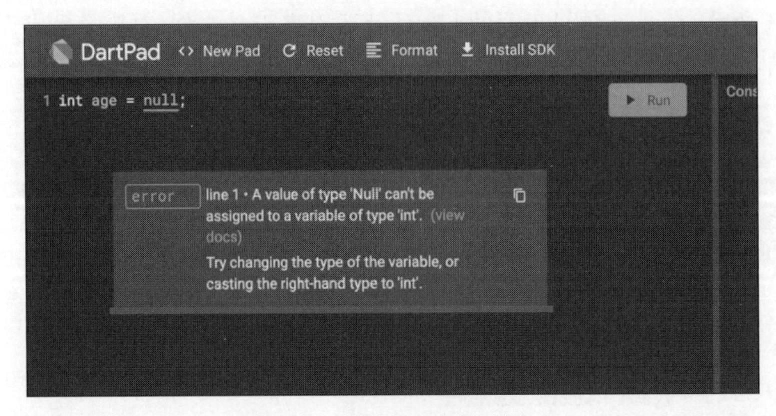

图 1-5 变量空安全提示图

需要注意的是，在 Dart 语言中有 var、object 及 dynamic 可声明弱类型数据，var、dynamic 在声明变量时，变量初始值可以为 null。

1.2.1 Dart 变量与方法

Dart 属于强类型语言，其中 var、dynamic 用于声明动态类型变量，对于 var，如果没有初始值，则可以变成任何类型，代码如下：

```
var testName;
//类型为字符串
testName = '张三';

//类型为整数
testName = 24;
```

如果有初始值，则其类型将会被锁定，例如下面语句将会报错，如图 1-6 所示。

```
//初始类型为数值
var testAge =12;
//修改类型会报错
testAge="张三";
```

```
A value of type 'String' can't be assigned to a variable of type 'int'. (Documentation)

Try changing the type of the variable, or casting the right-hand type to 'int'.

Add cast    More actions...
```

图 1-6 var 变量类型修改出错效果图

dynamic 可声明动态任意类型，编译阶段不检查类型初始值，并且不可为 null，代码如下：

```
//Dart 中声明字符串类型
  //动态声明任意类型,编译阶段不检查类型
  dynamic flag = "张三";
  //print 函数将结果输出在控制台
  //is 函数用来判断类型是否对应,如果对应,则返回 true
  //$ 在这里是向字符串中插入内容,结合{}大括号来使用
  print("flag 变量类型是 String ${flag is String}");

  //将类型修改成数值类型
  flag = 33;

  bool isBool = flag is int;
  //$ 在这里引入的是一个单变量,所以可以省略{}
  print("flag 变量类型是 bool $isBool ");
```

Dart 中数值类型分为 int(整数)和 double(浮点数),bool 用于声明是与否,其声明的变量可取值为 true、false,变量的基本声明方式如下:

```
//全局变量定义
//以下画线开头的变量只可以在文本文件中调用
int _testNumber = 10;
//全局变量定义,在项目工程中都可以引用
int testNumber = 100;
//定义浮点类型
double testDouble = 10.78;

class UserBean {
  //只可以在本类中访问外部,如果需要访问,则可以通过 get 方法使用
  int? _age;

  //定义 set 方法,在类外部为_age 属性赋值
  set age(int value) {
    _age = value;
  }

//定义 get 方法,在类外部获取 age 属性的值
  int? get getAge => _age;

  //未使用下画线开头定义的变量可直接通过类对象实例引用
  String? username;

  //静态变量可以直接使用类引用
  static String? userSchool;
}
```

在如下测试函数中使用 UserBean 中的变量，代码如下：

```
//代码清单 1-1
//代码路径 code1/dart_test_user_bean.dart
void main(List<String> arguments) {
  //无须创建 UserBean 对象,直接使用类名引用
  String? userSchool = UserBean.userSchool;

  //创建 UserBean 的实例对象
  UserBean userBean = new UserBean();
  //或者可以这样写
  UserBean userBean1 = UserBean();

  //使用类实例访问 username
  String? username = userBean.username;
  //使用类实例访问 _age 属性通过 get 方法得到 age
  //age 的值可能为 null,在使用时需要判断是否为 null
  int? age = userBean.getAge;
  //age2 为 null 时取默认值 0
  int age2 = userBean1.getAge ?? 0;

  //在使用时需要判断是否为 null
  if (age != null && age2 != null) {
    //计算
    int allAge = age + age2;
  }
}
```

在 Dart 语言中定义方法的基本格式如下：

```
返回值类型 方法名称(参数){

  return 返回值;
}
```

定义一个无参数无返回值的方法，代码如下：

```
//无参数方法无返回传值
void test(){

}

//返回值类型可以省略不写
test(){

}
```

定义一个求和的方法，代码如下：

```
//参数 name 必传
int add1(int value1, int value2) {
  return value1 + value2;
}

void main(List<String> arguments) {
  //调用方法求和
  int result = add1(1, 2);
  print("测试求和方法调用 $result");
}
```

上述 add1 求和方法中两个参数值必传，可以通过“[ ]”与“{}”来优化方法定义，基本使用代码如下：

```
//参数 value1 可选的位，参数不传时默认取值为 0
//参数 value2 可选的位，参数不传时默认取值为 0
int add2([int value1 = 0, int value2 = 0]) {
  return value1 + value2;
}

//参数 value1 为必须传参数
//参数 value2 为可选的位，参数不传时默认取值为 0
int add3({required int value1, int value2 = 0}) {
  return value1 + value2;
}

//测试函数调用
void main(List<String> arguments) {
  int result1 = add2(22);
  int result2 = add3(value1: 10, value2: 12);

}
```

闭包函数又叫匿名函数，就是没有函数名的函数，将闭包函数赋值给一个变量，通过变量名调用函数即可调用，基本使用代码如下：

```
//定义一个闭包函数，拥有 a 和 b 两个参数，这里忽略了参数类型
Function testAdd = (a, b) {
  //返回 bool 类型参数
  return a > b;
};
//上面例子改写成箭头函数，这两个函数是等价的
```

```
var testAdd2 = (a, b) => a > b;

void main(List<String> arguments) {
  //调用
  bool addRe = testAdd(2,3);

}
```

1.2.2 Map、List、Set 的基本使用概述

Map 用来存储对像类型的数据，List 与 Set 用来存储数组类型的数据，本节中描述 Map、List、Set 中数据的添加、修改、循环遍历查询。

Map 用来保存 key-value(键-值对)的数据集合，与 Object-C 中所讲的字典一致，分为无序的 HashMap、按 key 的插入顺序的 LinkedHashMap、按 key 的排序顺序的 SplayTreeMap。

在实际项目中结合数据创建 Map 实例，创建一个空的 Map，代码如下：

```
//创建一个 Map 实例，按插入顺序进行排列，默认无数据
  Map dic = new Map();
  print(dic);      //输出 {}
  //创建一个空的 Map，Map 允许 null 作为 key
  var dic5 = new Map.identity();
  print(dic5);    //{}
```

创建一个有初始值的 Map，代码如下：

```
//根据一个 Map 创建一个新的 Map，按插入顺序进行排列
var dic1 = new Map.from({'name': '张三'});
print(dic1); //{name: 张三}

//根据 List 创建 Map，按插入顺序进行排列
List<int> list = [1, 2, 3];
//使用默认方式，key 和 value 都是数组对应的元素
var dic2 = new Map.fromIterable(list);
print(dic2); //{1: 1, 2: 2, 3: 3}

//设置 key 和 value 的值
var dic3 = new Map.fromIterable(list,
    key: (item) => item.toString(), value: (item) => item * item);
print(dic3); //{1: 1, 2: 4, 3: 9}

//创建一个不可修改且基于哈希值的 Map
```

```
var dic6 = new Map.unmodifiable({'name': '张三'});
print(dic6); //{name: 张三}
```

根据 List 数据来创建 Map，代码如下：

```
//两个数组映射一个字典，按插入顺序进行排列
  List<String> keys = ['name', 'age'];
  var values = ['张三', 20];
  //如果有相同的 key 值，则后面的值会覆盖前面的值
  var dic4 = new Map.fromIterables(keys, values);
  print(dic4); //{name: 张三, age: 20}
```

对于 Map 来讲，初始化创建时可以赋值也可以是空的，当创建可变的 Map 数据集合时，在实际开发中往往会根据不同的操作来修改不同的数据，代码如下：

```
//根据一个 Map 创建一个新的 Map，按插入顺序进行排列
  //在这里通过泛型指定了 Map 中的 key 的类型为 String 类型，value 是动态的
  Map<String, dynamic> dic7 = new Map.from({'name': '张三'});
  print(dic7);          //{name: 张三}

  //修改 name 的值
  dic7['name'] = '李四';
  print(dic7);          //{name: 李四}

  //向 Map 中添加新的键 - 值对数据
  dic7['age'] = 23;
  print(dic7);          //{name: 李四, age: 23}
```

然后获取 Map 中的数据，代码如下：

```
//根据 key 获取对应的值
String name = dic7['name'];

//遍历获取 Map 中所有的数据
dic7.forEach((key, value) {
  print("${key} is ${value}");
});
```

对于 List 与 Set 来讲，都用来存储数组类型数据，区别是 Set 不可保存重复数据，也就是说 Set 中的数据具有唯一性，List 的基本使用代码如下：

```
//创建非固定长度的 Lsit，初始数据为空
  List testList = [];
```

```
print(testList.length);        //0
print(testList);               //[]

//向 Lsit 中添加数据
testList.add("hello");
testList.add(123);
print(testList.length);        //2
print(testList); //[hello, 123]

//创建元素类型固定的 List
List<String> typeList = []; //只能添加字符串类型的元素

typeList.add("张三");           //正确
//typeList.add(1);             //错误,类型不正确

//直接赋值创建 List
List<int> numList = [1, 2, 3];

//多类型数据的 List
List<dynamic> typeMulList = [22,"张三"];
```

对于获取 List 中数据的方法也比较多,基本获取 List 中单个元素值(获取 List 中指定位置的值)的方法的代码如下:

```
//直接根据索引获取 0 号位置上的数据
String value = list[0];
//等效于 elementAt 法获取
String value1 = list.elementAt(0);
```

查找 List 中的元素,代码如下:

```
List<String> list = ["test1", "xioming", "张三", "xioming", "张三", "李四"];

//从索引 0 处开始查找指定元素,返回指定元素的索引
int index = list.indexOf("张三");          //index 2

//
//从索引 0 处开始查找指定元素,如果存在,则返回元素索引,否则返回 -1
int index2 = list.indexOf("张三",3);     //4
//
//从后往前查找,返回查找到的第 1 个元素的索引
int index4 = list.lastIndexOf("张三");  //4
```

循环遍历 List 中的数据,代码如下:

```
//使用 List 创建测试
List<String> testList = ["test1", "xioming", "张三", "xiong", "张三", "李四"];

//方式一,遍历获取 List 中的所有数据
testList.forEach((value) {
  //value 就是 List 中对应的值
});

//方式二,遍历获取 List 中的所有的数据
for(int i = 0;i < testList.length;i++){
  //根据索引获取 List 中的数据
  var value = testList[i];
}

//方式三
//while + iterator 迭代器遍历,类似 Java 中的 iterator
while(testList.iterator.moveNext()) {
  //获取对应的值
  var value = testList.iterator.current;

}

//方式四,增强 for 循环
//for - in 遍历
for (var value in testList) {
  //value 就是 List 中对应的值
}
```

List 数据结构转 Map 数据结构，代码如下：

```
List<String> testList3 = ["test1", "xioming", "张三", "xioming", "张三", "李四"];
print(testList3); //[test1, xioming, 张三, xioming, 张三, 李四]
Map<int,String> map = testList3.asMap();
print(map); //{0: test1, 1: xioming, 2: 张三, 3: xioming, 4: 张三, 5: 李四}
```

随机排列 List 中的数据顺序，代码如下：

```
List<String> testLis4t = ["test1", "xioming", "张三", "xioming", "张三", "李四"];
print(testLis4t); //[test1, xioming, 张三, xioming, 张三, 李四]

//将 list 中数据重新随机排列
testLis4t.shuffle();
print(testLis4t); //[李四, test1, 张三, xioming, xioming, 张三]
```

Set 用来保存列表数据，相对于 List，Set 是无序的、唯一的（数据不重复），基本使用代码如下：

```
//测试数据列表
List testList = ['香蕉', '苹果', '西瓜', '香蕉', '苹果', '香蕉', '苹果'];
//创建 Set
Set testSet = new Set();
//将集合中的数据添加到 Set 中
testSet.addAll(testList);

print(testSet);          //{'香蕉','苹果','西瓜'}
print(testSet.toList());//['香蕉','苹果','西瓜']

//获取 Set 中的第 1 个数据
String first = testSet.first;
//获取 Set 中的最后一个数据
String last = testSet.last;

//遍历获取 Set 中的所有数据
testSet.forEach((element) {
  print("$element");
});

//遍历获取 Set 中的所有数据
for (var value in testSet) {
  print("$value");
}
```

1.2.3 Dart 中的流程控制

使用 for 循环,代码如下:

```
//定义集合
List<String> userList = ['张三', '李四'];
//可以用 var 或 int
for (var i = 0; i < userList.length; i++) {
  print(userList[i]);
}
```

使用 while 循环,代码如下:

```
List<String> userList = ['张三', '李四'];
var index = 0;
while (index < userList.length) {
  print(userList[index]);
  index++;
}
```

使用 do-while 循环，代码如下：

```
List<String> userList = ['张三', '李四'];
  var index2 = 0;
  do {
    print(userList[index2++]);
  } while (index2 < userList.length);
```

使用 if-else 逻辑判断，代码如下：

```
var numbers = [1, 2, 3, 4, 5, 6, 7];
for (var i = 0; i < numbers.length; i++) {
  if (numbers[i].isEven) {
    print('偶数: ${numbers[i]}');
  } else if (numbers[i].isOdd) {
    print('奇数: ${numbers[i]}');
  } else {
    print('非法数字');
  }
}
```

使用三元运算符（？：），代码如下：

```
int value1 = 90;
//如果 value1 == 9,testResult 的结果是 "正确"
String testResult = value1 == 9 ? "正确" : "错误";
```

使用 switch 语句，switch 语句的作用与 if 语句类似，用于检测各种条件是否成立，然后执行相应分支的代码，代码如下：

```
//条件状态变量
var command = 'close';

switch (command) {          //需要检测的变量
  case 'close':             //case 语句用于设置检测条件
    executeClosed();        //如果 command = 'close',则执行当前分支代码
    break;                  //结束当前分支执行逻辑
  case 'open':
    executeApproved();
    break;
  default:                  //如果上面的 case 语句都没有匹配成功,则执行 default 分支的逻辑
    executeUnknown();
}
```

1.2.4 Dart 异常(Exception)处理

Dart 提供了异常处理机制，主要有两种异常类型，即 Exception 和 Error，Exception 是开发者可以预知的错误，Error 是程序错误，例如，调用函数参数错误，代码如下：

```
void main(List<String> arguments) {

  //声明一个空变量
  int? a;
  //在这里通过"!"强制 a 为非空
  //实际上这是错误的操作
  var b = a! + 10;
  print("处理结果 $b");

}
```

如上所示，在这里声明的 null 变量 a，在使用时未校验是否为 null 而赋值，而使用“！”强制转换声明为非空，这样会出错，程序通过了编译，运行效果如图 1-7 所示，出现错误，程序抛出异常。

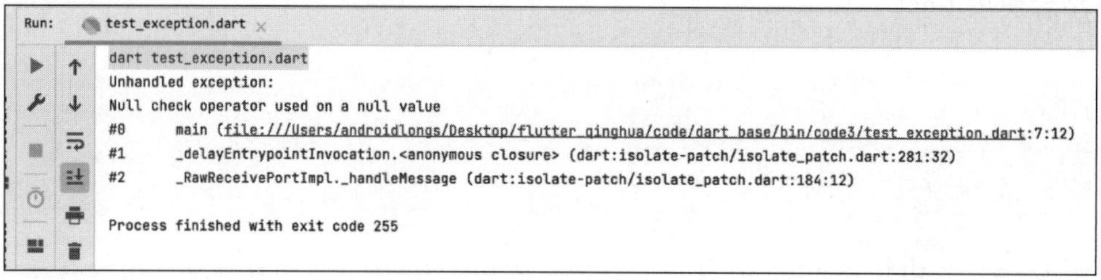

```
Run:    test_exception.dart
dart test_exception.dart
Unhandled exception:
Null check operator used on a null value
#0      main (file:///Users/androidlongs/Desktop/flutter_qinghua/code/dart_base/bin/code3/test_exception.dart:7:12)
#1      _delayEntrypointInvocation.<anonymous closure> (dart:isolate-patch/isolate_patch.dart:281:32)
#2      _RawReceivePortImpl._handleMessage (dart:isolate-patch/isolate_patch.dart:184:12)

Process finished with exit code 255
```

图 1-7 程序运行异常效果图

可以通过 try/catch 语句捕获异常，代码如下：

```
try {
  //业务代码
  var b = a! + 10;
  print("处理结果 $b");
} on Exception catch (e) {
  //通过 on 指定拦截的异常类型,通过 catch 捕获异常对象
  //拦截所有 Exception 和它的子类抛出的异常
  print('exception: $e');
} catch (e) {
  //没有指定异常类型,只是捕获异常对象,相当于捕获所有异常类型
  print('其他异常: $e');
}
```

使用 rethrow 重新抛出异常,代码如下:

```
try {
  dynamic foo = true;
  print(foo++); //这里会产生 Runtime error
} catch (e) {
  print('捕获到的异常类型 ${e.runtimeType}.');
  rethrow; //重新把异常抛出去
}
```

获取异常调用栈,捕获到异常的时候希望知道异常是从什么地方抛出来的,可以通过 catch 的第 2 个参数获取调用栈,代码如下:

```
try {
  //...
} catch (e, s) {
  print('异常信息:\n $e');
  print('调用栈信息:\n $s');
}
```

结合使用 finally,无论是否捕获到异常,都要执行 finally 中的内容,代码如下:

```
try {
  //业务代码
} catch (e) {
//处理异常
  print('Error: $e');
} finally {
    //无论是否捕获异常都会执行 finally 的代码
    //…
}
```

1.3 Flutter 项目创建与配置文件

Android Studio 是谷歌推出的非常强大的开发工具,本书中所有的代码都在 Android Studio 中进行开发调试,首先使用 Android Studio 来创建第 1 个 Flutter 项目 Hello World,创建过程与基本介绍可观看视频讲解【1.3 Flutter 项目的基本创建说明】。

4min

lib 目录下是编写 Dart 语言代码的目录空间,默认创建的 main.dart 文件就是 Flutter 项目的启动文件,main 函数是程序启动入口,默认生成的 Flutter 页面是一个单击累加的计数器,然后把默认生成的项目代码整理如下:

```
//代码清单 1-2
//代码路径 helloworld/lib/main.dart
import 'package:flutter/material.dart';
void main() {
  runApp(MyApp());
}

class MyApp extends StatelessWidget {
  const MyApp({Key? key}) : super(key: key);

  @override
  Widget build(BuildContext context) {
    //根视图
    return MaterialApp(
      //应用主题
      theme: ThemeData(
        primarySwatch: Colors.blue,
      ),
      //应用默认显示的页面
      home: const MyHomePage(),
    );
  }
}

//应用显示的第 1 个页面
class MyHomePage extends StatefulWidget {
  const MyHomePage({Key? key}) : super(key: key);
  @override
  _MyHomePageState createState() => _MyHomePageState();
}

class _MyHomePageState extends State<MyHomePage> {
  @override
  Widget build(BuildContext context) {
    //页面脚手架
    return Scaffold(
      //标题
      appBar: AppBar(
        title: const Text("第 1 个 Flutter 应用程序"),
      ),
      //中间显示的文本
      body: const Center(
        child: Text(
          "Hello World",
          //文本样式
```

```
            style: TextStyle(
              //文本大小
                fontSize: 33,
                //文本加粗
                fontWeight: FontWeight.bold),
          ),
        ),
      );
    }
  }
```

修改后的代码就是要创建生成的 Hello World，如图 1-8 所示，不用担心 Dart 语言的多重嵌套会导致代码阅读及维护性差，Android Studio 开发工具中提供了明显的代码结构层级效果图。

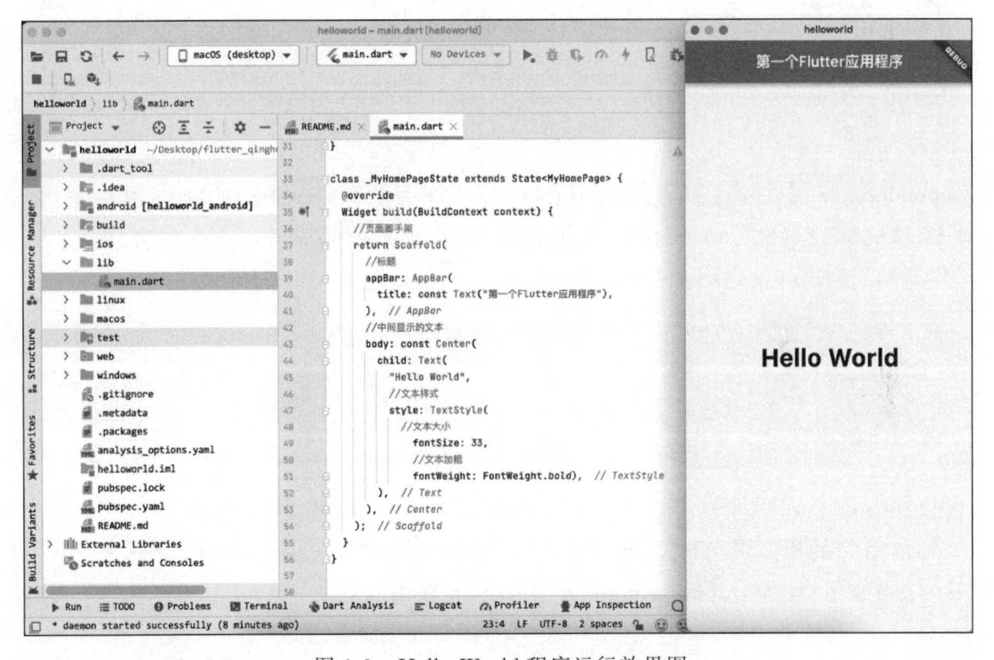

图 1-8 Hello World 程序运行效果图

1.3.1 pubspec 配置文件中依赖库引用说明

pubspec 配置文件，用来配置项目的名称、依赖 SDK 版本信息、依赖库、图片等资源文件，如图 1-9 所示，在 pubspec. yaml 文件中，name 属性是当前项目的项目名称，description 属性配置的是项目的说明，version 属性是配置的版本信息，包括两部分，如图 1-9 所示的 1.0.0+1，指的是版本号为 1.0.0，编译次数为 1，版本号对应 Android 原生中的 versionName，对应 iOS 原生中的 Version，编译次数对应 Android 原生中的 versionCode，对应 iOS 原生中的 Build，需要注意的是版本信息只能递增。

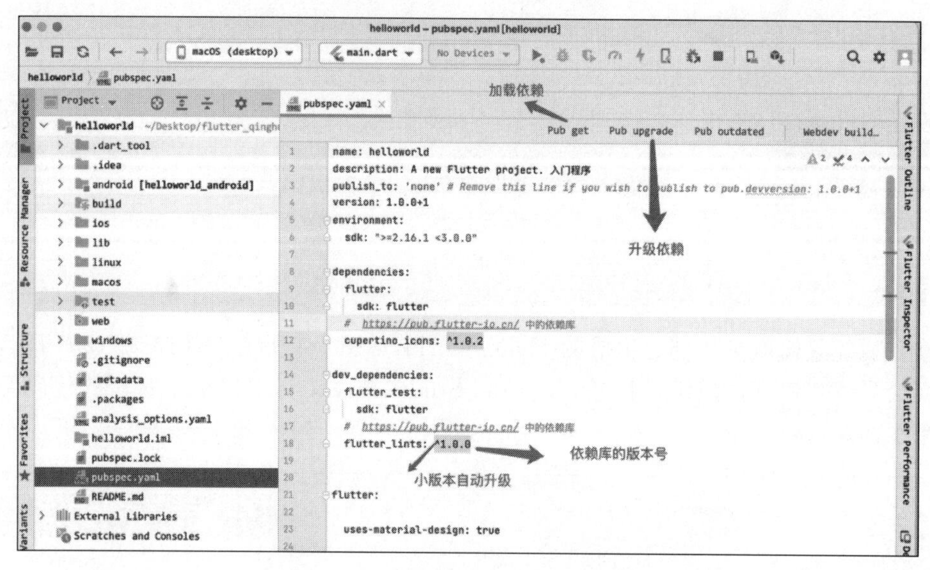

图 1-9　pubspec. yaml 文件基本说明图

dependencies 属性用于配置依赖库信息，依赖库有两种，一种是纯 Dart 语言库，另一种是插件类型的库，包含 Android、iOS 双平台的内容，配置依赖库有以下 3 种方式。

第 1 种是使用 pub 仓库，网址如下：

```
//国内
https://pub.flutter-io.cn/
//国外
https://pub.dev/
```

可以在 pub 仓库中搜索需要使用的依赖库或者插件，如图 1-10 所示。

添加依赖库的第 2 种方式是通过 git 形式，如这里添加 webview，可以修改为 GitHub 的依赖，代码如下（# 为 pubspec. yaml 文件中的注释内容）：

```
#URL 表示插件的网址,ref 表示插件的分支
  flutter_fai_webview:
    git:
      url: https://github.com/zhaolongs/Flutter_Fai_Webview.git
      ref: master
```

添加依赖库的第 3 种方式是通过本地路径依赖，使用 path 关键字，代码如下：

```
#抖动动画依赖库
  shake_animation_widget :
    path: /Volumes/code/ico/Desktop/… /shake_animation_widget
```

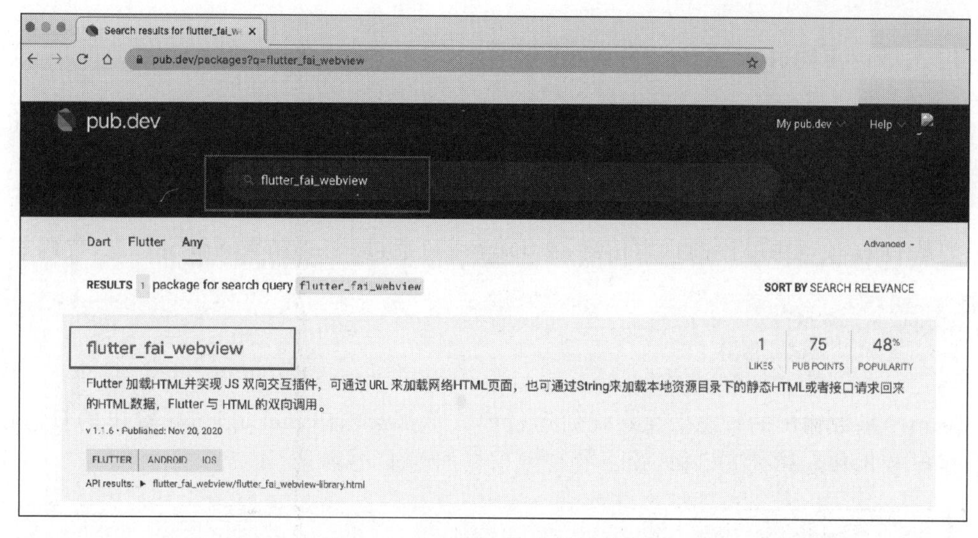

图 1-10　pub 仓库搜索插件效果图

在添加好依赖后，需要在 Terminal 命令工具中执行 flutter pub get 命令加载依赖，代码如下：

```
flutter packages get             //获取 pubspec.yaml 文件中所有的依赖关系
```

注：pub get 是 pub get packages 的简写，这是在 Dart 项目中下载包的方式，在 Flutter 项目中，flutter pub get 会自动更改为 flutter packages get。

1.3.2　图片等资源管理配置

自定义的静态资源目录如图 1-11 所示，一般在 App 中会使用图标、字体、动画、JSON、JS 文件等，这些文件会被打包到应用程序安装包内，assets 文件夹名称可以自定义，在这里根据 Android、iOS 原生习惯定义为 assets。

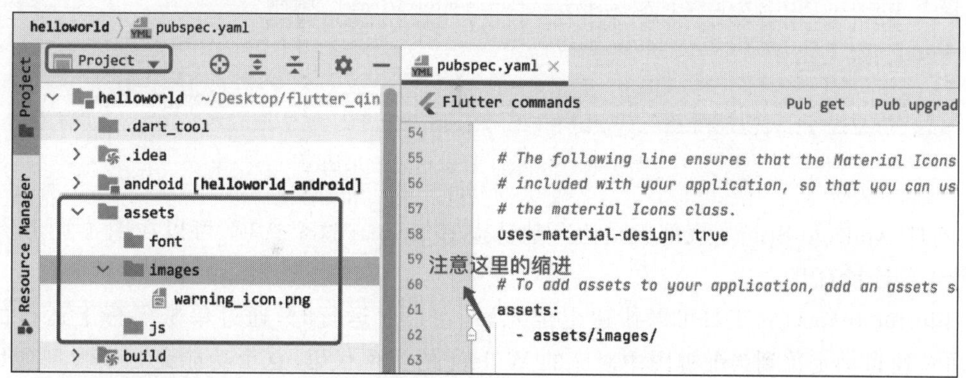

图 1-11　图片资源配置效果图

1.3.3 Flutter App 的调试技巧

合理地使用开发工具，可以有效地提升开发效率，本节介绍如何使用 Android Studio 日志调试、断点调试、Flutter Inspector 快速定位代码、Flutter Performance 内存跟踪调试。

Flutter 项目中的日志分为两类，一类是 Android 原生与 iOS 原生中输出的日志信息，另一类是 Dart 项目中输出的日志信息，在 Dart 中可通过 print 函数来输出日志，代码如下：

```
print("测试数据");
deBugPrint("测试数据");
```

print()函数输出的日志信息在任何运行模式下都会输出，deBugPrint 输出的日志信息只会在程序 deBug 模式下输出，输出的日志信息如图 1-12 所示。

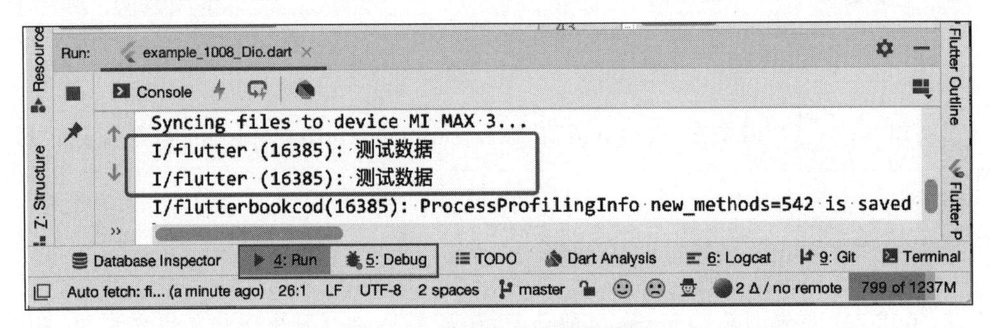

图 1-12　普通日志输出效果图

Logcat 也是 Android Studio 的一个强大的日志输出视图工具，如图 1-13 所示。

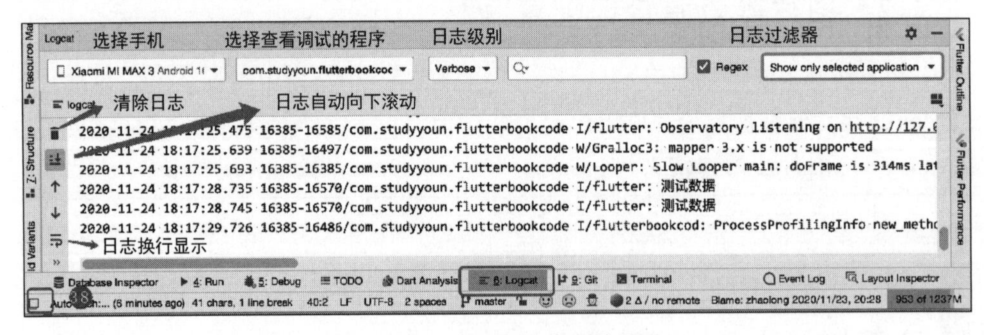

图 1-13　Logcat 日志信息效果图

有些 Android Studio 底部工具栏上默认没有 Logcat 这个选项，可以在图 1-13 所示的 38 号位置选择打开。

Flutter Inspector 工具可以快速定位元素，在程序运行时，通过单击屏幕上运行的视图，可快速自动定位到当前视图中对应的 Widget 的实现代码，这个功能在一个大型项目中非常适用，可以极大地节省开发者定位代码的时间，由原来的手动检索代码过渡到现在的自动定位代码。

如图 1-14 所示,Android Studio 右侧边栏的视图定位工具就是 Flutter Inspector,读者可观看视频讲解【1.3.3-1 Flutter Inspector 工具】【1.3.3-2 Flutter 项目在 Android 目录空间下的调试】【1.3.3-3 Android Studio Flutter 项目断点调试】和【1.3.3-4 Flutter 项目 Xcode 目录下的断点调试】。

图 1-14 Flutter Inspector 工具效果图

1.3.4 Flutter Widget 基本概述

在 Flutter 中,一切皆是 Widget,从构建页面使用的基础 Widget,如 Image、Text,到如手势姿势功能的 InkWell 等。

在 Flutter 程序中,程序入口 main()函数执行 runApp 加载根视图,一个简单程序的代码如下:

```
void main() => runApp(
      const MaterialApp(
        home: Scaffold(
          body: Text("早起的年轻人 = "),
        ),
      ),
    );
```

上述页面中包含 Scaffold 与 Text 两个 Widget,当 runApp()被调用时,程序会执行以下事件:

(1) Flutter 会构建包含这两个 Widget 的 Widgets 树。

(2) Flutter 遍历 Widget 树，然后根据其中的 Widget 调用 createElement()来创建相应的 Element 对象，最后将这些对象组建成 Element 树。

(3) 创建第 3 棵树 Render 树，这棵树中包含了与 Widget 对应的 Element 及通过 createRenderObject()创建的 RenderObject。

Element 是 Widget 的抽象，当一个 Widget 首次被创建的时候，这个 Widget 会通过 Widget. createElement 创建一个 Element，挂载到 Element Tree 遍历视图树，如图 1-15 所示，每个 Element 中都有着对应的 Widget 和 RenderObject 的引用，Element 是存在于可变 Widget 树和不可变 RenderObject 树之间的桥梁。

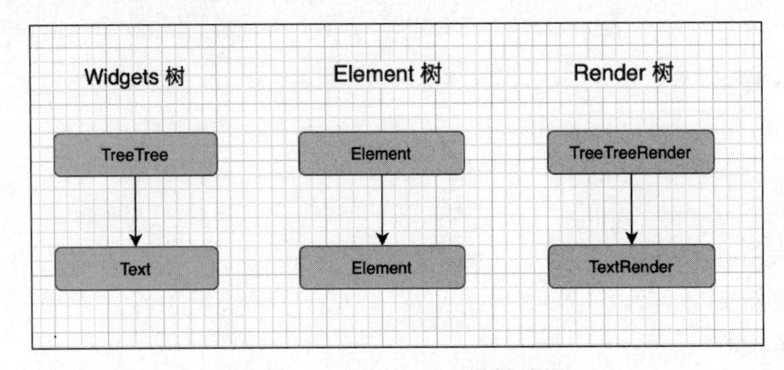

图 1-15 Flutter UI 三棵树结构

1.4 小结

本章为本书内容的开篇章节，概述了 Dart 语言的核心基础，这些核心基础在后续章节中会陆续使用到，创建了第 1 个 Flutter 项目工程，以及概述了 Android Studio 开发工具的调试技巧，应用调试技巧可以显著地提升开发效率。

第 2 章 Flutter 基础组件核心基础

CHAPTER 2

在 Flutter 中，从显示界面的 UI 组件，如 Text、Image 等，再到功能性的组件，如手势 InkWell 组件等，都是基于 Widget 构建的，StatefulWidget 和 StatelessWidget 是两个基本的 Widget，在进行 Flutter 项目开发时，基本的页面或者功能 Widget 一般继承于这两个 Widget 来使用，StatefulWidget 是有状态 Widget，StatelessWidget 是无状态 Widget，从另一个角度来讲，StatefulWidget 可用来构建数据变动的 UI 样式，而 StatelessWidget 构建的 UI 样式不能更新变动。

本章讲解核心 UI 组件 MaterialApp、Scaffold、AppBar、Text、TextField、Container、Image、Button 的基础与应用。本章对应的代码在本书配套源码中的 flutter_base_widget 项目中。

2.1 MaterialApp 用来搭建程序的入口

Widget 可分为 Material Design 风格的组件和 Cupertino 苹果风格的组件，MaterialApp 组件是 Material Design 设计风格在 Flutter 中的体现。使用 Material 组件，需要先引入：

```
import 'package:flutter/material.dart';
```

一般一个应用程序中可存在一个 MaterialApp 组件，所以 MaterialApp 组件可以理解为用来构建应用的根布局，Flutter 项目中默认配置的 main. dart 文件是项目的启动文件，在其中配置的程序应用入口 main()函数中通过 runApp()函数来加载应用程序，runApp 可以直接加载 MaterialApp，即作为第 1 个(根视图)来创建，对应的代码如下：

```
//代码清单 2-1
//代码路径 lib/code2/code201_MaterialApp.dart
//应用入口
void main() {
```

```
  //启动根目录
  runApp(MaterialApp(
    home: Example201(),
  ));
}

class Example201 extends StatelessWidget {
  @override
  Widget build(BuildContext context) {
    return MaterialApp(
      //安卓任务栏中的应用标题
      title: 'Flutter Demo',
      //使用的应用主题
      theme: ThemeData(
        primarySwatch: Colors.blue,
      ),
      //默认显示的首页面
      home: MyHomePage(title: 'Flutter Demo Home Page'),
    );
  }
}
```

2.1.1 路由配置

在 Flutter 中，打开并渲染一个新的页面有两种方式：一种是静态路由，另一种是动态路由，静态路由在 MaterialApp 组件中通过 routes 来配置跳转规则，由路由名称和指向的路由页面两部分构成，代码如下：

```
//代码清单 2-2
//代码路径 lib/code2/code202_MaterialApp_Router.dart
//静态路由配置
void main() => runApp(MyApp2());

//定义根目录 Widget
class MyApp2 extends StatelessWidget {
  @override
  Widget build(BuildContext context) {
    return MaterialApp(
      //配置路由规则
      routes: {
        //默认页面
        "/": (BuildContext context) => MyHomePage(),         //默认创建的启动页面
        "/first": (BuildContext context) => FirstPage(),     //自定义的页面
        "/second": (BuildContext context) => SecondPage(), //自定义的页面
```

```
      },
    );
  }
}
```

通过 home 属性或者 routes 中的配置符号 "/"来配置默认显示的页面，上述配置的 MyHomePage 就是当前 Flutter 应用启动加载的第 1 个 Flutter 页面。通过 Navigator 的 pushNamed(静态路由)的方法来打开对应的页面，代码如下：

```
Navigator.of(context).pushNamed('/second');
```

动态路由不通过 routes 配置，并且无指定的路由名字，在使用时直接使用 Navigator 来打开，代码如下：

```
Navigator.of(context).push(new MaterialPageRoute(builder: (_) {
//FirstPage 是要打开的目标页面
return FirstPage();
}),);
```

退出当前 Widget 页面有 3 种方式：

第 1 种是单击手机的物理返回按钮(安卓手机)，或者在全屏手机的左侧边缘或者右侧边缘手势滑动退出，在 iOS 手机上手势由左向右滑动退出。

第 2 种是单击页面 AppBar 中默认的左上角的返回箭头。

第 3 种是在页面中通过 Navigator 的 pop 方法主动调用退出页面 Widget 的方式。

调用 Navigator 的 pop()方法，可以理解为将页面退出路由栈，如果需要回传参数，则可以直接写在 pop()方法中，需要注意的是，在回传的数据与接收数据的地方数据类型应保持一致，在关闭当前页面时，最好判断一下当前页面是否是页面栈中的最后一个页面，也就是 Widgets 树中的最后一个 Widget，如果是，再强行执行 pop()方法，则栈中就没有 Widget 页面了，此时就会导致页面黑屏，所以安全退出的代码如下：

```
if (Navigator.of(context).canPop()) {
  Navigator.of(context).pop();
}
//或者
Navigator.of(context).maybePop();
```

Navigator 的 maybePop 方法在关闭页面时，会自动判断是否可关闭，也就是说判断当前要关闭的页面是否是 Widgets 树中的最后一个页面，如果不是，就正常通过 pop 关闭页面，否则就不进行操作。

一般在实际项目中会有这样的使用场景，即页面 A 将参数传递给页面 B，然后页面 B

在关闭时再将处理结果传递给页面A，代码如下：

```
//代码清单 2-3
//代码路径 lib/code2/code202_MaterialApp_Router.dart
//打开 B 页面并获取 B 页面回传的值
  void openBFunction() async{
    Map<String, String> map = new Map();
    map["name"] = "张三";
    //跳转第 2 个页面
    //[arguments]传到第 2 个页面的参数
    dynamic result = await Navigator.of(context)
                        .pushNamed("/second", arguments: map);
    //如果不需要 B 页面返回值，则在这里不处理 await 就可以了
//当 B 页面没有设置返回值时，B 页面关闭，程序会回调，这里 result 可能为 null
    if(result!= null&&result is Map){
      //获取回传的值
      String name = result["name"];
      print("页面 B 关闭 name $name");
    }
  }
```

在页面B中，需要获取A页面中传递的参数，获取代码如下：

```
//代码清单 2-4
//lib/code2/second_page.dart
String ? _title;
@override
void didChangeDependencies() {
  super.didChangeDependencies();

  ModalRoute? route = ModalRoute.of(context);
  if(route!= null){
    //是否是路由栈中的第 1 个页面
    bool? isFirst = route.isFirst;
    //当前手机屏幕上显示的是否是这个页面 Widget
    bool? isCurrent = route.isCurrent;
    //当前 Widget 是否是活跃可用的
    //当调用 pop 或者关闭当前 Widget 时 isActive 为 false
    bool isActive = route.isActive;
    if (isActive) {
      //获取路由信息
      RouteSettings routeSettings = route.settings;
      //获取传递的参数
      Object? arguments = routeSettings.arguments;
      //安全判断
      if(arguments!= null&&arguments is Map<String, String>){
```

```
        Map<String, String> map = arguments;
        String? title = map["name"];
        //变量赋值
        _title = title;
      }

    }
  }
}
```

在动态路由中传值比较方便，可直接通过构造函数传值，代码如下：

```
//代码清单 2-5
//代码路径 lib/code2/code202_MaterialApp_Router.dart
//以动态路由方式打开 B 页面
void openB() {
  //跳转第 2 个页面
  //在这里通过 then 函数获取 A 页面关闭时回传的参数
  Navigator.of(context).push(new MaterialPageRoute(builder: (_) {
    //直接通过构建函数来传参数
    return new FirstPage(title: "这是传递的参数");
  })).then((value) {
    if (value) {
      Map<String, String> resultMap = value;
      print("第 2 个页面回传的数据是 ${resultMap['name']}");
    }
  });
}
```

在对应的 B 页面中，获取 A 页面中传递的参数，代码如下：

```
//代码清单 2-6
//代码路径 lib/code2/second_page.dart
class FirstPage extends StatefulWidget {

  final String? title;

  const FirstPage({Key? key, this.title}) : super(key: key);

  @override
  State<StatefulWidget> createState() {
    return FirstPageState();
  }
}
```

```
class FirstPageState extends State<FirstPage> {
  @override
  Widget build(BuildContext context) {
    return Scaffold(
          ...

          //使用上一个页面传递的参数
          Text("${widget.title}"),

          ...
    );
  }
}
```

2.1.2 语言环境配置

在默认情况下，创建的 Flutter 应用程序文本输入框 TextField 长按时复制与粘贴、日期控件显示的是英文，当手机系统设置的是中文语言环境时，显示的仍然是英文，实际需求是让这些显示英文的情况变成中文，可通过配置应用的多语言国际化功能来解决。在配置文件 pubspec.yaml 中添加 localizations 多语言环境支持，代码如下：

```
dependencies:
  flutter:
    sdk: flutter
#在这里配置，注意空格
  flutter_localizations:
    sdk: flutter
```

然后在根布局视图下 MaterialApp 组件中配置中文语言环境，代码如下：

```
//代码清单 2-7
//代码路径 lib/code2/code203_MaterialApp_Lan.dart
//定义根目录 Widget
MaterialApp(
  localizationsDelegates: const [
    //初始化默认的 Material 组件
    GlobalMaterialLocalizations.delegate,
    //初始化默认的通用 Widget 组件
    GlobalWidgetsLocalizations.delegate,
    GlobalCupertinoLocalizations.delegate,
  ],
  //当前区域，如果为 null，则使用系统区域，一般用于语言切换
  //传入两个参数，即语言代码和国家代码
```

```
    //这里配置为中国
    locale: const Locale('zh', 'CN'),
    //定义当前应用程序所支持的语言环境
    supportedLocales: const [
      Locale('en', 'US'),     //英文
      Locale('zh', 'CN'),     //中文
    ],

  );
```

2.2 Scaffold 用来搭建页面主体

Scaffold 在英文中释意为"脚手架、建筑架",在 Flutter 应用程序中,可称为脚手架,在 Flutter 开发中,Scaffold 可以用来搭建页面的主结构。

2.2.1 AppBar 用来配置页面的标题

通过 Scaffold 组件的 appBar 属性来配置 AppBar 组件,用来显示页面的标题部分,如图 2-1 所示,基本使用代码如下:

```
//代码清单 2-8 Scaffold 中 AppBar 基本使用
//代码路径 lib/code2/code204_Scaffold_AppBar.dart
class Exam204HomePage extends StatefulWidget {
  @override
  State<Exam204HomePage> createState() => _Exam204HomePageState();
}

class _Exam204HomePageState extends State<Exam204HomePage> {
  @override
  Widget build(BuildContext context) {
    //Scaffold 用来搭建页面的主体结构
    return Scaffold(
      //页面的头部
      appBar: AppBar(
        //左侧按钮会覆盖默认的左侧回退按钮
        leading: IconButton(
          icon: const Icon(Icons.close),
          onPressed: () {},
        ),
        //中间显示的内容
        title: const Text("这里是 title"),
        //title 内容居中
```

```
          centerTitle: true,
          //右侧显示的内容
          actions: [
            IconButton(
              icon: Icon(Icons.share),
              onPressed: () {},
            ),
            IconButton(
              icon: Icon(Icons.message),
              onPressed: () {},
            ),
            IconButton(
              icon: Icon(Icons.more_horiz_outlined),
              onPressed: () {},
            ),
          ],
        ),
        //页面的主内容区
        body: Center(child: Text("测试页面")),
      );
    }
  }
```

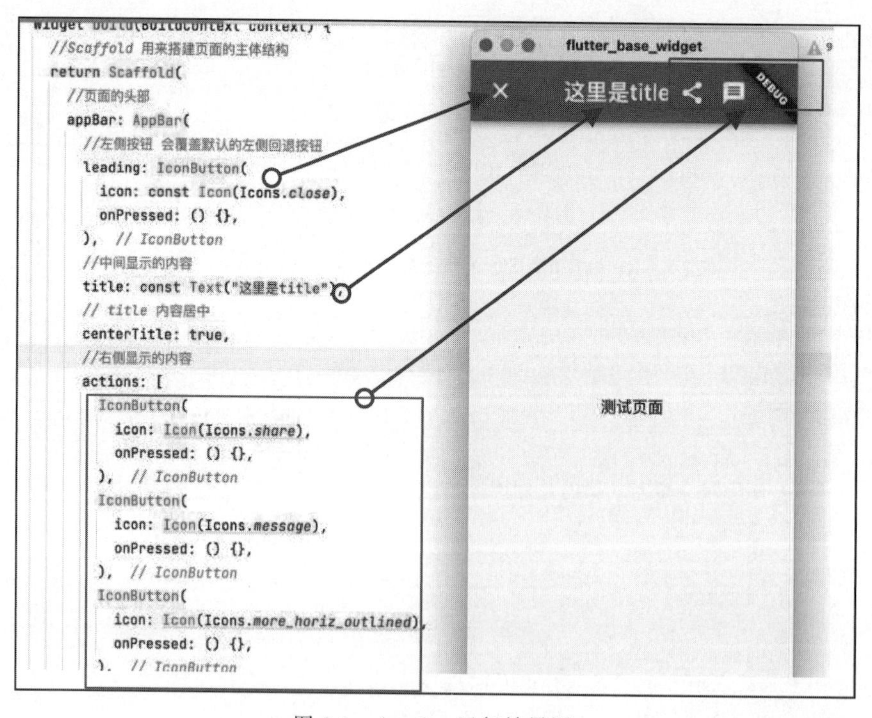

图 2-1　AppBar 运行效果图

AppBar 组件的 title 属性是一个 Widget 类型，这就意味着可以定义任意的 Widget，在这里结合 TabBar 与 TabBarView 实现一个可滑动的标签栏切换页面显示效果，如图 2-2 所示，代码如下：

```
//代码清单 2-9 Scaffold 中 AppBar 中配置可滑动的标签
//代码路径 lib/code2/code205_Scaffold_AppBar.dart
class Exam205HomePage extends StatefulWidget {
  @override
  State<Exam205HomePage> createState() => _Exam205HomePageState();
}

class _Exam205HomePageState extends State<Exam205HomePage>
    with SingleTickerProviderStateMixin {
  //控制器使用 late 关键字来修饰，说明这个变量不为 null
  late TabController _tabController;
  //Tab 集合
  List<Tab> tabs = <Tab>[];
  //主体页面的集合
  List<Widget> bodyList = [];

  @override
  void initState() {
    super.initState();
    //初始化 Tab，如新闻信息类型 App 的分类
    tabs = <Tab>[
      Tab(text: "Tab0"), Tab(text: "Tab1"),
      Tab(text: "Tab3"), Tab(text: "Tab4"),
    ];
    //创建模拟页面，如新闻信息类型 App 的分类列表
    for (int i = 0; i < tabs.length; i++) {
      bodyList.add(ItemPage(i));
    }
    //initialIndex 初始选中第几个
    //[length] 标签的个数
    //[vsync]动画同步依赖
    _tabController =
        TabController(initialIndex: 0, length: tabs.length, vsync: this);
  }

  @override
  Widget build(BuildContext context) {
    //Scaffold 用来搭建页面的主体结构
    return Scaffold(
      //标题
      appBar: AppBar(
```

```
        //配置 TabBar
        title: TabBar(
          //可以和 TabBarView 关联使用同一个 TabController
          controller: _tabController,
          //子 Tab
          tabs: tabs,
          //为 false 时,标签平均分配宽度,不可滑动为 true 时,标签包裹,过多标签,可滑动
          isScrollable: false,
        ),
        //标题居中
        centerTitle: true,
      ),
      //页面的主内容区
      body: TabBarView(
        //联动控制器
        controller: _tabController,
        //所有的子页面
        children: bodyList,
      ),
    );
  }
}
```

TabBar 用来实现标签栏功能，在 TabBar 中配置 Tab，默认情况下 TabBar 的属性 isScrollable 为 false，TabBar 中的 Tab 被平均分配在水平方向的空间中，不可滑动，这种一般应用于数量固定只有少量 Tab 的情况；当配置为 true 时，每个 Tab 的宽度就是包裹其内容的宽度，并且可滑动。

将在上述代码中构建的 TabBar 配置在 AppBar 的 bottom 属性上，实现效果如图 2-3 所示，这种效果是比较常见，如新闻信息类应用的分类浏览、商品交易应用中的订单列表页面等。

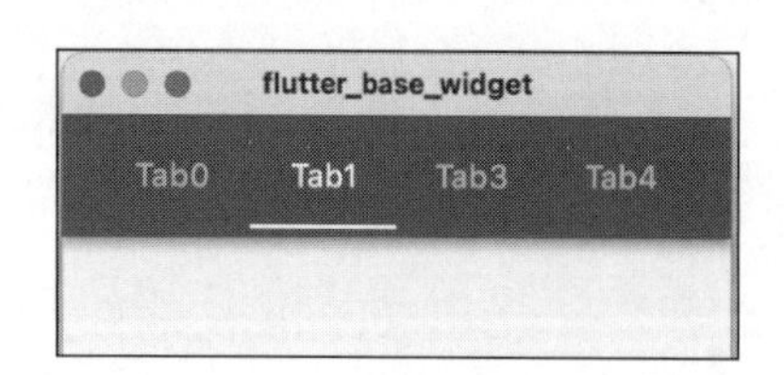

图 2-2　AppBar 中配置可滑动的标签页面

图 2-3　AppBar bottom 中配置可滑动的标签页面

代码如下：

```
//代码清单 2 - 10 Scaffold 中在 AppBar 中配置可滑动的标签
//代码路径 lib/code2/code206_Scaffold_AppBar.dart
```

```
AppBar(
  //配置 TabBar
  title: Text("这里是标题"),
  //标题居中
  centerTitle: true,
  bottom: TabBar(
    //可以和 TabBarView 关联使用同一个 TabController
    controller: _tabController,
    //子 Tab
    tabs: tabs,
    //为 false 时,标签平均分配宽度,不可滑动为 true 时,标签包裹,过多标签,可滑动
    isScrollable: false,
  ),
)
```

AppBar 的 bottom 属性可配置使用 PreferredSizeWidget 组件的子类,AppBar、TabBar 都继承于此,在实际应用场景中,如果需要配置其他组件,则可通过 PreferredSize 组件来结合使用,代码如下:

```
//代码清单 2-11 Scaffold 中 AppBar
//代码路径 lib/code2/code207_Scaffold_AppBar.dart
PreferredSize buildAppBarTitle() {
  //PreferredSize
  return PreferredSize(
    //bottom 的高度设置
    preferredSize: Size.fromHeight(44),
    //bottom 的子 widget
    child: Container(
      color: Colors.white,
      height: 44,
      child: Row(
        //子布局居中
        mainAxisAlignment: MainAxisAlignment.center,
        //两个子 Widget
        children: const [
          Text("测试标签 1"),
          //SizedBox 这里可理解为用来填充空间
          SizedBox(width: 20),
          Text("测试标签 2"),
        ],
      ),
    ),
  );
}
```

2.2.2 FloatingActionButton 悬浮按钮效果

FloatingActionButton 悬浮按钮，简称 FAB(下文中会使用简称)，Scaffold 组件通过属性 floatingActionButton 来配置页面右下角的悬浮按钮功能，代码如下：

```
//代码清单 2-12 Scaffold 中悬浮按钮
//代码路径 lib/code2/code208_Scaffold_FAB.dart.dart
class Exam208HomePage extends StatefulWidget {
  @override
  State<Exam208HomePage> createState() => _Exam208HomePageState();
}
class _Exam208HomePageState extends State<Exam208HomePage>{
  @override
  Widget build(BuildContext context) {
    //Scaffold 用来搭建页面的主体结构
    return Scaffold(
      //页面的头部
      appBar: AppBar(title: Text("标题"),),
      //页面的主内容区
      //可以是单独的 StatefulWidget ,也可以是当前页面构建的,如 Text 文本组件
      body: Center(child: Text("显示内容"),),
      //悬浮按钮
      floatingActionButton: FloatingActionButton(
        //建议使用 Icon
        child: Icon(Icons.add), onPressed: () {
        print("单击了悬浮按钮");
      },
      //长按在按钮上的提示文本
      tooltip: "这里是 FAB!",
      //背景色为灰色
      backgroundColor: Colors.grey,
      //单击水波纹,颜色为黄色
      splashColor: Colors.yellow,
      //前景色为紫色
      foregroundColor: Colors.deepPurple,
    ),);
  }
}
```

如图 2-4 所示，FAB 有 3 种类型：regular、mini、extended，由于 mini 类型是缩小版本的，所以将 FAB 的 mini 属性配置为 true 可实现，默认创建的 FAB 是 regular 类型的，创建的 FAB 的 width 和 height 都为 56.0，mini 类型创建的 FAB 的 width 和 height 都为 46.0。

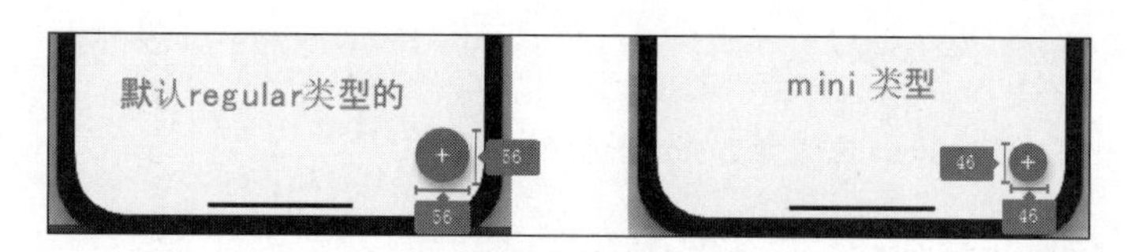

图 2-4　FAB 按钮大小说明

2.2.3　侧拉页面 Drawer

Scaffold 的 drawer 属性用来配置页面左侧侧拉的页面，属性 endDrawe 用来配置右侧侧拉的页面，在实际应用场景中，侧拉的页面一般是一个 ListView 或是一个 Column 线性布局排列的多个条目，如图 2-5 所示，当配置了 Scaffold 的 drawer 内容时，会在 AppBar 位置的左侧多出一个菜单按钮，当单击这个按钮或者从左侧手机边缘向右滑动时，可触发这个页面，代码如下：

```
//代码清单 2 - 13 Scaffold 中 drawer 侧拉页面
//代码路径 lib/code2/code209_Scaffold_Dwaber.dart
class Exam209HomePage extends StatefulWidget {
  @override
  State<Exam209HomePage> createState() => _Exam209HomePageState();
}
class _Exam209HomePageState extends State<Exam209HomePage>{

  @override
  Widget build(BuildContext context) {
    //Scaffold 用来搭建页面的主体结构
    return Scaffold(
      //左侧侧拉页面
      drawer:buildDrawer() ,
      //右侧侧拉页面
      endDrawer: buildDrawer(),
      //页面的头部
      appBar: AppBar(title: Text("标题"),),
      //页面的主内容区
      //可以是单独的 StatefulWidget,也可以是当前页面构建的,如 Text 文本组件
      body: Center(child: Text("body 内容区域"),),
    );
  }

  //以封装方法来构建 Widget,代码块可以是单独的一个 StatefulWidget 页面
  Container buildDrawer(){
    //Container 可看作一个容器,用来包裹一些 Widget
    return Container(
      //背景颜色
```

```
        color: Colors.white,
        width: 200,
    //Column 可以让子 Widget 在垂直方向线性排列
        child: Column(
          children: <Widget>[
            Container(
              color: Colors.blue,
              height: 200,
              child: Text("这是一个 Text"),),
            Container(
              color: Colors.red,
              height: 200,
              child: Text("这是一个 Text2"),),
          ],
        ),
      );
    }
  }
```

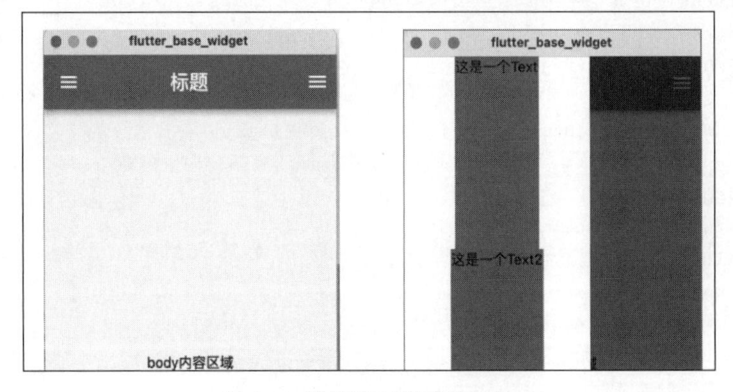

图 2-5　侧拉页面配置效果图

在实际业务开发中，通常的需求是在页面中的其他位置（例如单击按钮），触发侧拉页面主动显示，在与 Scaffold 同级页面中，需要使用 Builder 组件来结合按钮使用，代码如下：

```
//代码清单 2-14 Scaffold 中 drawer 侧拉页面
//代码路径 lib/code2/code209_Scaffold_Dwaber.dart
Builder buildCustomBuilder() {
  return Builder(
    //使用 Builder 包裹后，Scaffold.of 中使用 Builder 中回调的 context
    //这个 context 的父级是 ScaffoldState
    builder: (BuildContext context) {
      return IconButton(
        icon: Icon(Icons.access_alarm),
```

```
          //单击事件
          onPressed: () {
            //打开左侧拉页面
            Scaffold.of(context).openDrawer();
            //打开右侧拉页面
            Scaffold.of(context).openEndDrawer();
          },
        );
      },
    );
}
```

2.2.4　常用底部导航菜单栏

在 Scaffold 中，通过 bottomNavigationBar 属性来配置页面底部导航栏，基本使用代码如下：

```
//代码清单 2-15 Scaffold 中 BottomNavigation 底部导航栏
//代码路径 lib/code2/code210_Scaffold_BottomNavigation.dart
class Exam210HomePage extends StatefulWidget {
  const Exam210HomePage({Key? key}) : super(key: key);
  @override
  State<Exam210HomePage> createState() => _Exam210HomePageState();
}

class _Exam210HomePageState extends State<Exam210HomePage> {
  @override
  Widget build(BuildContext context) {
    //Scaffold 用来搭建页面的主体结构
    return Scaffold(
      //页面的头部
      appBar: AppBar(title: const Text("标题")),
      //页面的主内容区
      //可以是单独的 StatefulWidget,也可以是当前页面构建的,如 Text 文本组件
      //可以结合 TabBarView 和 PageView 实现标签面的切换
      body: const Center(child: Text("body 内容区域")),
      bottomNavigationBar: buildBottomNavigation(),
    );
  }

  //选中当前标签的索引
  int _tabIndex = 0;
  //构建底部导航栏
  BottomNavigationBar buildBottomNavigation() {
```

```
    //创建一个 BottomNavigationBar
    return BottomNavigationBar(
      items: const <BottomNavigationBarItem>[
        //icon 用来配置显示的图标,读者可以在这里使用其他图片替换
        BottomNavigationBarItem(icon: Icon(Icons.home), label: "首页"),
        BottomNavigationBarItem(icon: Icon(Icons.message), label: "消息"),
        BottomNavigationBarItem(icon: Icon(Icons.people), label: "我的"),
      ],
      type: BottomNavigationBarType.fixed,//显示效果
      currentIndex: _tabIndex,//当前选中的页面
      backgroundColor: Colors.white, //导航栏的背景颜色
//fixedColor: Colors.deepPurple,//选中时图标与文字的颜色
      selectedItemColor: Colors.blue,//选中时图标与文字的颜色
      unselectedItemColor: Colors.grey, //未选中时图标与文字的颜色
      iconSize: 24.0,//图标的大小
      //单击事件
      onTap: (index) {
        setState(() {
          _tabIndex = index;
        });
      },
    );
  }
```

运行结果如图 2-6 所示。

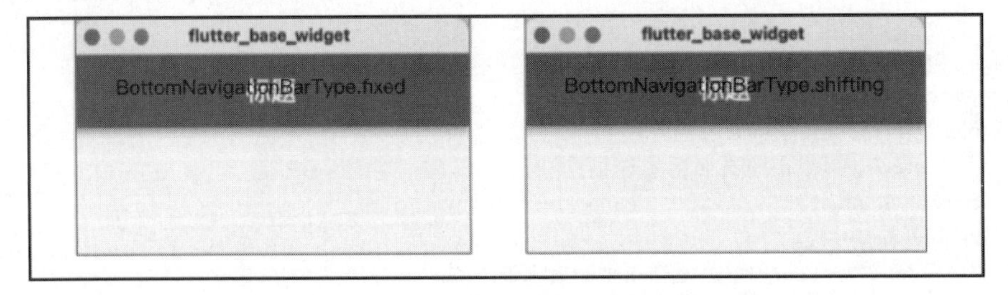

图 2-6 BottomNavigationBar 底部菜单栏效果图

BottomAppBar 结合 Scaffold 的 bottomNavigationBar 属性也可以实现自定义底部导航栏效果,如底部导航栏小图标上面需要显示小红点提示,BottomAppBar 的基本使用代码如下:

```
//代码清单 2-16 01 Scaffold 中 BottomAppBar 底部导航栏
//代码路径 lib/code2/code211_Scaffold_BottomAppBar.dart
BottomAppBar buildBottomAppBar() {
  //创建一个 BottomAppBar
  return BottomAppBar(
```

```
    child: Theme(
      data: ThemeData(
        //单击时的高亮颜色
        highlightColor: Colors.blueGrey[600],
        //水波纹颜色
        splashColor: Colors.grey,
      ),
      //用来配置 TabBar 的背景颜色
      child: Material(
        color: Colors.grey[300],
        child: TabBar(
          labelColor: Colors.blue,          //选中的 Tab 图标与文字的颜色
          //其他未选中的 Tab 图标与文字的颜色
          unselectedLabelColor: Colors.blueGrey,
          tabs: tabWidgetlist,              //所有的 Tab
          controller: _tabController,       //联动控制器
          //下画线的颜色
          indicatorColor: Colors.grey[300],
          //下画线的高度
          indicatorWeight: 1.0,
        ),
      ),
    ),
  );
}
```

运行结果如图 2-7 所示。

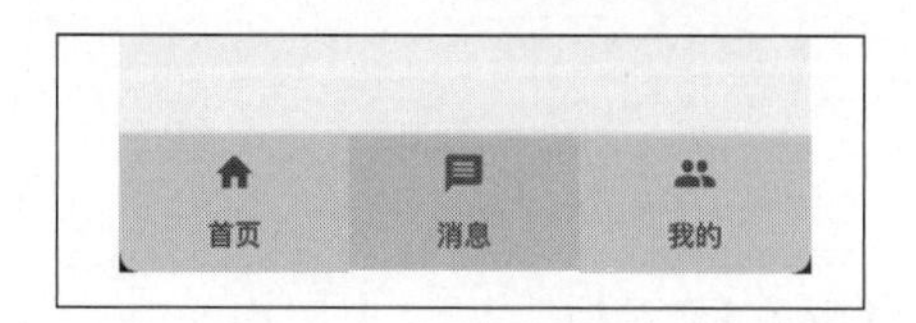

图 2-7 BottomAppBar 底部菜单栏效果图

在本实例中，通过 BottomAppBar 结合 TabBarView 实现标签页面切换，两者通过 TabController 实现联动，代码如下：

```
//代码清单 2-16 Scaffold 中 BottomAppBar 底部导航栏
//代码路径 lib/code2/code211_Scaffold_BottomAppBar.dart
class Exam211HomePage extends StatefulWidget {
  const Exam211HomePage({Key? key}) : super(key: key);
  @override
  State<Exam211HomePage> createState() => _Exam211HomePageState();
}
```

```
class _Exam211HomePageState extends State<Exam211HomePage>
    with SingleTickerProviderStateMixin {
  //底部导航栏使用数据
  List<Tab> tabWidgetlist = [
    const Tab(text: "首页", icon: Icon(Icons.home)),
    const Tab(text: "消息", icon: Icon(Icons.message)),
    const Tab(text: "我的", icon: Icon(Icons.people)),
  ];
  //子页面
  List<Widget> bodyWidgetList = [
    ItemPage(0),
    ItemPage(1),
    ItemPage(2),
  ];

  //创建控制器,late 关键字声明非 null
  late TabController _tabController;

  @override
  void initState() {
    super.initState();

    //创建控制器
    _tabController = TabController(
        initialIndex: 0,          //初始化显示的页面
        length: 3,                //页面个数
        vsync: this);
  }

  @override
  Widget build(BuildContext context) {
    //Scaffold 用来搭建页面的主体结构
    return Scaffold(
      //页面的头部
      appBar: AppBar(title: const Text("标题")),
      //页面的主内容区
      //可以结合 TabBarView 和 PageView 实现标签面的切换
      body: TabBarView(
        controller: _tabController,
        children: bodyWidgetList,
      ),
      bottomNavigationBar: buildBottomAppBar(),
    );
  }
}
```

对于 BottomAppBar 的更多应用，读者可查看本书 11.3.6 节与 12.2.1 节的应用篇内容。

2.2.5 小提示框 SnackBar

SnackBar 用于在屏幕的底部显示便捷提示消息，代码如下：

```
//代码路径 lib/code2/code219_Button.dart
ElevatedButton buildElevatedButton(BuildContext context) {
  return ElevatedButton(
    onPressed: () {
      final snackBar = SnackBar(
        content: Text('提示内容 '),
        backgroundColor: Colors.black54, //背景颜色
        elevation: 10, //阴影高度
        shape: RoundedRectangleBorder(//背景样式
          borderRadius: BorderRadius.circular(10),
        ),
        behavior: SnackBarBehavior.floating, //显示模式
        action: SnackBarAction(//交互操作按钮
          label: '知道了',
          onPressed: () {
            //移动显示
            ScaffoldMessenger.of(context).removeCurrentSnackBar();
          },
        ),
        duration: Duration(seconds: 5),
        onVisible: () {
          //显示出来的回调
        },
      );

      //Find the ScaffoldMessenger in the widget tree
      //and use it to show a SnackBar.
      ScaffoldMessenger.of(context).showSnackBar(snackBar);
    },
    child: const Text('单击显示 SnackBar'),
  );
}
```

2.3 Text 用来显示文件段落

在 Flutter 中，Text 组件用于显示文本，类似 Android 中的 TextView 和 iOS 中的 UILabel，代码如下：

```
new Text('这里是文本')
```

2.3.1 Text 文本的常用属性配置

在 Text 组件中通过 textAlign 来配置文字的对齐方式，取值类型为 TextAlign 枚举类型，可取值如表 2-1 所示。

表 2-1　文本对齐方式取值简述

类　别	描　述	类　别	描　述
TextAlign. left	左对齐	TextAlign. right	右对齐
TextAlign. center	居中对齐	TextAlign. start	开始方向对齐
TextAlign. justify	单行文字占满空间	TextAlign. end	结束方向对齐

可以通过设置 Text 组件的 softWrap 属性来控制其是否自动换行，当 softWrap 为 false 且 Text 中的文本超出父组件配置的宽度时，不会自动换行，而会直接被裁剪掉，如图 2-8 所示，对应代码如下：

```
Text(
  "早起的年轻人 for your whole life, it's right there by your side all along.",
  style: TextStyle(fontSize: 25),//字体样式
  //设置是否自动换行默认值为 true;自动换行
  //设置为 false,不会自动换行
  softWrap:true,
maxLines: 1,//设置最大显示行数
overflow: TextOverflow.ellipsis,//超出一行后显示省略号
)
```

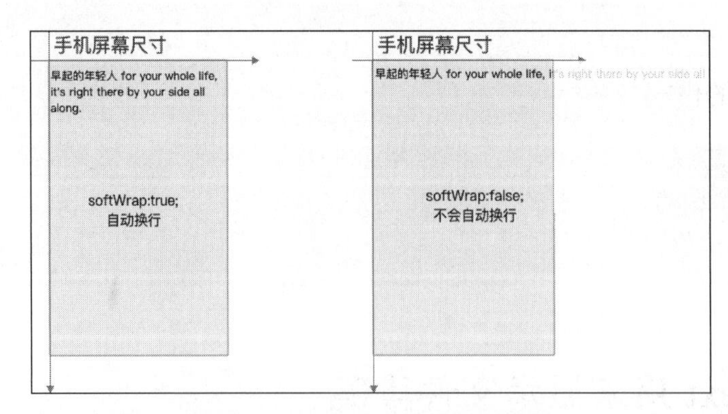

图 2-8　Text 文本自动换行效果图

Text 的 maxLines 可用来配置最多显示的行数，overflow 属性用来限制超出部分的显示模式，可取值描述如下：

```
enum TextOverflow {
  //Clip the overflowing text to fix its container.
//超出宽度的文本部分直接裁剪
  clip,
  //Fade the overflowing text to transparent.
//超出宽度的文本部分以透明渐变方式结尾
  fade,
  //Use an ellipsis to indicate that the text has overflowed.
//超出宽度的文本部分以省略号的方式结尾
  ellipsis,
  //Render overflowing text outside of its container.
//超出的部分照常显示,效果与clip的效果差不多,只不过是clip直接裁剪掉了多余的部分
visible,
}
```

2.3.2 TextStyle用来配置文本显示样式

在文本显示组件Text中,通过style属性来配置显示文本样式,style的取值为TextStyle类型,TextStyle的常用类型配置如表2-2表示。

表2-2 TextStyle文本样式配置概述表

类　别	描　述	类　别	描　述
color	文本的颜色	backgroundColor	Text文本的背景色
fontSize	字体大小配置	fontWeight	字体粗细配置
fontStyle	配置常规体、斜体	letterSpacing	字条之间的间隔
wordSpacing	单词之间的间隔	shadows	字体阴影设置
decoration	如下画线、删除线	decorationThickness	装饰线的粗细
fontFamily	自定义字体		

TextStyle的代码如下:

```
//代码清单 2-17-1 Text 文本的基本使用
//代码路径 lib/code2/code212_Text.dart
Text(
    "早起的年轻人 ",
    style: TextStyle(
          fontSize: 25,//文字大小
          //将文字的粗细规则设置为常规体
          fontWeight: FontWeight.normal,
          //斜体 FontStyle.italic ;常规 FontStyle.normal
          fontStyle: FontStyle.italic,
          //字符与字符之间的间隔
```

```
            letterSpacing: 12.0,
            //单词之间的间隔
            wordSpacing: 20.0,
  ),
  )
```

Text 通过 TextStyle 设置文本删除线，代码如下：

```
//代码清单 2-17-2 Text 删除线
//代码路径 lib/code2/code212_Text.dart
Text(
  "早起的年轻人奋斗青年",
  style: TextStyle(
    fontSize: 16,
    //设置中间删除装饰样式
    decoration: TextDecoration.lineThrough,
    //将删除线配置为红色
    decorationColor: Colors.red,
    //双线样式
    decorationStyle: TextDecorationStyle.double,
    //加粗
    decorationThickness: 2,
  ),
)
```

AnimatedDefaultTextStyle 动画样式切换，通过动画过渡的方式来切换文本的显示样式，旨在提升用户视觉体验效果，代码如下：

```
//代码清单 2-17-3 Text 动画样式使用
  //代码路径 lib/code2/code212_Text.dart
  bool _isSelected = false;
  //构建动画样式组件
  Widget buildAnimatedDefaultTextStyle() {
    //例如文本的颜色发生变化，这种颜色变化是通过动画效果来渐变的
    return AnimatedDefaultTextStyle(
      //设置 Text 中的文本样式
      //当样式有改变时会以动画的方式过渡切换
      style: _isSelected
          ? const TextStyle( color: Colors.red)
          : const TextStyle(color: Colors.black),
      //动画切换的时间
      duration: const Duration(milliseconds: 200),
      curve: Curves.bounceInOut,//动画执行插值器
      textAlign: TextAlign.start,//文本对齐方式
      softWrap: true, //文本是否应该在软换行符处换行
```

```
        //超过文本行数区域的裁剪方式
        overflow: TextOverflow.ellipsis,//设置为省略号
        maxLines: 1, //最大显示行数
        //当样式有修改触发动画时
        //动画执行结束的回调
        onEnd: () {
          print("动画执行结束");
        },
        //文本组件
        child: const Text("Hello, Flutter"),
      );
    }
```

2.3.3 RichText 实现多种文本风格组合显示

RichText 结合 TextSpan 可以实现文本段落中的文字多样式显示，结合使用 WidgetSpan 可以将更为复杂的 Widget 展示在段落中间，代码如下：

```
//代码清单 2-18 RichText 文本的基本使用
//代码路径 lib/code2/code213_RichText.dart
Widget buildRichText() {
  return RichText(
    //文字区域
    text: TextSpan(
      text: "登录即代表同意",
      style: const TextStyle(color: Colors.grey),
      children: [
        //文本
        TextSpan(
            text: "《用户注册协议》",
            style: const TextStyle(color: Colors.blue),
            //单击事件
            recognizer: TapGestureRecognizer()
              ..onTap = () {
                print("单击用户协议");
              }),
        //文本
        const TextSpan(
          text: "与",
          style: TextStyle(color: Colors.grey),
        ),
        //其他 Widget,这里使用的是图片 Image
        WidgetSpan(
          //对齐方式
```

```
            alignment: PlaceholderAlignment.middle,
            //这里是中间显示的图片，也可以是其他任意的 Widget
            child: Image.asset(
              "assets/images/warning_icon.png",
              width: 20,
              height: 20,
            ),
          ),
        ],
      ),
    );
  }
```

运行效果如图 2-9 所示。

登录即代表同意《用户注册协议》与

图 2-9　RichText 文本使用效果图

2.3.4　SelectableText 实现文件显示

使用 Text、RichText 可以便捷地显示文本，但是它们不允许用户选择文本，SelectableText 可以用来显示文本，也可以用来使用用户单击快速选择的文本单词，代码如下：

```
SelectableText(
    'Flutter 早起的年轻人正在实践',
    style: TextStyle(color: Colors.blue, fontSize: 40),
    textAlign: TextAlign.center,
)
```

默认情况下，当用户单击文本时，它不会显示光标。将 showCursor 的值设置为 true，用来显示光标，也可以自定义显示光标的颜色等，代码如下：

```
SelectableText(
    'Flutter 早起的年轻人正在实践',

    showCursor: true,
    cursorWidth: 5,
    cursorColor: Colors.red,
    cursorRadius: Radius.circular(5),
)
```

默认情况下，当选择激活时，它会显示“复制”和“选择所有”选项，可以使用toolbarOptions属性来传递一个toolbarOptions的实例并以此修改，代码如下：

```
SelectableText(
    'Flutter 早起的年轻人正在实践',
    onTap: () => print('The text is tapped'),//文本单击事件
    toolbarOptions: ToolbarOptions(copy: true, selectAll: false),
)
```

SelectableText也支持RichText显示多样式文本，代码如下：

```
SelectableText.rich(
    TextSpan(
      children: <TextSpan>[
        TextSpan(text: 'hello', style: TextStyle(color: Colors.blue)),
        TextSpan(text: '早起的年轻人', style: TextStyle(color: Colors.black)),
        TextSpan(text: '听说你在研究, style: TextStyle(color: Colors.red)),
      ],
    ),
    style: TextStyle(
        color: Colors.blue, fontWeight: FontWeight.bold, fontSize: 48),
    textAlign: TextAlign.center,
  )
```

2.4 TextField用来实现文本输入功能

在Flutter中组件TextField用来输入文本，例如用户输入用户名与密码实现登录页面，TextField的代码如下：

```
new TextField()
```

运行效果如图2-10所示，默认有一条下画线，并且默认情况下不获取输入焦点，下画线为灰色，获取输入焦点后下画线变为高亮蓝色。

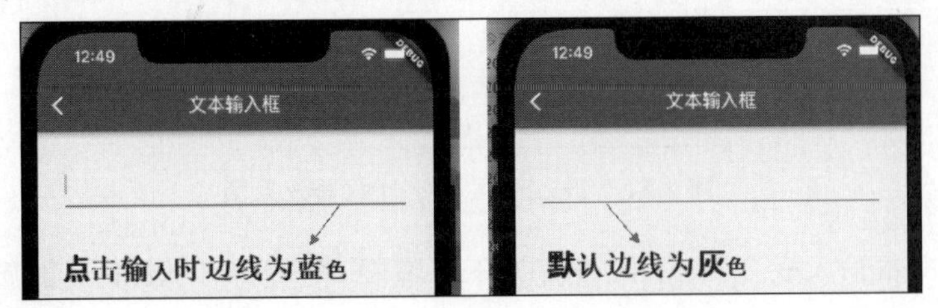

图2-10 TextField输入文本框使用效果图

2.4.1 TextField 文本输入的常用属性配置

通过将属性 autofocus 的值配置为 true，使用 TextField 默认的自动获取输入焦点，通过将属性 obscureText 配置为 true，可以隐藏输入的文本，通常在输入密码选项时使用，此时需要将 TextField 的最大行数 maxLines 的值设置为 1。

可通过 maxLength 属性来配置文本框最多可输入的字符数，当设置了 maxLength 属性时，在文本输入框 TextField 的右下角默认会出现输入文字计数器，如图 2-11 所示，有些使用场景往往不显示右下角的文本输入计数器，但是依然需要限制输入的字数，代码如下：

```
//代码清单 2-19 TextField 输入限制
//代码路径 lib/code2/code214_TextField.dart
Widget buildTextField4() {
  //表情内容
  const String REGEX_EMOJI =
      "[^\\u0020-\\u007E\\u00A0-\\u00BE\\u2E80-\\uA4CF\\uF900-\\uFAFF\\uFE30-\\uFE4F\\uFF00-\\uFFEF\\u0080-\\u009F\\u2000-\\u201f\r\n]";
  return TextField(
    //设置键盘的类型
    keyboardType: TextInputType.phone,
    //输入文本格式过滤
    inputFormatters: [
      //输入的内容长度为 11 位
      LengthLimitingTextInputFormatter(11),
      //禁止表情输入
      FilteringTextInputFormatter.deny(RegExp(REGEX_EMOJI)),
      //只能输入数字
      FilteringTextInputFormatter.digitsOnly,                  //数字,只能是整数
      //只允许输入数字正则
      FilteringTextInputFormatter.allow(RegExp("[0-9.]")),     //数字包括小数
      FilteringTextInputFormatter.allow(RegExp("[a-zA-Z]")),   //只允许输入字母
    ],
  );
}
```

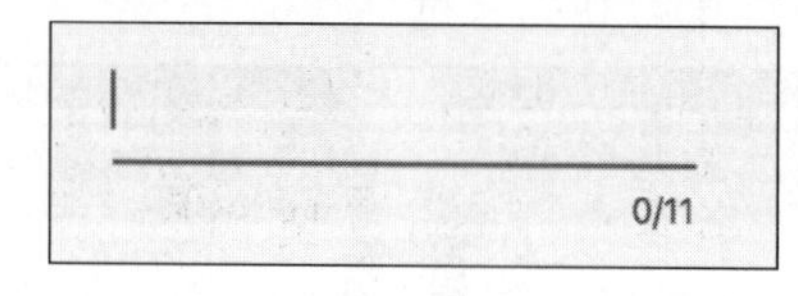

图 2-11 TextField 文字计数器效果图

如果想将文本输入框 TextField 设置为不可编辑，则可将 TextField 的属性 enabled 的值设置为 false，也可将 TextField 的只读属性 readOnly 设置为 true，两者的区别是使用的

边框样式不一样。

2.4.2 文本输入框的边框配置

TextFiled 默认情况下有一个底部边框效果，在实际项目开发中，会有很多情况无法满足设计要求，通过属性 decoration 配置 InputDecoration 可以定义任意想要的效果，InputDecoration 中的 border 用来设置边框的样式，可取值包括 InputBorder. none（无边框）、OutlineInputBorder（圆角边框）、UnderlineInputBorder（下边框默认使用），如实现去掉底部边框线，代码如下：

```
TextField(
  //边框样式设置
  decoration: InputDecoration(
    //设置无边框
    border: InputBorder.none,
  ),
);
```

默认的 OutlineInputBorder 创建的四周的四个角的弧度为 4.0，如图 2-12 所示，可配置修改四个角的弧度，代码如下：

```
TextField(
  //边框样式设置
  decoration: InputDecoration(
    //设置上下左右都有边框
    border: OutlineInputBorder(
      //设置边框四个角的弧度
      borderRadius: BorderRadius.all(Radius.circular(40)),
    ),
  ),
)
```

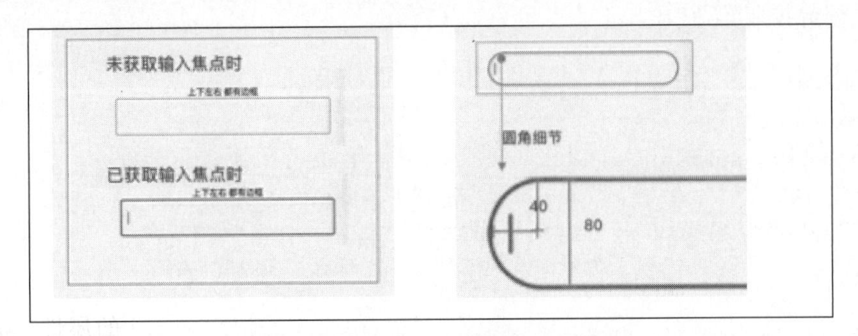

图 2-12　TextField 边框设置效果图

TextField未获取焦点时是灰色的边框，获取焦点时是蓝色的边框，此时使用的是程序默认主题中配置的颜色，在实际应用中可通过InputDecoration中相关border属性修改边框的样式，如表2-3所示。

表2-3 InputDecoration边框样式概述表

类　　别	描　　述
border	通用样式
enabledBorder	TextField属性enabled为true时，表示当前的输入框为激活状态时使用的边框样式
disabledBorder	TextField属性enabled为true时，表示当前的输入框为未激活状态时的边框样式
errorBorder	输入框错误提示边框样式
fontFamily	输入框获取输入焦点时的边框样式

TextField的decoration属性中的InputDecoration中的hintText用来配置输入提示语，当输入框中有内容时，显示输入内容；当输入框中无内容时，显示hintText配置的文本，代码如下：

```
TextField(
    decoration: InputDecoration(
        //提示文本
        hintText: "请输入姓名",
        //提示文本样式
    hintStyle: TextStyle(color: Colors.grey,fontSize: 12.0)
    ),
),
```

2.4.3 TextField输入内容的监听与获取

通过TextEditingController可实现为绑定的输入框TextField预设内容、获取TextField中输入的内容、监听TextField中文字输入变化等。

首先创建控制器，代码如下：

```
//创建文本控制器实例
//创建方式一
final TextEditingController _editingController = TextEditingController();
//创建方式二,其中 text 是预设置的内容,输入光标在最前面
final TextEditingController _controller2 =
                            TextEditingController(text: "初始化内容");
```

上述创建的控制器预设显示的文本内容在输入框中，输入框的光标显示在文本最前面，为提高用户体验，需要将光标显示在文本最后，代码如下：

```
void setControText(TextEditingController controller, String flagText) {
  //用来设置初始化时显示
  TextEditingValue value = TextEditingValue(
    //用来设置文本
    text: flagText,
    //设置光标的位置
    selection: TextSelection.fromPosition(
      //用来设置文本的位置
      TextPosition(
          affinity: TextAffinity.downstream,
          //光标向后移动的长度
          offset: flagText.length),
    ),
  );
  //控制初始化时光标保持在文字最后
  controller.value = value;
}
```

然后在 TextField 中使用，代码如下：

```
new TextField(
    //绑定控制器
    controller: _editingController,
)
```

最后通过控制器获取 TextField 中输入的内容，代码如下：

```
String result = _editingController.text;
```

在实际项目中用户的操作习惯是当输入键盘是弹出状态时，用户单击屏幕的空白处，键盘要隐藏，实现思路是在页面根视图添加一个单击事件监听，在事件监听回调中隐藏键盘，同时使输入框失去焦点，对应代码如下：

```
//代码清单 2-20 TextField FocusNode 输入焦点控制
//代码路径 lib/code2/code217_TextField.dart
class Exam217HomePage extends StatefulWidget {
  const Exam217HomePage({Key? key}) : super(key: key);

  @override
  State<Exam217HomePage> createState() => _Exam217HomePageState();
}

class _Exam217HomePageState extends State<Exam217HomePage> {
  //创建 FocusNode 对象实例
```

```
  final FocusNode _focusNode = FocusNode();
  @override
  void initState() {
    super.initState();
    //添加 listener 监听可以在这里做输入数据的校验
    //对应的 TextField 在失去或者获取焦点时都会回调此监听
    _focusNode.addListener(() {
      if (_focusNode.hasFocus) {
        print('得到焦点');
      } else {
        print('失去焦点');
      }
    });
  }

  @override
  Widget build(BuildContext context) {
    return GestureDetector(
      onTap: () {
        //隐藏键盘
        SystemChannels.textInput.invokeMethod('TextInput.hide');
      },
      child: Scaffold(
        //页面的头部
        appBar: AppBar(title: const Text("标题")),
        //页面的主内容区
        body: Center(
          child: SizedBox(
            width: 224,
            child: TextField(
              focusNode: _focusNode,
              onEditingComplete: (){
                //单击 Enter 键时的回调
              },
            ),
          ),
        ),
      ),
    );
  }

//页面销毁
  @override
  void dispose() {
    super.dispose();
    //释放
```

```
    _focusNode.dispose();
  }
}
```

例如单击按钮时，可以通过 FocusNode 来使输入框 TextField 主动获取输入焦点，代码如下：

```
//获取焦点
void getFocusFunction(BuildContext context){
  FocusScope.of(context).requestFocus(focusNode);
}

//失去焦点
void unFocusFunction(){
  focusNode.unfocus();
}
//隐藏键盘而不丢失文本字段焦点
void hideKeyBoard(){
  SystemChannels.textInput.invokeMethod('TextInput.hide');
}
```

2.5 按钮实现用户单击事件

Container 容器及其他显示类的 Widget 结合手势处理 Widget 可以实现单击事件功能，Button 是 Flutter 中专门用来实现单击事件的组件，本节概述 Flutter 基础按钮 MaterialButton、TextButton、ElevatedButton、ShakeAnimationWidget 抖动动画按钮、ActionChip 胶囊组合按钮、DraggableFloatingActionButton 悬浮拖动按钮。

2.5.1 常用按钮 Button 概述

在 Flutter 中，Button 常用于处理用户的单击事件，常用的按钮有 MaterialButton、TextButton、ElevatedButton。

ElevatedButton 创建出的按钮，默认情况下会带有圆角边框背景，当单击时，会有水波纹颜色与阴影，如图 2-13 所示；TextButton 创建出的按钮默认无背景填充，也无阴影高度，是文本类型按钮。

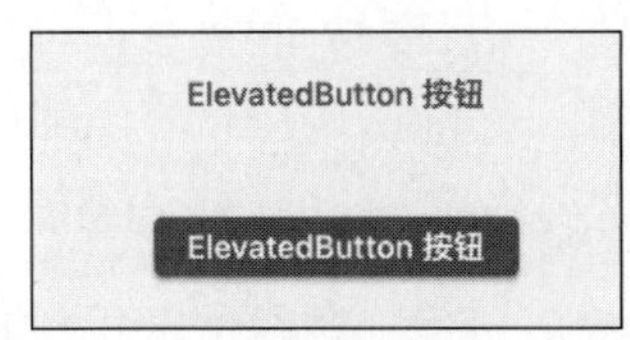

图 2-13 基本 Button 使用效果图

代码如下：

```
//竖直方向排开子 Widget
Column(
  children: [

    TextButton(
      child: const Text("ElevatedButton 按钮"),
      onPressed: () {
        //单击事件
      },
    ),
    const SizedBox(height: 44, ),

    ElevatedButton(
      child: const Text("ElevatedButton 按钮"),
      onPressed: () {
        //单击事件
      },
    )
  ],
)
```

ElevatedButton 与 TextButton 自定义样式，如按钮的背景颜色、字体等，可通过属性 style 来配置 ButtonStyle 实现，ButtonStyle 的代码如下：

```
//代码清单 2 - 22 ButtonStyle 样式
//代码路径 lib/code2/code219_Button.dart
ButtonStyle buileButtonStyle() {
  return ButtonStyle(
    //定义文本的样式,这里设置的颜色不起作用
    textStyle:
        MaterialStateProperty.all(TextStyle(fontSize: 18, color: Colors.red)),
    //设置按钮上字体与图标的颜色
    //foregroundColor: MaterialStateProperty.all(Colors.deepPurple),
    //以更优美的方式设置
    foregroundColor: MaterialStateProperty.resolveWith(
      (states) {
        if (states.contains(MaterialState.focused) &&
            !states.contains(MaterialState.pressed)) {
          //获取焦点时的颜色
          return Colors.blue;
        } else if (states.contains(MaterialState.pressed)) {
          //按下时的颜色
          return Colors.deepPurple;
```

```
          }
          //默认状态使用灰色
          return Colors.grey;
        },
      ),
      //背景颜色
      backgroundColor: MaterialStateProperty.resolveWith((states) {
        //设置按下时的背景颜色
        if (states.contains(MaterialState.pressed)) {
          return Colors.blue[200];
        }
        //默认不使用背景颜色
        return null;
      }),
      //设置水波纹颜色
      overlayColor: MaterialStateProperty.all(Colors.yellow),
      //设置阴影不适用于这里的 TextButton
      elevation: MaterialStateProperty.all(0),
      //设置按钮内边距
      padding: MaterialStateProperty.all(EdgeInsets.all(10)),
      //设置按钮的大小
      minimumSize: MaterialStateProperty.all(Size(200, 100)),

      //设置边框
      side: MaterialStateProperty.all(BorderSide(color: Colors.grey, width: 1)),
      //外边框装饰会覆盖 side 配置的样式
      shape: MaterialStateProperty.all(StadiumBorder()),
    );
  }
```

MaterialButton 是 Material Design 风格按钮，需要在 MaterialApp 组件中使用，代码如下：

```
  //代码清单 2-23
  //代码路径 lib/code2/code220_MaterialButton.dart.dart
  MaterialButton builеButton() {
    return MaterialButton(
      //按钮的背景
      color: Colors.blue,
      //按钮上显示的文字
      child: const Text('登录'),
      //配置按钮上文本的颜色
      textColor: Colors.deepOrange,
      //未设置单击时的背景颜色，也就是当默认 onPressed: null 时显示这种颜色
      disabledColor: Colors.yellow,
```

```
      //按钮单击下的颜色
      highlightColor: Colors.deepPurple,
      //水波纹的颜色
      splashColor: Colors.green,
      //按钮的阴影
      elevation: 10,
      //按钮按下时的阴影高度
      highlightElevation: 20,
      //未设置单击时的阴影高度
      disabledElevation: 5.0,
      //按钮的单击事件
      onPressed: () {
        print("单击了按钮");
      },
    );
  }
```

MaterialButton 的 shape 属性用来配置边框样式，可通过 RoundedRectangleBorder 来配置圆角矩形边框，代码如下：

```
MaterialButton(
  …

  //用来设置圆角矩形边框
  shape: const RoundedRectangleBorder(
      //边框样式
      side: BorderSide(
          color: Colors.deepOrange, width: 2.5, style: BorderStyle.solid),
      //边框的圆角
      borderRadius: BorderRadius.all(Radius.circular(20))),
);
```

2.5.2 抖动按钮

插件 shake_animation_widget 是笔者在大量项目实战中集成的几个特殊动画效果合集，如抖动组件 ShakeAnimationWidget、向上弹出的动画菜单、抖动的文本、进度提交的按钮、类似开源中国底部圆形弹出的动画菜单等，首先需要在项目的配置文件 pubspec.yaml 中添加依赖，代码如下：

```
dependencies:
#效果图查看 pub 仓库说明文档
shake_animation_widget: ^3.0.1
```

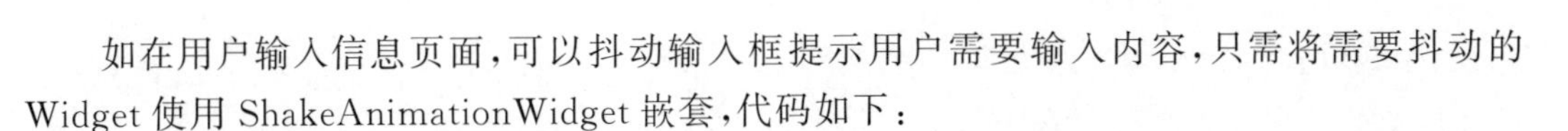

如在用户输入信息页面，可以抖动输入框提示用户需要输入内容，只需将需要抖动的Widget使用ShakeAnimationWidget嵌套，代码如下：

```
//代码清单 2-25-1 抖动组件
//抖动动画控制器
final ShakeAnimationController _shakeAnimationController =
    ShakeAnimationController();

//构建抖动效果
ShakeAnimationWidget buildShakeAnimationWidget() {
  return ShakeAnimationWidget(
    //抖动控制器
    shakeAnimationController: _shakeAnimationController,
    //微旋转的抖动
    shakeAnimationType: ShakeAnimationType.SkewShake,
    //设置不开启抖动
    isForward: false,
    //默认为 0 无限执行
    shakeCount: 0,
    //抖动的幅度,取值范围为[0,1]
    shakeRange: 0.2,
    //执行抖动动画的子 Widget
    child: RaisedButton(
      child: const Text(
        '测试',
        style: TextStyle(color: Colors.white),
      ),
      onPressed: () {
        //判断抖动动画是否正在执行
        if (_shakeAnimationController.animationRunging) {
          //停止抖动动画
          _shakeAnimationController.stop();
        } else {
          //开启抖动动画
          //参数 shakeCount 用来配置抖动次数
          //通过 controller start()方法,默认为 1
          _shakeAnimationController.start(shakeCount: 1);
        }
      },
    ),
  );
}
```

可以通过ShakeTextAnimationWidget组件实现在一行文本中每个文字都在随机抖动的功能，代码如下：

```
//代码清单 2-25-2 抖动的文本
buildTextAnimationWidget() {
  return ShakeTextAnimationWidget(
    //需要设置抖动效果的文本
    animationString: "这里是文字的抖动",
    space: 1.0,               //字符间距
    runSpace: 10,             //行间距
    //文字的样式
    textStyle: const TextStyle(
      //文字的大小
      fontSize: 25,
    ),
    //抖动次数
    shakeCount: 0,
  );
}
```

一个表单提交场景，如用户的登录页面，单击“登录”按钮，按钮动画转换为一个小圆圈进度，如果提交失败，则再转换回按钮样式，可以通过 AnimatedStatusButton 实现，代码如下：

```
//代码清单 2-25-3 动画过渡按钮
AnimatedStatusController animatedStatusController =
                              AnimatedStatusController();
//切换样式的动画按钮
Widget buildAnimatedStatusButton() {
  return AnimatedStatusButton(
    animatedStatusController: animatedStatusController,     //控制器
    width: 220.0,                                            //显示按钮的宽度
    height: 40,                                              //显示按钮的高度
    milliseconds: 1000,                                      //动画交互时间
    buttonText: '提交',
    backgroundNormalColor: Colors.white,                     //背景颜色
    borderNormalColor: Colors.deepOrange,                    //边框颜色
    textNormalCcolor: Colors.deepOrange,                     //文字颜色
    //单击回调
    clickCallback: () async {
      print("单击事件回调");
      //模拟耗时操作
      await Future.delayed(Duration(milliseconds: 4000));
      //返回值为 false,会一直在转圈圈
      //返回值为 true,会回到默认显示样式
      return Future.value(false);
    },
  );
}
```

可以通过 BottomRoundFlowMenu 实现当单击悬浮按钮时显示一个底部圆形导航栏菜单选项，代码如下：

```
//代码清单 2-25-4 底部弹出的圆形导航栏菜单
class _ExampleState extends State<Example309> {
  //构建菜单所使用的图标
  List<Icon> iconList = [
    Icon(Icons.android, color: Colors.blue, size: 18),
    Icon(Icons.image, color: Colors.red, size: 18),
    Icon(Icons.find_in_page, color: Colors.orange, size: 18),
    Icon(Icons.add, color: Colors.lightGreenAccent, size: 28),
  ];

  @override
  Widget build(BuildContext context) {
    return Scaffold(
      backgroundColor: Colors.grey[200],
      body: Container(
        //文字标签流式布局
        child: BottomRoundFlowMenu(
          //图标使用的背景
          defaultBackgroundColor: Colors.white,
          //菜单所有的图标
          iconList: iconList,
          //对应菜单项单击事件回调
          clickCallBack: (int index) {
            print("单击了 $index");
          },
        ),
      ),
    );
  }
}
```

2.5.3 ActionChip 胶囊组合按钮

在业务开发中常见的一小卡片形状如图 2-14 所示。在 Flutter 中，可通过 ActionChip 组件来便捷实现，代码如下：

```
//代码清单 2-26 ActionChip
//代码路径 lib/code2/code224_ActionChip.dart
Widget buildActionChip() {
  return ActionChip(
      label: Text('删除'),
```

```
        labelStyle: TextStyle(color: Colors.black),
        labelPadding: EdgeInsets.all(10.0),
        avatar: Icon(Icons.delete),
        onPressed: () {
          print('Processing to delete item');
        });
}
```

图 2-14　ActionChip 使用效果图

2.5.4　悬浮拖动按钮

Scaffold 组件的 floatingActionButton 属性可配置 FloatingActionButton 按钮悬浮在页面右下角，在业务开发中，时常会使用悬浮按钮，悬浮按钮在页面上可拖动移动，本节中使用 DraggableFloatingActionButton 实现，如图 2-15 所示。

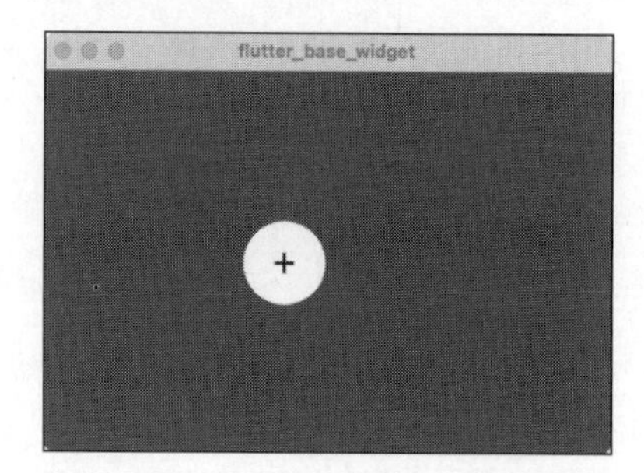

图 2-15　页面中可拖动的悬浮按钮使用效果图

代码如下：

```
//代码清单 2-27 可拖动的悬浮按钮
//代码路径 lib/code2/code225_FAB.dart
class _Exam223HomePageState extends State<Exam223HomePage> {
  //Stack 使用的 Key
  final GlobalKey _parentKey = GlobalKey();

  @override
  Widget build(BuildContext context) {
    return Scaffold(
```

```
      body: SizedBox(
        width: double.infinity,
        height: double.infinity,
        child: Stack(
          key: _parentKey,
          children: [
            Container(color: Colors.blueGrey),

            DraggableFloatingActionButton(
              child: Container(
                width: 60,
                height: 60,
                decoration: const ShapeDecoration(
                  shape: CircleBorder(),
                  color: Colors.white,
                ),
                child: const Icon(Icons.add),
              ),
              initialOffset: const Offset(120, 70),
              parentKey: _parentKey,
              onPressed: () {},
            ),
          ],
        ),
      ),
    );
  }
}
```

笔者已将 DraggableFloatingActionButton 提交到插件 shake_animation_widget 中，读者可以直接使用，本书中概述一下实现的知识点。在这里使用了父组件的 GlobalKey，主要是为了防止拖动按钮时按扭被拖出边界，根据这个键，从 currentContext 属性得到 RenderBox，通过 findRenderObject()方法可以从 RenderBox 的 size 属性中获取父组件的大小，findRenderObject()方法必须在构建树之后调用，所以需要使用 addPostFrameCallback 控件绑定后调用它，代码如下：

```
//代码清单 2-27-1 可拖动的悬浮按钮
//代码路径 lib/code2/code225_FAB.dart
class DraggableFloatingActionButton extends StatefulWidget {
  final Widget child;
  final Offset initialOffset;
  final VoidCallback onPressed;
  GlobalKey<State<StatefulWidget>> parentKey;
  DraggableFloatingActionButton({
```

```
    required this.child,
    required this.initialOffset,
    required this.onPressed,
    required this.parentKey,
  });

  @override
  State<StatefulWidget> createState() => _DraggableFloatingActionButtonState();
}

class _DraggableFloatingActionButtonState
    extends State<DraggableFloatingActionButton> {
  //拖动按钮使用的 Key
  final GlobalKey _key = GlobalKey();
  bool _isDragging = false;
  late Offset _offset;
  late Offset _minOffset;
  late Offset _maxOffset;

  @override
  void initState() {
    super.initState();
    //拖动按钮的初始位置
    _offset = widget.initialOffset;
    //添加视图监听
    WidgetsBinding.instance?.addPostFrameCallback(_initBoundary);
  }
  //页面第一帧绘制完成后调用
  void _initBoundary(_) {
    //获取组件的 RenderBox
    final RenderBox parentRenderBox =
        widget.parentKey.currentContext?.findRenderObject() as RenderBox;
    //获取拖动按钮组件的 RenderBox
    final RenderBox renderBox =
        _key.currentContext?.findRenderObject() as RenderBox;

    try {
      //分别获取两者的大小,从而计算边界
      final Size parentSize = parentRenderBox.size;
      final Size size = renderBox.size;
      setState(() {
        _minOffset = const Offset(0, 0);
        _maxOffset = Offset(
            parentSize.width - size.width, parentSize.height - size.height);
      });
    } catch (e) {
```

```
            print('catch: $e');
        }
    }
}
```

使用 Listener 监听手势滑动，Listener 的 onPointerMove 用于实现当手指在页面上滑动时实时回调，返回手指位置信息，onPointerUp 用于实现当手指抬起时回调，在这种方法中当滑动操作结束时，需要更新滑动结束标识，如果是单击事件，就回调单击事件，代码如下：

```
//代码清单 2-27-2 可拖动的悬浮按钮
@override
Widget build(BuildContext context) {
  return Positioned(
    left: _offset.dx,
    top: _offset.dy,
    child: Listener(
      onPointerMove: (PointerMoveEvent pointerMoveEvent) {
        //更新位置
        _updatePosition(pointerMoveEvent);
        setState(() {
          _isDragging = true;
        });
      },
      onPointerUp: (PointerUpEvent pointerUpEvent) {
        print('onPointerUp');
        if (_isDragging) {
          setState(() {
            _isDragging = false;
          });
        } else {
          widget.onPressed();
        }
      },
      child: Container(
        key: _key,
        child: widget.child,
      ),
    ),
  );
}
```

手指在拖动过程中，需要实时计算相对位置，以获得拖动按钮的位置，代码如下：

```
//代码清单 2-27-3 计算按钮位置
void _updatePosition(PointerMoveEvent pointerMoveEvent) {
  double newOffsetX = _offset.dx + pointerMoveEvent.delta.dx;
  double newOffsetY = _offset.dy + pointerMoveEvent.delta.dy;

  if (newOffsetX < _minOffset.dx) {
    newOffsetX = _minOffset.dx;
  } else if (newOffsetX > _maxOffset.dx) {
    newOffsetX = _maxOffset.dx;
  }

  if (newOffsetY < _minOffset.dy) {
    newOffsetY = _minOffset.dy;
  } else if (newOffsetY > _maxOffset.dy) {
    newOffsetY = _maxOffset.dy;
  }

  setState(() {
    _offset = Offset(newOffsetX, newOffsetY);
  });
}
```

2.6 其他常用 Widget 概述

MaterialApp 是程序的根视图，Scaffold 是页面的基础结构，Text 用来显示文件，TextField 用来输入文本，Button 用来实现按钮单击，这些是最基本的 Widget，也是应用程序开发起步构建基础页面的基础元素，Image 也是最常用的一个 Widget，用来加载显示图片。

页面展示内容通常是丰富多彩的，Container 可以用来嵌套各种 Widget，为其提供边框、背景、外边距、内边距、大小限制、对齐方式等样式的装饰；SizedBox 可以提供限制大小的功能；Padding 可以提供设置内边距功能；Align 可以提供对齐功能设置，Center 可以使子 Widget 居中。

2.6.1 Image 用来显示图片

在应用开发中，图片资源一般来自网络、应用程序内 asset 资源目录、手机磁盘上存储的图片、SDK 提供的 Icon 及相册等。

通过 Image.network 来加载网络图片，只需传入对应的网络图片链接，代码如下：

```
//代码清单 2-24
//代码路径 lib/code2/code221_Image.dart
String imageUrl =
    "https://img-blog.csdnimg.cn/20201031094959816.gif";
Widget buildImage() {
  return Image.network(
    imageUrl,
    //图片的填充模式
    fit: BoxFit.fill,
    //图片的宽和高
    width: 100,
    height: 100,
    //加载中的占位
    loadingBuilder: (
        BuildContext context,
        Widget child,
        loadingProgress,
        ) {
      return const Text("加载中");
    },
    //加载出错
    errorBuilder: (
        BuildContext context,
        Object error,
        stackTrace,
        ) {
      return const Text("加载出错");
    },
  );
}
```

也可以通过 NetworkImage 来配合 Image 加载网络图片，代码如下：

```
Widget buildImage2() {
    return Image(
      fit: BoxFit.fill,
      width: 100,
      height: 100,
      image: NetworkImage(imageUrl),
    );
}
```

FadeInImage 可以实现图片加载，加载时，显示一个加载动画，属性 placeholder 用于设置加载中的占位图片，可以是.gif 动图，加载成功后，以透明渐变的方式显示图片，代码如下：

```
FadeInImage.assetNetwork(
  placeholder: 'assets/loading.gif',
  image: '图片地址',
)
```

CachedNetworkImage 可以实现图片缓存及将一个 Widget 设置为加载中占位,需要添加依赖,代码如下:

```
cached_network_image: ^3.2.0
```

基本使用代码如下:

```
CachedNetworkImage(
  placeholder: (context, url) => const CircularProgressIndicator(),
  imageUrl: '',
),
```

在 Flutter 中通过 Image.asset 来加载资源目录下的图片(在 1.3.2 节中有配置图片资源目录路径),代码如下:

```
Widget buildImage3() {
  return Image.asset(
    "assets/images/banner_mang.png",
  );
}

//或者
Widget buildImage4() {
  return const Image(
    image: AssetImage(
      "assets/images/banner_mang.png",
    ),
  );
}
```

在 Image 中,通过 fit 属性来配置图片的填充模式,如表 2-4 所示。

表 2-4　Image 图片填充模式概述表

类　别	描　述
BoxFit.fill	充满父容器,并不按照比例来伸缩
BoxFit.fitHeigh	等比例地缩放,图片填满高度,宽度可能会被截断
BoxFit.fitWidth	等比例地缩放,图片填满宽度,高度可能会被截断

续表

类　别	描　述
BoxFit. cover	等比例地缩放，直到图片的宽和高都充满整个控件为止，图片可以超出控件的范围，显示不完整
BoxFit. contain	图片成比例地伸缩，直到图片的高或者宽填满控件为止
BoxFit. scaleDown	图片可以完整显示，但是可能不能填满
BoxFit. none	不进行任何缩放，保持原生态。当控件过小时会造成显示不完整。默认会居中显示图片

2.6.2 Container 用来设置边框样式

Container 组件是用来放置 Widget 的容器，可以设置 padding、margin、位置、大小、边框和阴影等参数，Container 可理解为 Flutter 中的盒子模型，代码如下：

```
Container(
  width: 220,//宽
  height: 120,//高
  padding: const EdgeInsets.only(left: 12, right: 24),//内边距
  margin: const EdgeInsets.all(40),//外边距
  child: const Text("早起的年轻人"),
)
```

Container 可实现多种多样的装饰效果，如圆角边框、渐变背景、体育场背景等，通过 decoration 属性来配置使用，color 属性与 decoration 不能同时配置，否则会有冲突，decoration 最常用的是 BoxDecoration 盒模型装饰，其常用属性说明如表 2-5 所示。

表 2-5　BoxDecoration 属性概述表

类　别	描　述	类　别	描　述
border	配置边框、颜色及线的样式	borderRadius	边框四个角的圆角度
color	Container 的填充色	boxShadow	阴影
gradient	渐变过渡色填充		

通过 Container 的 BoxDecoration 实现圆角边框，如图 2-16 所示，对应的实现代码如下：

```
//代码清单 2 - 21 Container 实现圆角边框
//代码路径 lib/code2/code218_Container.dart
Container buildBorderAndColor() {
  return Container(
    //子 Widget 居中对齐
    alignment: Alignment.center,
```

```
      width: 200,
      height: 100,
      //主要是这里
      decoration: BoxDecoration(
        //边框颜色与宽度
        border: Border.all(color: const Color(0x2A00FFFF), width: 3.5),
        //边框圆角四个角的全部配置如表 2-5 所示
        borderRadius: const BorderRadius.all(
          Radius.circular(20),
        ),
        //填充色
        color: Color(0x559E9E9E),
      ),
      child: Text("配置边框与填充色"),
    );
  }
```

图 2-16 圆角边框效果图

通过 Container 的 BoxDecoration 的 boxShadows 属性来配置阴影，代码如下：

```
BoxDecoration(
  color: Colors.blue,
  //可配置多组阴影效果
  boxShadow: [
    //配置阴影
    BoxShadow(
      color: Colors.red,              //阴影的颜色
      offset: Offset(10.0, 10.0),     //阴影在 x 轴与 y 轴上的偏移量
      blurRadius: 20.0,               //模糊半径
      spreadRadius: 1,                //阴影的延伸量
    )
  ],
)
```

通过 Contrainer 的 BoxDecoration 的 gradient 属性来配置渐变过渡的样式，包括线性渐变（LinearGradient）、扫描渐变（SweepGradient）、中心辐射（RadialGradient），代码如下：

```
BoxDecoration(
  //线性渐变
  gradient: LinearGradient(
    //渐变过渡的颜色体系
    colors: [Colors.blue, Colors.yellow, Colors.red],
    //过渡的开始位置
    begin: FractionalOffset(0, 0.5),
    //过渡的结束位置
    end: FractionalOffset(1, 0.5),
  ),)
```

添加扫描渐变(SweepGradient),代码如下:

```
BoxDecoration(
  //扫描渐变
  gradient: SweepGradient(
    //渐变过渡的颜色体系
    colors: [Colors.blue, Colors.yellow, Colors.red],
    //过渡的开始角度,默认为 0
    startAngle: 0.0,
    //过渡的结束角度,默认为 1×3.1415926,也就是在默认情况下为 0~360°
    endAngle: 1 * 3.14,
    //中心点默认就是当前 Container 的中心
    //也可以通过 Alignment.center 来配置
    center: FractionalOffset(0.5, 0.5),
  ),
)
```

添加中心辐射(RadialGradient),代码如下:

```
BoxDecoration(
    //环形渐变
    gradient: RadialGradient(
      //渐变过渡的颜色体系
      colors: [Colors.red,Colors.yellow, Colors.black, ],
      //渐变颜色的过渡半径
      //由 Container 的短边决定
      radius: 0.5,
      //过渡半径之外的颜色填充模式
      //默认为 clamp,直接使用 colors 中配置的最后一种颜色填充
      //如这里配置的黑色
      tileMode: TileMode.clamp,
      //中心点默认为当前 Container 的中心
      //也可以通过 Alignment.center 配置
      center: FractionalOffset(0.5, 0.5),
```

```
    ),
  ),
  child: Text("配置渐变颜色"),
)
```

2.6.3 ShaderMask 实现利用轮廓

如需要使用一个比较复杂的图形轮廓生成一个背景轮廓，则可以使用 ShaderMask 嵌套图形，如图 2-17 所示。

图 2-17 ShaderMask 使用效果图

代码如下：

```
//代码清单 2-28
ShaderMask(
  blendMode: BlendMode.srcATop,
  shaderCallback: (Rect bounds) {
    return const LinearGradient(
      colors: [Colors.orange, Colors.yellow],
    ).createShader(bounds);
  },
  child: Image.asset(
    'assets/images/qq_icon.png',
    width: 40,
  ),
)
```

2.6.4 SizedBox 用来限制子 Widget 大小

SizedBox 可以用来设置子 Widget 大小，代码如下：

```
//代码清单 2-29-1
//lib/code2/code227_SizedOverflowBox.dart
```

```
class _Exam220HomePageState extends State<Exam220HomePage> {
  @override
  Widget build(BuildContext context) {
    return Scaffold(
      body: Center(
        child: SizedBox(
          width: 100,
          height: 100,
          child: Container(
            height: 50.0,
            width: 150.0,
            color: Colors.teal,
          ),
        ),
      ),
    );
  }
}
```

如图 2-18 所示，SizedBox 限制大小时，当子 Widget 尺寸大于 SizedBox 限制的尺寸时，子 Widget 尺寸会被限制，甚至出现越界异常，如果使用 SizedOverflowBox 来限制大小，则允许子 Widget 越界显示。

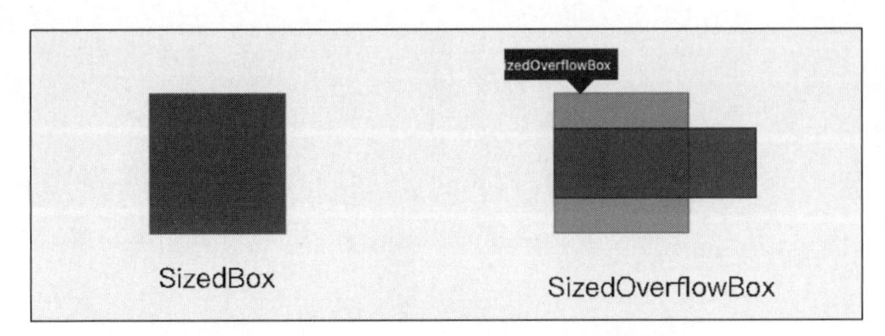

图 2-18　组件尺寸限制使用效果图

代码如下：

```
//代码清单 2-29-2
//lib/code2/code227_SizedOverflowBox.dart
class _Exam220HomePageState extends State<Exam220HomePage> {
  @override
  Widget build(BuildContext context) {
    return Scaffold(
      body: Center(
        child: SizedOverflowBox(
          size: const Size(100, 100),
          child: Container(height: 50.0, width: 150.0, color: Colors.teal),
```

```
            alignment: Alignment.centerLeft,
          ),
        ),
      );
    }
  }
```

2.7 小结

本章概述了 Flutter 项目开发中 MaterialApp、Scaffold、AppBar、Text、Image、Container、TextField、Button 等基础 Widget 组件，通过这些基础组件就可以构建基本的应用程序，本章中概述的这些基础组件的属性为在实际开发中比较常用的，在第 3 章中，通过 UI 排版组件，就可以将本章中基础组件应用起来，从而开发完整的 UI 页面。

第3章 CHAPTER 3

Flutter UI布局排版组件核心基础

在Flutter中，应用于页面布局排版方式的组件Widget有线性排列(Column、Row)、层叠排列(Stack)、流式排列(Flow、Wrap)等。

3.1 Column与Row实现线性排列

线性布局指页面布局中的控件摆放方式是以线性的方式摆放的，线性排列分为纵向(垂直方向)和横向(水平方向)，如图3-1所示。

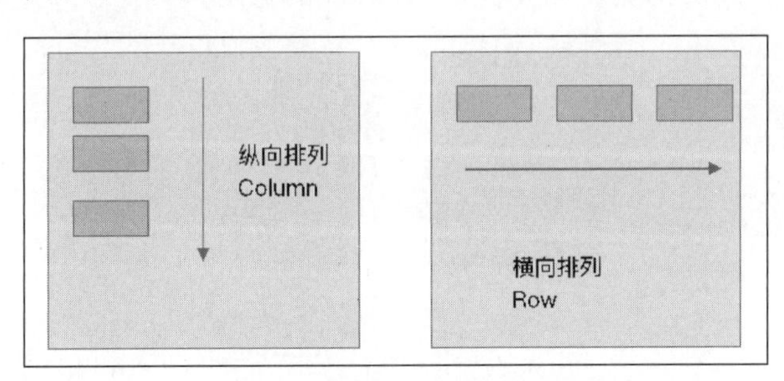

图3-1 线性布局排列效果图

3.1.1 Column用来实现竖直方向线性排列

Column组件的主要功能是处理垂直方向的布局，Column可以将子组件在竖直方向线性排列，如图3-2所示，代码如下：

```
//代码清单 3-1 Column的基本使用
//代码路径 lib/code3/code301_Column.dart
class Exam220HomePage extends StatelessWidget {
  const Exam220HomePage({Key? key}) : super(key: key);
  @override
```

```
    Widget build(BuildContext context) {
      return Scaffold(
        //页面的头部
        appBar: AppBar(title: const Text("标题")),
        body: SizedBox(
          //宽度填充屏幕
          width: double.infinity,
          child: Column(
            children: const [
              Text("测试数据一"),
              Text("测试数据二"),
              Text("测试数据三"),
            ],
          ),
        ),
      );
    }
  }
```

对于 Column 来讲，在默认情况下，将竖直方向称为主轴方向，将水平方向称为交叉轴方向，Column 默认在主轴方向填充，在交叉轴方向包裹，在上述代码中，通过组件 SizeBox 来设定填充宽度，Column 的宽度也会填充 SizeBox 的宽度。

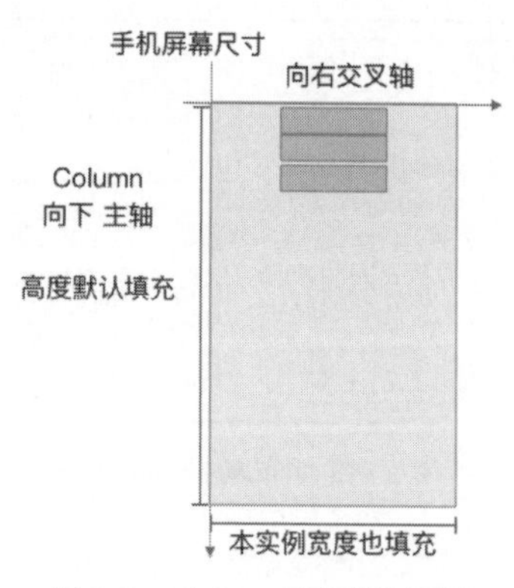

图 3-2　Column 排列效果图

将 Column 在主轴方向设置为包裹，以及将交叉轴对齐方式设置为左对齐，如图 3-3 所示，代码如下：

```
Column(
  //主轴方向包裹
  mainAxisSize: MainAxisSize.min,
  //主轴方向顶部对齐,默认方式
  mainAxisAlignment: MainAxisAlignment.center,
```

```
  //交叉轴方向左对齐,默认居中
  crossAxisAlignment: CrossAxisAlignment.start,
  children: const [
    Text("测试数据一"),
    Text("测试数据二"),
    Text("测试数据三"),
  ],
)
```

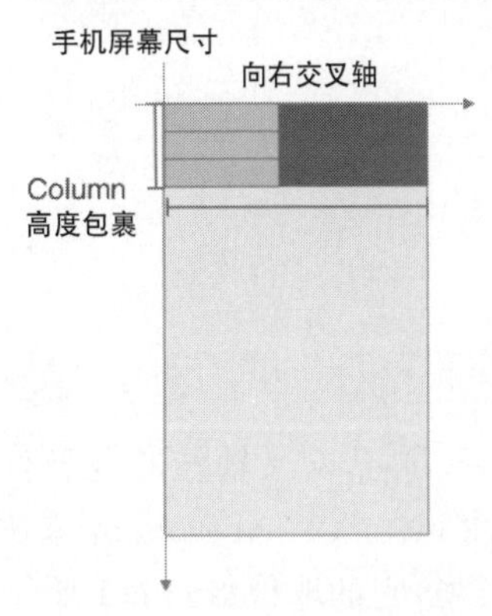

图 3-3 Column 左对齐效果图

3.1.2 Row 用来实现水平方向线性排列

Row 组件的主要功能是处理水平方向的布局,如图 3-4 所示,通过 Row 实现子组件的水平线性排列。

图 3-4 Row 基本使用

代码如下:

```
//代码清单 3-2 Row 的基本使用
//代码路径 lib/code3/code302_Row.dart
class Exam302HomePage extends StatelessWidget {
  const Exam302HomePage({Key? key}) : super(key: key);
  @override
  Widget build(BuildContext context) {
    return Scaffold(
```

```
        //页面的头部
        appBar: AppBar(title: const Text("标题")),
        body: Row(
          //主轴方向包裹
          //mainAxisSize: MainAxisSize.min,
          //主轴方向左对齐,默认方式
          //mainAxisAlignment: MainAxisAlignment.center,
          //交叉轴方向顶部对齐,默认居中
          //crossAxisAlignment: CrossAxisAlignment.start,
          children: const [
            Text("测试数据一"),
            Text("测试数据二"),
            Text("测试数据三"),
          ],
        ),
      );
    }
  }
```

对于 Row,其主轴指的是水平方向,交叉轴指的是竖直方向,在默认情况下,Row 在主轴方向上填充,主轴方向通过属性 mainAxisAlignment 来决定子 Widget 的对齐方式,配置的是 MainAxisAlignment 枚举类型,它的取值如表 3-1 所示。

表 3-1 主轴 alignment 对齐方式

类　别	描　述
MainAxisAlignment. start	沿着主轴方向开始位置对齐
MainAxisAlignment. end	沿着主轴方向结束位置对齐
MainAxisAlignment. center	沿着主轴方向居中对齐
MainAxisAlignment. spaceBetween	沿着主轴方向平分剩余空间
MainAxisAlignment. spaceAround	把剩余空间平分成 n 份,n 是子 widget 的数量,然后把其中一份空间分成两份,放在第 1 个 child 的前面和最后一个 child 的后面
MainAxisAlignment. spaceEvenly	把剩余空间平分成 $n+1$ 份,然后平分所有的空间

交叉轴方向通过属性 crossAxisAlignment 来决定子 Widget 的对齐方式,配置的是 CrossAxisAlignment 枚举类型,它的取值如表 3-2 所示。

表 3-2 交叉轴 alignment 对齐方式

类　别	描　述
CrossAxisAlignment. start	交叉轴方向开始位置对齐
CrossAxisAlignment. end	交叉轴方向结束位置对齐
CrossAxisAlignment. center	交叉轴方向居中位置对齐
CrossAxisAlignment. stretch	拉伸,使用子 Widget 填充交叉轴
CrossAxisAlignment. baseline	与基线相匹配(不常用)

3.1.3 Column 与 Row 中子 Widget 按比例权重布局

如图 3-5 所示，线性布局 Row 中两个按钮水平排列，默认按钮包裹子文本的大小，结合 Expanded 组件使用后，第 2 个 Button 填充了水平方向剩余的所有空白区域，代码如下：

```
//代码清单 3-3 Row 权重适配
//代码路径 lib/code3/code303_Row.dart
class Exam302HomePage extends StatelessWidget {
  const Exam302HomePage({Key? key}) : super(key: key);

  @override
  Widget build(BuildContext context) {
    return Scaffold(
      //页面的头部
      appBar: AppBar(title: const Text("标题")),
      body: Row(
        //主轴方向居中对齐(对于 Row 来讲就是水平方向)
        mainAxisAlignment: MainAxisAlignment.center,
        children: [
          ElevatedButton(
            onPressed: () {},
            child: Text("A"),
          ),
          //填充
          Expanded(
            child: ElevatedButton(
              onPressed: () {},
              child: Text("B"),
            ),
          ),
        ],
      ),
    );
  }
}
```

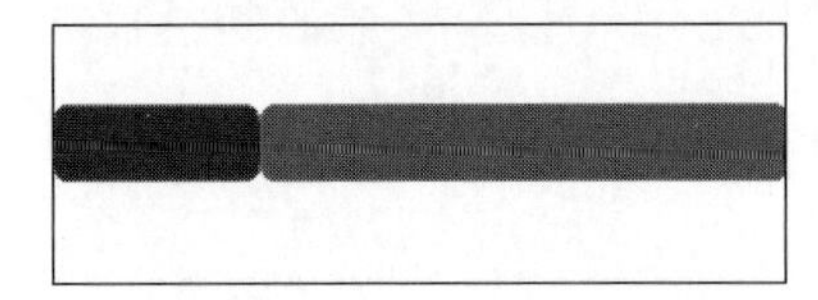

图 3-5 Row 中权重适配

如果期望 Row 中的子 Widget 平均分配宽度，如图 3-6 所示，则可以将 Row 中的子 Widget 分别使用 Expanded 来包裹，代码如下：

```
Row(
  //主轴方向居中对齐(对于 Row 来讲就是水平方向)
  mainAxisAlignment: MainAxisAlignment.center,
  children: [
Expanded(
flex: 1,
      child: ElevatedButton(
        onPressed: () {},
        child: Text("A"),
      ),
    ),
    //填充
Expanded(
flex: 1,
      child: ElevatedButton(
        onPressed: () {},
        child: Text("B"),
      ),
    ),
  ],
)
```

在这里通过 Expanded 中配置的 flex 值，来决定当前 Expanded 的子 Widget 占用的 height，如在这里的 3 个区域内容中，Expanded 分别将 flex 配置为 1，也就是将当前的 Column 的高度 height 平均分成 3 份，然后每个子 Widget 占用一份 height，也就实现了等比分布。同理应用在 Column 中，便可在竖直方向按比例分布。

3.2 非线性布局综合概述

3.2.1 Stack 用来实现层叠布局

Stack 是将子 Widget 重叠在一起，如图 3-6 所示，小点浮在图片上层。

图 3-6 Stack 的基本使用

图 3-6 中的小点可以使用 Container 结合裁剪组件 ClipOval 实现，Stack 中结合 Positioned 组合实现子 Widget 的上、下、左、右、居中排列，代码如下：

```
//代码清单 3-4 Stack 层叠布局
//代码路径 lib/code3/code304_Stack.dart
buildStack() {
  return SizedBox(
    width: 40, height: 40,//限定大小
    child: Stack(
      alignment: Alignment.center,//子组合居中
      children: [
        Positioned(
          width: 33, height: 33,//设定子组件大小
          child: Image.asset(
            "assets/images/warning_icon.png",
          ),
        ),
        Positioned(
          top: 0,right: 0,//设定子组件右上角对齐
          child: ClipOval(//裁剪
            child: Container(
              width: 10, height: 10,
              color: Colors.red,
            ),
          ),
        ),
      ],
    ),
  );
}
```

对于 alignment 属性配置的值，只对没有配置定位或部分定位的子组件起作用，如本实例中的图片没有设置对齐方式，默认使用 Stack 的 alignment 属性配置的对齐方式。

3.2.2 Wrap 用来实现层叠布局

在使用线性布局 Row 和 Colum 时，如果子 Widget 的宽度或者高度超出屏幕范围，则会报溢出错误，通过 Wrap 来支持流式布局，溢出部分则会自动折行，如图 3-7 所示。

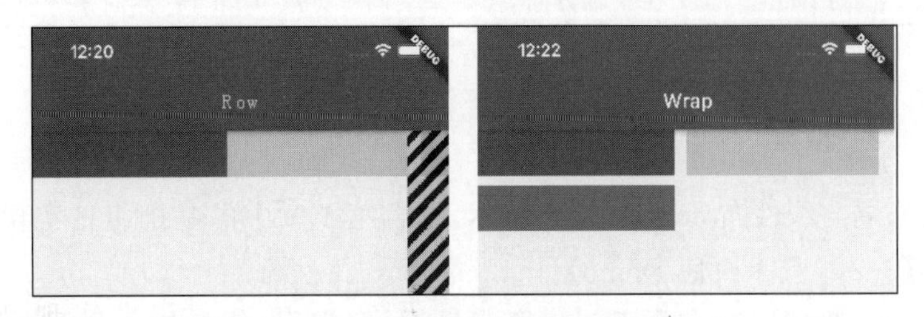

图 3-7 Row 与 Wrap 排列对比图

Wrap 流式布局的代码如下：

```
//代码清单 3-5 Wrap 流式布局
//代码路径 lib/code3/code305_Wrap.dart
Widget buildWrap(){
  return Wrap(
    //包裹的子 view
    children: [
      Container(color: Colors.blue,width: 200,height: 45,),
      Container(color: Colors.yellow,width: 200,height: 45,),
      Container(color: Colors.grey,width: 200,height: 45,),
    ],
    direction: Axis.horizontal,       //水平排列，默认此方式
    spacing: 12,                      //主轴方向上的两个 Widget 之间的间距
    runSpacing: 10,                   //行与行之前的间隔
    alignment: WrapAlignment.start,   //主轴方向的 Widget 的对齐方式
    runAlignment: WrapAlignment.start,//次轴方向上的对齐方式
  );
}
```

Wrap 的 alignment 属性用来配置主轴方向上子 Widget 的对齐方式，如这里配置为水平方向，它的取值如表 3-3 所示。

表 3-3　Wrap 的 alignment 取值概述

类　别	描　述
WrapAlignment.start	子组件沿开始方向对齐
WrapAlignment.end	子组件沿结束方向对齐
WrapAlignment.center	子组件居中对齐
WrapAlignment.spaceBetween	使子组件在主轴方向平均分配未占用的空间，两端对齐
WrapAlignment.spaceAround	在主轴方向，把剩余空间平分成 n 份，n 是子 widget 的数量，然后把其中一份空间分成两份，放在第 1 个 child 的前面和最后一个 child 的后面
WrapAlignment.spaceEvenly	在主轴方向，把剩余空间平分成 $n+1$ 份，然后平分所有的空间

3.2.3　实现登录页面

24min

使用层叠布局与线性布局，如图 3-8 所示，结合第 1 章中的基础组件实现登录页面，读者可观看视频【3.2.3 登录页面 Demo】。

页面主结构是通过层叠布局 Stack 将背景层、模糊层、信息输入层展示在一起，代码如下：

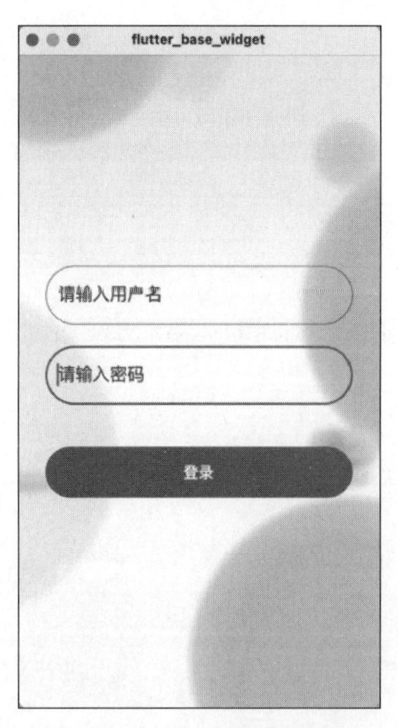

图 3-8 登录页面显示效果图

```
//代码清单 3-6 登录页面 Demo
//代码路径 lib/code3/code305_Wrap.dart
class Exam306HomePage extends StatefulWidget {
  const Exam306HomePage({Key? key}) : super(key: key);
  @override
  State<Exam306HomePage> createState() => _MyHomePageState();
}

class _MyHomePageState extends State<Exam306HomePage> {

  @override
  Widget build(BuildContext context) {
    return Scaffold(
      body: SizedBox(
        width: double.infinity,
        height: double.infinity,
        child: Stack(
          children: [
            //第一层背景图片
            buildFunction1(),
            //第二层高斯模糊
            buildFunction2(),
            //第三层登录输入层
            buildFunction3(),
```

```
            ],
          ),
        ),
      );
    }
  …
  }
```

第一层通过 Image 加载需要显示的图片，第二层通过 BackdropFilter 结合 ImageFilter 实现高斯模糊效果，代码如下：

```
buildFunction2() {
  return Positioned.fill(
    child: BackdropFilter(
      filter: ImageFilter.blur(sigmaX: 3, sigmaY: 3),
      child: Container(
        color: Colors.white.withOpacity(0.4),
      ),
    ),
  );
}
```

本实例的完整源码可查看本书配套源码 flutter_base_widget 中的代码清单 3-6。

3.3 弹框用于提示用户信息

弹框又可称为对话框，用于提示用户，如软件应用升级信息弹框、优惠券活动提示、提交修改删除操作给用户的确认弹框等，本节内容概述 Material 风格与苹果风格的弹框使用、弹框中内容刷新问题。

一个基本 Material 风格弹框 AlertDialog 的内容分布区域效果图，如图 3-9 所示。

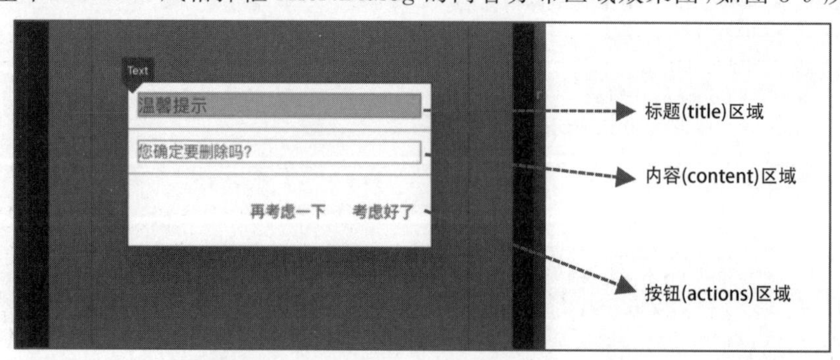

图 3-9　Material 风格 AlertDialog 效果图

3.3.1 showDialog 显示基本弹框

showDialog()方法是 Material 组件库提供的一个用于弹出 Material 风格弹框的方法，代码如下：

```
//代码清单 3-7 Dialog 的基本使用 showDialog 运行效果如图 3-9 所示
//lib/code3/example_307_showDialog.dart
void showDialogFunction() async {
  bool? isSelect = await showDialog<bool>(
    context: context,
    builder: (context) {
      return AlertDialog(
        title: const Text("温馨提示"),
        //title 的内边距,默认 left: 24.0,top: 24.0, right 24.0
        //默认底部边距,如果 content 不为 null,则底部内边距为 0
        //如果 content 为 null,则底部内边距为 20
        titlePadding: EdgeInsets.all(10),
        //标题文本样式
        titleTextStyle: TextStyle(color: Colors.black87, fontSize: 16),
        //中间显示的内容
        content: const Text("你确定要删除吗?"),
        //中间显示的内容边距
        //默认 EdgeInsets.fromLTRB(24.0, 20.0, 24.0, 24.0)
        contentPadding: EdgeInsets.all(10),
        //中间显示内容的文本样式
        contentTextStyle: TextStyle(color: Colors.black54, fontSize: 14),
        //底部按钮区域
        actions: <Widget>[
          TextButton(
            child: const Text("再考虑一下"),
            onPressed: () {
              //关闭返回 false
              Navigator.of(context).pop(false);
            },
          ),
          FlatButton(
            child: const Text("考虑好了"),
            onPressed: () {
              //关闭返回值为 true
              Navigator.of(context).pop(true);
            },
          ),
        ],
      );
    },
  );

  print("弹框关闭 $ isSelect");
}
```

3.3.2 showCupertinoDialog 显示苹果风格弹框

showCupertinoDialog 用于弹出苹果风格的弹框，如图 3-10 所示。

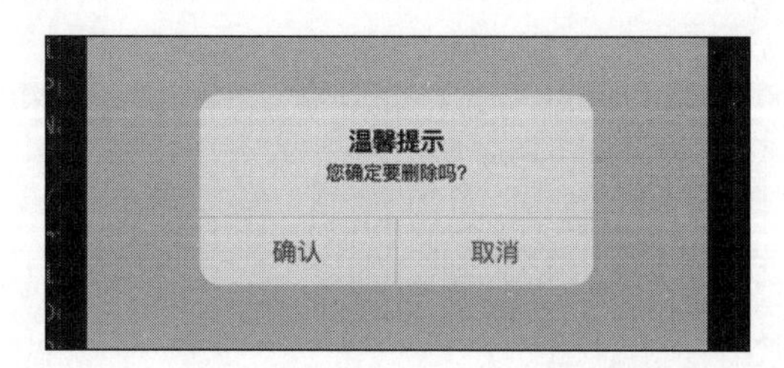

图 3-10　showCupertinoDialog 苹果风格弹框效果图

代码如下：

```
//代码清单 3-7-1 运行效果如图 3-10 所示
//lib/code3/example_307_showDialog.dart
void showCupertinoDialogFunction(BuildContext context) async {
  bool isSelect = await showCupertinoDialog(
    //单击背景弹框是否消失
    barrierDismissible: true,
    context: context,
    builder: (context) {
      return CupertinoAlertDialog(
        title: Text('温馨提示'),
        //中间显示的内容
        content: Text("你确定要删除吗?"),
        //底部按钮区域
        actions: [
          CupertinoDialogAction(
            child: Text('确认'),
            onPressed: () {
              Navigator.of(context).pop();
            },
          ),
          CupertinoDialogAction(
            child: Text('取消'),
            isDestructiveAction: true,
            onPressed: () {
              Navigator.of(context).pop();
            },
          ),
        ],
      );
    },
  );
}
```

3.3.3 showBottomSheet 底部显示弹框

showBottomSheet 用来在视图底部弹出一个 Material Design 风格对话框，如图 3-11 所示。这种业务应用场景也比较多，如页面中的分享面板等。

图 3-11 showBottomSheet 底部弹框效果图

代码如下：

```
//代码清单 3-7-2 showBottomSheet
//lib/code3/example_307_showDialog.dart
void showBottomSheetFunction(BuildContext context) async {
  showBottomSheet(
    context: context,
    builder: (BuildContext context) {
      return buildContainer(context);
    },
  );
}

Container buildContainer(BuildContext context) {
  return Container(
    color: Colors.white,
    height: 240,
    width: double.infinity,
    child: Column(
      children: <Widget>[
        Container(
          alignment: Alignment.center,
          height: 44,
          child: Text(
            "温馨提示",
            style: TextStyle(fontSize: 16, fontWeight: FontWeight.w500),
          ),
        ),
        Expanded(
```

```
            child: Text("这里是内容区域"),
          ),
          Container(
            height: 1,
            color: Colors.grey[200],
          ),
          Container(
            height: 64,
            child: Row(
              children: [
                Expanded(
                  child: TextButton(
                    child: Text("再考虑一下"),
                    onPressed: () {
                      //关闭后返回值为 false
                      Navigator.of(context).pop(false);
                    },
                  ),
                ),
                Container(
                  width: 1,
                  color: Colors.grey[200],
                ),
                Expanded(
                  child: FlatButton(
                    child: Text("考虑好了"),
                    onPressed: () {
                      //关闭后返回值为 true
                      Navigator.of(context).pop(true);
                    },
                  ),
                ),
              ],
            ),
          ),
        ],
      ),
    );
  }
```

在使用 showBottomSheet 时，可能会出现异常，异常信息如下：

```
[VERBOSE-2:ui_dart_state.cc(177)] Unhandled Exception: No Scaffold widget found.
Example701 widgets require a Scaffold widget ancestor.
The specific widget that could not find a Scaffold ancestor was:
...
```

这是因为在调用 showBottomSheet 时使用的 Context 不是 Scaffold 对应的 Context，可以考虑使用 Builder 组件来包裹，如在单击按钮时调用显示底部弹框，结合 Builder 的代码如下：

```
Builder(
  builder: (BuildContext context) {
    return ElevatedButton(
      child: Text("BottomSheet "),
      onPressed: () {
        showBottomSheetFunction(context);
      },
    );
  },
),
```

showBottomSheet()方法相当于调用了 Scaffold 的 showBottomSheet()方法，相当于在当前视图中插入显示的一个布局视图。

3.3.4 showModalBottomSheet 底部弹出对话框

showModalBottomSheet 用来在视图底部弹出一个 Modal Material Design 风格对话框，是菜单或对话框的替代品，可以防止用户与应用程序的其余部分进行交互，showBottomSheet 创建的底部弹窗视图，不阻止用户与应用程序交互基本使用，代码如下：

```
//代码清单 3-7-3 showModalBottomSheet
//lib/code3/example_307_showDialog.dart
void showModalBottomSheetFunction(BuildContext context) async {
  showModalBottomSheet(
    context: context,
    //背景颜色
    backgroundColor: Colors.grey,
    //阴影颜色
    barrierColor: Color(0x30000000),
    //单击背景消失
    isDismissible: true,
    //下滑消失
    enableDrag: true,
    builder: (BuildContext context) {
      //代码清单 3-7-2 中定义的视图布局
      return buildContainer(context);
```

```
    },
  );
}
```

showCupertinoModalPopup 用来快速构建并弹出 iOS 风格的底部弹框，代码如下：

```
//代码清单 3-7-4 showCupertinoModalPopup
  //lib/code3/example_307_showDialog.dart
  void showCupertinoModalFunction(BuildContext context) async {
    showCupertinoModalPopup<int>(
      context: context,
      builder: (cxt) {
        CupertinoActionSheet dialog = CupertinoActionSheet(
          title: Text("温馨提示"),
          message: Text('请选择分享的平台'),
          //取消按钮
          cancelButton: CupertinoActionSheetAction(
            onPressed: () {},
            child: Text("取消"),
          ),
          actions: <Widget>[
            CupertinoActionSheetAction(
                onPressed: () {
                  Navigator.pop(cxt, 1);
                },
                child: Text('QQ')),
            CupertinoActionSheetAction(
                onPressed: () {
                  Navigator.pop(cxt, 2);
                },
                child: Text('微信')),
            CupertinoActionSheetAction(
                onPressed: () {
                  Navigator.pop(cxt, 3);
                },
                child: Text('系统分享')),
          ],
        );
        return dialog;
      },
    );
  }
```

运行效果如图 3-12 所示。

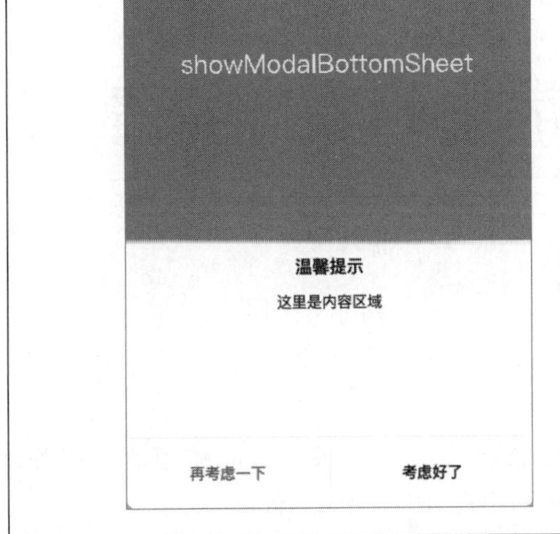

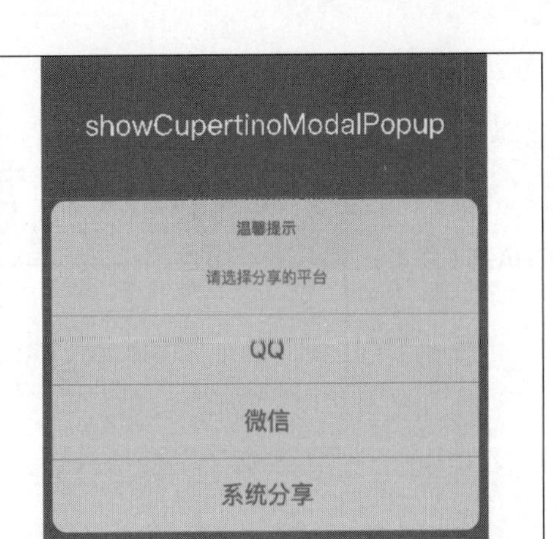

图 3-12 showModalBottomSheet 运行效果图

3.4 小结

本章概述了线性布局 Column、Row、层叠布局 Stack、流式布局 Wrap 的基本排版 Widget，用这些排版 Widget 再结合第 1 章中的基础组件综合使用，就可以构建出基本的应用程序，这样便进入了 Flutter 开发的初级阶段。

第4章 Flutter 常用组件核心基础

CHAPTER 4

本章概述 App 开发中常用的功能性 Widget，如页面加载过渡时使用的进度条，以及 App 设置中常用的单选与复选、手势识别、裁剪组件、阴影模糊组件等。

4.1 加载过渡指示器

如进入页面时请求网络加载数据，这里最好在页面上显示一个加载中的小圆圈。

4.1.1 线性指示器 LinearProgressIndicator

LinearProgressIndicator 是一个线性进度指示器 Widget，当 value 为 null 时，进度条是一个线性的循环模式，表示正在加载；当 value 不为空时，可设置为 0.0～1.0 的非空值，用来表示加载进度，如图 4-1 所示，代码如下：

```
//代码清单 4-1-1 线性指示器 LinearProgressIndicator
//code4/example_401_progress_page.dart
Widget buildLinearProgress() {
  //Container 用来约束大小
  return const SizedBox(
    width: 300, height: 10,     //覆盖进度条的 minHeight
    child: LinearProgressIndicator(
      //value: 0.3,
      //进度高亮颜色
      valueColor: AlwaysStoppedAnimation<Color>(Colors.blue),
      //总进度的颜色
      backgroundColor: Color(0xff00ff00),
      //设置进度条的高度
      minHeight: 10,
    ),
  );
}
```

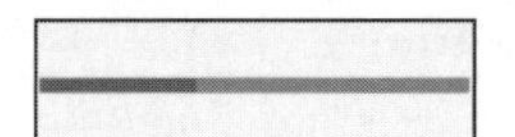

图 4-1 线性指示器

4.1.2 圆形指示器 CircularProgressIndicator

Material 风格的圆形小圆圈，当未设置指定值时，一直顺时针旋转，设置值后，可用来指定具体的进度，效果如图 4-2 所示，代码如下：

```
//代码清单 4-1-2 圆形指示器 LinearProgressIndicator
//code4/example_401_progress_page.dart
Widget buildCircularProgress() {
  //通过 Container 或者 SizeBox 来限制大小
  return const SizedBox(
    width: 55, height: 55,
    child: CircularProgressIndicator(
      //value: 0.3,
      //进度高亮颜色
      valueColor: AlwaysStoppedAnimation<Color>(Colors.blue),
      //总进度的颜色
      backgroundColor: Color(0xff00ff00),
      strokeWidth: 6.0,      //圆圈的厚度
    ),
  );
}
```

图 4-2 圆形指示器

4.1.3 小菊花 CupertinoActivityIndicator

CupertinoActivityIndicator 是一个顺时针旋转的 iOS 风格的活动指示器，效果如图 4-3 所示，代码如下：

```
//代码清单 4-1-3 苹果风格圆形指示器
//code4/example_401_progress_page.dart
Widget buildCupertinoIndicator() {
  //通过 Container 或者 SizeBox 来限制大小
  return const SizedBox(
    width: 85, height: 85,
```

```
    child: CupertinoActivityIndicator(
      radius: 30, //半径
      color: Colors.red,//颜色
      animating: true,//是否转动,默认为 true 开启转动
    ),
  );
}
```

图 4-3 小菊花指示器

4.2 单选框 Radio、复选框 CheckBox、开关 Switch

单选框、复选框、开关常用在配置型业务中，如用户是否打开消息提示配置，就可以使用本节的开关效果来实现。

4.2.1 单选框 Radio 组件

在 Flutter 中通过 Radio 组件实现单选框效果，通常 Radio 不单独使用，常常在有多组数据时使用，只能选择其中之一的情景中，如图 4-4 所示。

图 4-4 单选框 Radio 效果图

代码如下：

```
//代码清单 4-2-1 单选框 Radio 组件的基本使用
//lib/code4/example_404_radio_page.dart
//默认选中的单选框的值
int _groupValue = 0;
//单选框组件的使用
Row buildRadioGroupWidget() {
  return Row(children: [
    Radio(
      value: 0,                    //此单选框绑定的值为必选参数
      groupValue: _groupValue,     //当前组中选定的值为必选参数
      activeColor:Colors.red,      //选中时填充颜色
```

```
      //内边距,默认最小单击区域为 48 × 48
      //MaterialTapTargetSize.shrinkWrap 为组件实际大小
      materialTapTargetSize: MaterialTapTargetSize.shrinkWrap,
      //单击状态改变时的回调的必选参数
      onChanged: < int >(v) {
        setState(() {
          _groupValue = v;
        });
      },
    ),
    Radio(
      value: 1,                  //此单选框绑定的值为必选参数
      groupValue: _groupValue,  //当前组中选定的值为必选参数
      //单击状态改变时的回调的必选参数
      onChanged: < int >(v) {
        setState(() {
          _groupValue = v;
        });
      },
    ),
  ]);
}
```

4.2.2 单选框 RadioListTile

如图 4-5 所示，RadioListTile 是一个用于便捷、快速构建列表样式的组件，与 SwitchListTile、CheckboxListTile、ListTile 类似，代码如下：

```
//代码清单 4-2-2 单选框 RadioListTile 组件的基本使用
  //llib/code4/example_404_radio_page.dart
  Widget buildRadioListTile() {
    return RadioListTile(
      value: 0,//当前对应的单选框的标识
      //是否选中发生变化时的回调,回调的 bool 值是否被选中 , true 是选中
      onChanged: < int >(value) {
        setState(() {
          _groupValue = value;
        });
      },
      //选中时 Radio 的填充颜色
      activeColor: Colors.red,
      //标题,selected 如果是 true
      //如果不设置 text 的 color
      //text 的颜色使用 activeColor
```

```
      title: const Text("标题"),
      //副标题(在 title 下面)
      subtitle: const Text("副标题"),
      //是否是三行文本
      //如果是 true,则副标题不能为 null
      //如果是 false 且没有副标题
      //就只有一行,如果有副标题,就只有两行
      isThreeLine: true,
      dense: false,//是否密集垂直
      //左边的控件
      secondary: Image.asset(
        "assets/images/warning_icon.png",
        fit: BoxFit.fill,
      ),
      //text 和 icon 的 color 是否是 activeColor 的颜色
      selected: false,
      //方向模型
      controlAffinity: ListTileControlAffinity.trailing,
      groupValue: _groupValue,
    );
  }
```

图 4-5　单选框 RadioListTile 效果图

4.2.3　复选框 CheckBox

复选框 CheckBox 的效果图如图 4-6 所示。

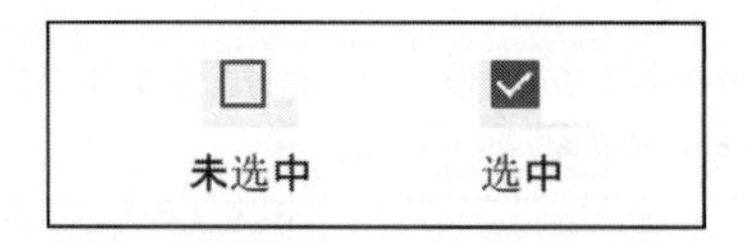

图 4-6　复选框 CheckBox 效果图

代码如下：

```
//代码清单 4-3 复选框 CheckBox 的基本使用
  //lib/code4/example_405_checkbox_page.dart
  //默认选中的单选框的值
  bool checkIsSelect = false;
//复选框 Checkbox 的基本使用
```

```
Checkbox buildCheckBox() {
  return Checkbox(
    //单击选择时的回调
    onChanged: <bool>(value) {
      setState(() {
        checkIsSelect = value;
      });
    },
    //为 true 时 Checkbos 是选中状态
    //为 false 时 Checkbos 是未选中状态
    value: checkIsSelect,
    //选中时的填充颜色
    activeColor: Colors.blue,
    //选中时小对勾的颜色
    checkColor: Colors.white,
  );
}
```

4.2.4 复选框 CheckboxListTile

CheckboxListTile 组件用来快速构建列表样式的组件布局，与 SwitchListTile、RadioListTile、CheckboxListTile、ListTile 类似，如图 4-7 所示，包括左侧显示的图片 Widget、中间显示的文字区域和右侧的复选框。

图 4-7 复选框 CheckBox 效果图

代码如下：

```
//代码清单 4-4 CheckboxListTile 的基本使用
//lib/code4/example_406_checkbox_page.dart
//默认选中的值
bool checkIsSelect = false;

Widget buildCheckBox() {
  return CheckboxListTile(
    //当前对应的复选框是否选中
    value: checkIsSelect,
    //是否选中发生变化时的回调，回调的 bool 值为是否选中，true 是选中
    onChanged: <int>(value) {
      setState(() {
        checkIsSelect = value;
```

```
      });
    },
    //选中时 checkbox 的填充颜色
    activeColor: Colors.red,
    //标题,selected 如果是 true
    //如果不设置 text 的 color
    //text 的颜色使用 activeColor
    title: Text("标题"),
    //副标题(在 title 下面)
    subtitle: Text("副标题"),
    //是否是三行文本
    //如果是 true,则副标题不能为 null
    //如果是 false 且没有副标题
    //就只有一行,如果有副标题,就只有两行
    isThreeLine: true,
    //是否密集垂直
    dense: false,
    //左边的控件
    secondary: Image.asset(
      "assets/images/warning_icon.png",
      height: 32,
      fit: BoxFit.fill,
),
    //text 和 icon 的 color 是否是 activeColor 的颜色
    selected: false,
    //方向模型
    controlAffinity: ListTileControlAffinity.trailing,
  );
}
```

4.2.5 开关 Switch

Switch 组件是 Material Design 设计风格的开关,CupertinoSwitch 是苹果设计风格的开关,Switch 的基本使用代码如下:

```
//代码清单 4-5
//lib/code4/example_407_switch_page.dart
//记录开关的状态
bool switchValue = false;
//开关[Switch]的基本使用
Widget buildSwitchWidget() {
  return Switch(
```

```
    //开关状态改变时的回调
    onChanged: (bool value) {
      setState(() {
        switchValue = value;
      });
    },
    value: switchValue,                //当前开关的状态
    activeColor: Colors.blue,          //选中时小圆滑块的颜色
    activeTrackColor: Colors.yellow,   //选中时底部的颜色
    //未选中时小圆滑块的颜色
    inactiveThumbColor: Colors.deepPurple,
    //未选中时底部的颜色
    inactiveTrackColor: Colors.redAccent,
  );
}
```

运行效果如图 4-8 所示。

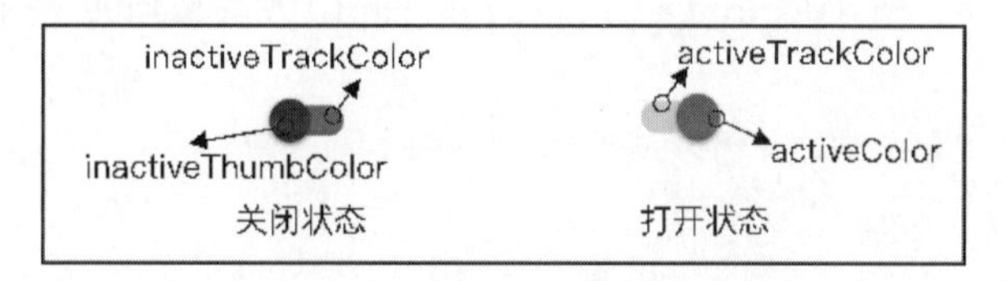

图 4-8　开关 Switch 效果图

开关 CupertinoSwitch 组件的效果如图 4-9 所示,代码如下:

```
//代码清单 4-5-2
  //lib/code4/example_407_switch_page.dart
  //苹果风格的形状小圆圈一直是白色的
  Widget buildCupertinoSwitch() {
    return CupertinoSwitch(
      //开关状态改变时的回调
      onChanged: (bool value) {
        setState(() {
          switchValue = value;
        });
      },
      value: switchValue,        //当前开关的状态
      activeColor: Colors.blue, //选中时底部的颜色
      trackColor: Colors.grey,  //未选中时底部的颜色
    );
  }
```

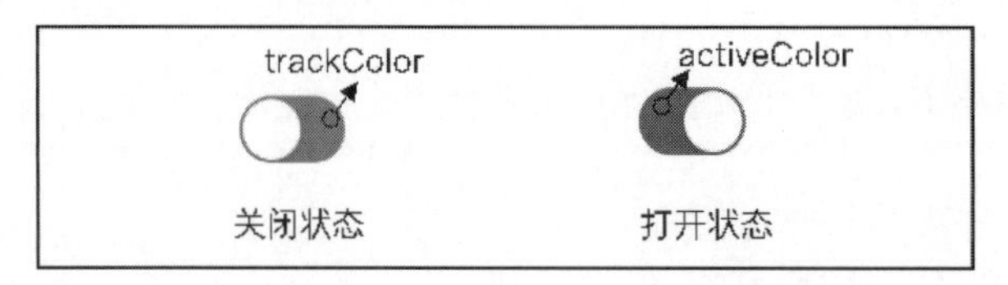

图 4-9　开关 CupertinoSwitch 效果图

4.3　手势处理

在 Flutter 中，GestureDetector 或者 InkWell 用来为不具备事件响应的组件（如图片 Image、文本 Text）添加单击事件监听，如单击事件、触摸与滑动事件监听。

4.3.1　GestureDetector 手势监听

在项目开发中，如果需要对某张图片添加一个单击事件监听，则可通过 GestureDetector 组件实现，GestureDetector 的 onTap()方法监听的是手指抬起时的回调，代码如下：

```
//代码清单 4-6
//lib/code4/example_408_gesture_page.dart
//手势识别[GestureDetector]的基本使用
Widget buildGestureDetector() {
  return GestureDetector(
    //手指抬起时的回调
    onTap: () {
      print("单击了图片");
    },
    child: Container(
      width: 200,
      height: 100,
      child: Image.asset(
        "assets/images/warning_icon.png",
      ),
    ),
  );
}
```

小球跟随手指移动，拖动手势主要由 onPanDown（手指按下）、onPanUpdate（手指滑动）、onPanEnd（滑动结束）构成，代码如下：

```
//代码清单 4-7-1
//lib/code4/example_408_gesture_drag_page.dart
```

```
double _top = 0;
double _left = 0;
Widget buildDragStack(){
  return Stack(
    children: <Widget>[
      Positioned(
        top: _top,
        left: _left,
        child: GestureDetector(
          onPanDown: (DragDownDetails e) {
            //打印手指按下的位置
            print("手指按下: ${e.globalPosition}");
          },
          //手指滑动
          onPanUpdate: (DragUpdateDetails e) {
            setState(() {
              _left += e.delta.dx;
              _top += e.delta.dy;
            });
          },
          onPanEnd: (DragEndDetails e) {
            //打印滑动结束时在 x 轴和 y 轴上的速度
            print(e.velocity);
          },

          child: Container(
            width: 72,
            height: 72,
            decoration: BoxDecoration(
                color: Colors.blueAccent,
                borderRadius: BorderRadius.circular(36)
            ),
          ),
        ),
      )
    ],
  );
}
```

在 GestureDetector 的 onScaleUpdate 中实现缩放手势功能,代码如下:

```
//代码清单 4-7-2 缩放
//lib/code4/example_408_gesture_drag_page.dart
double _imageWidth = 100;
Widget buildScaleImage() {
```

```
    return Center(
      child: GestureDetector(
        child: OverflowBox(
          maxWidth: 2000.0,
          child: Image.asset("assets/images/loginbg.png", width: _imageWidth),
        ),
        onScaleUpdate: (ScaleUpdateDetails e) {
          setState(() {
            //缩放倍数在 0.8 到 10 倍之间
            _imageWidth = 200 * e.scale.clamp(0.8, 10);
          });
        },
      ),
    );
  }
```

4.3.2 Ink 与 InkWell

复杂点的动作(如拖动、旋转、缩放手势等)可使用 GestureDetector 实现,通过 InkWell 可以实现单击事件监听,会有单击高亮及水波纹效果,而通过 GestureDetector 实现的单击无单击效果出现,所以在实际项目开发中,如果只是普通的单击事件,则建议使用 InkWell 组件,代码如下:

```
  //代码清单 4-8-1
  //lib/code4/example_409_gesture_page.dart
  //手势识别[InkWell]的基本使用
  Widget buildInkWell() {
    return InkWell(
      //radius: 30,//水波纹执行的半径
      borderRadius: BorderRadius.all(Radius.circular(20)),
      splashColor: Colors.grey,//水波纹颜色配置
      //单击事件回调
      onTap: () {
        print("onTap 单击回调");
      },
      //需要设置单击事件的子 Widget
      child: Container(
        color: Colors.black38,
        width: 100,
        height: 100,
      ),
    );
  }
```

Ink 常与 InkWell 和 InkResponse 一起使用，用来配置单击效果的水波纹与高亮的样式，Ink 必须在 Material Desin 风格下使用，所以一般使用 Material 组件配合 Ink 来使用，代码如下：

```
//代码清单 4-8-2
//lib/code4/example_409_gesture_page.dart
//通过 [InkWell] 为 [Container]设置单击事件
//通过 [Ink] 来添加背景样式
Widget buildInkWellContainer() {
  return Material(
    child: Ink(
      color: Colors.blue,
      child: InkWell(
        //单击事件回调
        onTap: () {},
        //不要在这里设置背景色,否则会遮挡水波纹效果
        child: buildContainer(),
      ),
    ),
  );
}
```

4.4 其他常用交互视觉类型组件

在实际应用开发中，Clip 系列的裁剪 Widget，可以实现任意的个性化视图；卡片形式的布局也是常用的布局，通过 Card 实现，CupertinoSegmentedControl 可以用来实现分组标签选择效果。

4.4.1 Clip 裁剪系列组件

Clip 裁剪系列组件包括 ClipOval、ClipRect、ClipRRect、ClipPath。

ClipOval 用于裁剪椭圆，如果 child 为正方形，则剪裁之后是圆形；如果 child 为矩形，则剪裁之后为椭圆形，代码如下：

```
ClipOval(
  child: Image.asset(
    "assets/images/loginbg.png",
    width: 100,
    height: 100,
  ),
)
```

ClipRect 将子 child 裁剪为矩形，代码如下：

```
//ClipRect 将 child 剪裁为给定的矩形大小
ClipRect(
  //自定义裁剪路径
  clipper: ClipperPath(),
  child: Container(
    width: 100,
    height: 100,
    color: Colors.blue,
  ),
)
```

通过 CustomClipper 可以自定义裁剪图形，可以实现任意的裁剪效果，ClipperPath 的定义如下：

```
//代码清单 4-9-1
//lib/code/code4/example_410_imageshow_page.dart
class ClipperPath extends CustomClipper<Rect> {
  @override
  Rect getClip(Size size) {
    //自定义一个矩形
    return new Rect.fromLTRB(
      100, 10,
      size.width, size.height,
    );
  }

  @override
  bool shouldReclip(CustomClipper<Rect> oldClipper) {
    return true;
  }
}
```

ClipRRect 可以将任意的子 Widget 裁剪为圆角矩形，代码如下：

```
ClipRRect(
  clipBehavior: Clip.antiAlias,//抗锯齿
  //对应的圆角
  borderRadius: const BorderRadius.all(Radius.circular(10)),
  child: Container(
    width: 100,
    height: 100,
    color: Colors.blue,
  ),
)
```

通过 Path(在后续绘图章节中会有定义)可以定义任意的不规则的图形,ClipPath 可以结合 Path 将子 Widget 裁剪为任意想要的效果,如这里裁剪一个三角形,代码如下:

```
ClipPath(
  clipper: TriangleClipper(),
  child: Image.asset(
    "assets/images/loginbg.png",
    width: 100,
    height: 100,
  ),
)
```

TriangleClipper 是自定义的一个 CustomClipper,在其中通过 Path 构建三角形,代码如下:

```
//代码清单 4-9-2
//lib/code/code4/example_410_imageshow_page.dart
class TriangleClipper extends CustomClipper<Path> {
  @override
  Path getClip(Size size) {
    final path = Path()
      ..moveTo(0.0, size.height)
      ..lineTo(size.width, size.height)
      ..lineTo(size.width / 2, size.height / 2)
      ..close();
    return path;
  }

  @override
  bool shouldReclip(CustomClipper<Path> oldClipper) => false;
}
```

ClipOval、ClipRect、ClipRRect、ClipPath 都具有裁剪功能,所以都涉及边缘抗锯齿问题,通过属性 clipBehavior 可以配置裁剪的方式,不能为 null,可选值包括 none、hardEdge、antiAlias、antiAliasWithSaveLayer,详细描述如表 4-1 所示。

表 4-1　Clip 取值概述

类　别	描　述
Clip.none	不裁剪
Clip.hardEdge	裁剪但不应用抗锯齿,裁剪速度比 none 模式慢一点,但比其他方式快
Clip.antiAlias	裁剪而且抗锯齿,以实现更平滑的外观。裁剪速度比 antiAliasWithSaveLayer 快,但比 hardEdge 慢
Clip.antiAliasWithSaveLayer	带有抗锯齿的剪辑,并在剪辑之后立即保存 saveLayer

4.4.2 Card 用来实现卡片效果

Card 是 Flutter 提供的一个卡片组件，提供了圆角和阴影，如图 4-10 所示，代码如下：

```
//代码清单 4-10
//lib/code4/example_413_card_page.dart
Card(
  color: Colors.blue,                          //背景色
  shadowColor: Colors.black,                   //阴影颜色
  elevation: 20,                               //阴影高度
  borderOnForeground: true,                    //是否在 child 前绘制 border，默认为 true
  margin: EdgeInsets.fromLTRB(0, 50, 0, 30),//外边距
  child: Container(
    width: 200,
    height: 100,
    alignment: Alignment.center,
    child: Text("卡片效果", style: TextStyle(color: Colors.white),),
  ),
)
```

图 4-10 Card 卡片效果图

4.4.3 分段选择标签效果

在 Flutter 中通过 CupertinoSegmentedControl 组件实现，代码如下：

```
//代码清单 4-11
//lib/code4/example_411_segmented_page.dart
CupertinoSegmentedControl(
  //子标签
  children: const <int, Widget>{
    0: Text("全部"),
    1: Text("收入"),
    2: Text("支出 "),
  },
  groupValue: _currentIndex,                   //当前选中的索引
  onValueChanged: (int index) {                //单击回调
```

```
      print("当前选中 $ index");
      setState(() {
        _currentIndex = index;
      });
    },
    selectedColor: Colors.blue,                    //选中的背景颜色
    unselectedColor: Colors.white,                 //未选中的背景颜色
    borderColor: Colors.blue,                      //边框颜色
    pressedColor: Colors.blue.withOpacity(0.4),   //按下的颜色
  )
```

运行效果如图 4-11 所示。

图 4-11 分段组件效果图

4.4.4 手机横屏与竖屏处理

应用程序有时会强制用户在竖屏下使用,这就需要应用全局禁止横屏,可以在程序入口 main 函数()中处理,代码如下:

```
import 'package:flutter/services.dart';
//应用入口
void main() {
  //强制横屏
  SystemChrome.setPreferredOrientations([
    DeviceOrientation.landscapeLeft,
    DeviceOrientation.landscapeRight
  ]);
  //启动根目录
  runApp(BaseApp(
    homeWidget: Example410(),
  ));
}
```

在某些情况下,进入页面后允许横屏显示,退出页面后禁止横屏显示,可以这样来操作:

```
@override
  void initState() {
    super.initState();
    SystemChrome.setPreferredOrientations([
```

```
        DeviceOrientation.landscapeLeft,
        DeviceOrientation.landscapeRight,
        DeviceOrientation.portraitUp,
        DeviceOrientation.portraitDown,
      ]);
    }


    @override
    void dispose() {
      SystemChrome.setPreferredOrientations([
        DeviceOrientation.portraitUp,
        DeviceOrientation.portraitDown,
      ]);
      super.dispose();
    }
```

在某些情况下，当用户将屏幕从纵向模式旋转到横向模式时，需要更新应用程序的显示，在 Flutter 中可以通过 OrientationBuilder 实现，如本实例实现的是手机坚屏时显示 2 列，横屏时显示 3 列，代码如下：

```
//代码清单 4-12
//lib/code4/example_416.dart
class _ExampleState extends State<Example410> {
  @override
  Widget build(BuildContext context) {
    return Scaffold(
      body: OrientationBuilder(
        builder: (context, orientation) {
          return GridView.count(
            //坚屏时显示 2 列，横屏时显示 3 列
            crossAxisCount: orientation == Orientation.portrait ? 2 : 3,
            //List for 循环模拟 100 条数据
            children: List.generate(100, (index) {
              return Center(
                child: Text(
                  '测试使用数据 $ index',
                ),
              );
            }),
          );
        },
      ),
    );
  }
}
```

4.5 小结

本章概述了一些常用的功能性的组件，进度指示器用于处理 App 与用户交互方面的内容，单选框、复选框、开关常用于设置中心或者一些筛选编辑应用场景，手势处理则可以让应用程序中的任何一个 Widget 都设置事件监听。

第5章 滑动视图

CHAPTER 5

在 Flutter 中，当内容超过显示视图时，如果没有特殊处理，Flutter 则会提示 Overflow 错误，一般当内容体超过显示视图时，采用滑动（滚动视图）来处理。

在 Flutter 中通过 ScrollView 组件实现滑动视图效果，当 ScrollView 的内容大于它本身的大小时，ScrollView 会自动添加滚动条，并可以竖直滑动，如 Android 中的 ScrollView、iOS 中的 UIScrollView。

在 Flutter 中常用的滚动视图有 SingleChildScrollView、NestedScrollView、CustomScrollView、Scrollable、ListView、GridView、TabBarView、PageView。

5.1 长页面滑动视图

Flutter 中 SingleChildScrollView、NestedScrollView、CustomScrollView 用来处理长页面滑动效果。

默认情况下滑动视图可在竖直方向上下滑动，在 Android 平台中只是一个滑动，当滑动到边缘时会有浅水波纹拉伸效果，在 iOS 平台中，有回弹效果，可通过 physics 属性修改为指定的效果，physics 属性的取值如表 5-1 所示。

表 5-1 滑动回弹效果概述

类 别	描 述
BouncingScrollPhysics	可滑动，当滑动到边界时有回弹效果，iOS 平台默认使用
ClampingScrollPhysics	可滑动，当滑动到边界时有水波回弹效果，Android 平台默认使用
AlwaysScrollableScrollPhysics	列表总是可滚动的，在 iOS 上会有回弹效果，在 Android 上不会回弹
PageScrollPhysics	一般用于 PageView 的滑动效果，如果将 ListView 设置为滑动到末尾，则会有个比较大的弹起和回弹
FixedExtentScrollPhysics	一般用于 ListWheelScrollViews
NeverScrollableScrollPhysics	不可滑动

scrollDirection 属性用来设置滚动方向（滑动组件的通用配置），默认为垂直，也就是在

竖直方向上下滚动，可取值包括 Axis. vertical（竖直方向）、Axis. horizontal（水平方向）。

5.1.1 滑动组件 SingleChildScrollView

SingleChildScrollView 适用于简单滑动视图处理，如 App 中常见的商品详情页面、订单详情页面（笔者建议在实际应用开发中常用的详情页面使用这个组件），代码如下：

```
//代码清单 5-1-1 SingleChildScrollView 的基本使用
//lib/code5/example_501_SingleChildScrollView.dart
class Example501 extends StatefulWidget {
  const Example501({Key? key}) : super(key: key);
  @override
  State<StatefulWidget> createState() {
    return _ExampleState();
  }
}

class _ExampleState extends State<Example501> {
  //滑动控制器
  final ScrollController _scrollController = ScrollController();
  //文本的关键 Key,用于获取文本的位置信息
  final GlobalKey _globalKey = GlobalKey();

  @override
  Widget build(BuildContext context) {
    return Scaffold(
      appBar: AppBar(title: Text("滑动")),
      body: SingleChildScrollView(
        //设置内边距
        padding: const EdgeInsets.all(20),
        physics: const BouncingScrollPhysics(),
        controller: _scrollController,  //配置滑动控制器
        //子 Widget,通常是 UI 布局系列的 Column、Stack、Row
        child: Container(
          alignment: Alignment.center,
          color: Colors.grey,
          height: 1600,
          child: Text("测试数据",key: _globalKey,),
        ),
      ),
      floatingActionButton: FloatingActionButton(
        onPressed: () {
          //scrollToTop();
          scrollToWidgetPostion(_globalKey);
        },
```

```
            child: Icon(Icons.arrow_upward),
          ),
        );
      }

      …

    }
```

ScrollController是滑动视图使用的控制器，在ScrollController中可添加监听，实时获取滚动的距离，代码如下：

```
//代码清单 5-1-2 ScrollController 的监听
//lib/code5/example_501_SingleChildScrollView.dart
@override
void initState() {
  super.initState();
  //添加滚动监听
  _scrollController.addListener(() {
    //滚动时会实时回调这里
    //获取滚动的距离
    double offsetValue = _scrollController.offset;
    //ScrollView 最大可滑动的距离
    double max = _scrollController.position.maxScrollExtent;
    if (offsetValue <= 0) {
      //如果有回弹效果,则 offsetValue 会出现负值
      print("滚动到了顶部");
    } else if (offsetValue >= max) {
      //如果有回弹效果,则 offsetValue 的值可能大于 max
      print("滚动到了底部");
    } else {
      print("滑动的距离 offsetValue $offsetValue max $max");
    }
  });
}
```

可通过ScrollController的animateTo()方法滑动到指定位置，如这里滑动到视图中的文本位置，通过绑定的GlobalKey获取对应文本的位置，代码如下：

```
//代码清单 5-1-3 ScrollController 的监听
//lib/code5/example_501_SingleChildScrollView.dart
//通过 Widget 绑定的 GlobalKey 获取位置信息
void scrollToWidgetPostion(GlobalKey key) {
  //根据 key 获取上下文对象,即获取 Element 信息
```

```
  BuildContext ? stackContext = key.currentContext;
  if (stackContext != null) {
    //获取对应的 RenderObj 对象
    RenderObject? renderObject = stackContext.findRenderObject();
    if (renderObject != null) {
      //获取指定的 Widget 的位置信息
      Size size = renderObject.paintBounds.size;
      //获取矩阵
      Matrix4 matrix4 = renderObject.getTransformTo(null);
      var translation = matrix4.getTranslation();
      //距离手机屏幕顶部的距离
      double? top = translation.y;
      //获取状态栏高度
      final double statusBarHeight = MediaQuery.of(context).padding.top;
      //获取 AppBar 的高度
      final double appBarHeight = 76;
      //滑动到这个 Widget 的位置
      double tagHeight = top - statusBarHeight - appBarHeight;
scrollOffset(tagHeight);
    }
  }
}
```

scrollOffset 对应的滑动方法如下：

```
//代码清单 5-1-4
//lib/code5/example_501_SingleChildScrollView.dart
//滚动到指定的位置
void scrollOffset(double offset) {
  //返回顶部指定位置
  _scrollController.animateTo(
    offset,
    //返回顶部的过程中执行一个滚动动画,动画持续时间是 200ms
    duration: const Duration(milliseconds: 200),
    //动画曲线是 Curves.ease
    curve: Curves.ease,
  );
}
```

Demo 运行调试图如图 5-1 所示，SingleChildScrollView 的顶部是状态栏与标题栏的高度，此高度为 79 像素，SingleChildScrollView 中的子组件 TextView 距离滑动组件的顶部为 790 像素，在代码中通过 GlobalKey 获取的 top 值为 866＝76＋790。

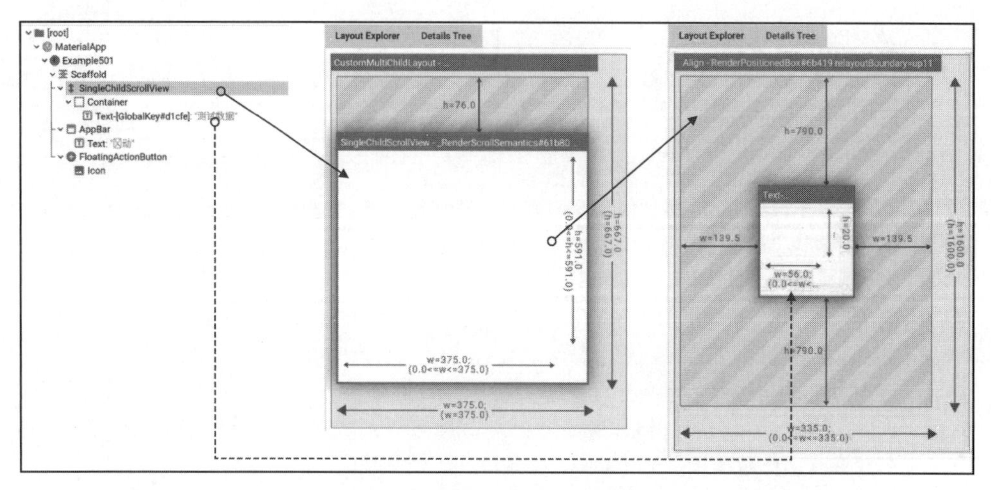

图 5-1　滑动视图构建效果图

5.1.2　滑动布局 NestedScrollView 与 SliverAppBar

1min

NestedScrollView 继承于 CustomScrollView，它比 SingleChildScrollView 更强大，可以用来实现诸如滑动折叠头部的功能，如图 5-2 所示。读者可观看视频讲解【5.1.2 滑动布局 NestedScrollView 与 SliverAppBar】。

图 5-2　滑动折叠页面显示效果图

页面主视图通过 NestedScrollView 来搭建，代码如下：

```
//代码清单 5-2-1 NestedScrollView 的基本使用
//lib/code/code5/example_503_NestedScrollView.dart
class Example503 extends StatefulWidget {
  @override
  State<StatefulWidget> createState() {
    return _ExampleState();
  }
}
class _ExampleState extends State<Example503>
    with SingleTickerProviderStateMixin {

  @override
  Widget build(BuildContext context) {
    return Scaffold(
      body: NestedScrollView(
        //配置可折叠的头布局
headerSliverBuilder: (BuildContext context, bool innerBoxIsScrolled) {
          //回调参数 innerBoxIsScrolled,当折叠头部隐藏时为 true
          //当折叠头部显示时为 false
          print("innerBoxIsScrolled $ innerBoxIsScrolled");
          return [buildSliverAppBar()];
        },
        //超出显示内容区域的 Widget 可以是一个列表、一个滑动视图
        //也可以是一个 TabBarView 来结合 TabBar 使用
        body: buildBodyWidget(),
      ),
    );
  } //S1
…
}
```

在构建函数 headerSliverBuilder 中可使用 Sliver 家族的组件，在这里使用 SliverAppBar 组件，SliverAppBar 与 AppBar 类似，代码如下：

```
//代码清单 5-2-2 SliverAppBar 的详细配置
//lib/code/code5/example_503_NestedScrollView.dart
buildSliverAppBar() {
  return SliverAppBar(
    backgroundColor: Colors.white,
    title: const Text("这里是标题",style: TextStyle(color: Colors.blue),),
    //标题居中
    centerTitle: true,
    //当此值为 true 时 SliverAppBar title 会固定在页面顶部
```

```
    //当此值为 false 时 SliverAppBar title 会随着滑动向上滑动
    pinned: true,

    //当 pinned 属性值为 true 时才会起作用
    //当 floating 为 true 且滑动到顶部时 title 会隐藏
    //当为 false 时 title 不会隐藏
    floating: false,

    //当 snap 配置为 true 时,向下滑动页面,SliverAppBar(
    //及其中配置的 flexibleSpace 内容)会立即显示出来

    //反之当 snap 配置为 false 且向下滑动时,
    //只有当 ListView 的数据滑动到顶部时,SliverAppBar 才会被下拉显示出来
    snap: false,
    //展开的高度
    expandedHeight: 200,
    //AppBar 下的内容区域
    flexibleSpace: FlexibleSpaceBar(
      //背景
      //在这里直接使用的是一张图片
      background: Image.asset(
        "assets/images/banner_icon.jpg",
        height: 200,
        width: MediaQuery.of(context).size.width,
        fit: BoxFit.fill,
      ),
    ),
    bottom: TabBar(
      labelColor: Colors.blue,
      unselectedLabelColor: Colors.grey,
      controller: tabController,
      tabs: const <Widget>[
        Tab(text: "标签一"), Tab(text: "标签二"), Tab(text: "标签三"),
      ],
    ),
  );
}
```

页面的主体部分通过 Scaffold 的 body 属性来配置,可以通过 TabBarView 结合上述 SliverAppBar 中的 bottom 属性配置的 TabBar 实现标签页面的切换,代码如下:

```
//因为页面主体使用了 TabBar,所以用到了控制器
late TabController tabController;
@override
void initState() {
```

```
    super.initState();
    //初始化控制器
    tabController = TabController(length: 3, vsync: this);
  }
  buildBodyWidget() {
    return TabBarView(
      controller: tabController,
      children: [
        ItemPage(),
        ItemPage(),
        ItemPage(),
      ],
    );
  }
```

5.1.3 滑动组件 CustomScrollView

当一个页面中,既有九宫格布局 GridView,又有列表 ListView,由于二者有各自的滑动区域,所以不能进行统一滑动,可通过 CustomScrollView 将二者结合起来,也就是可将 CustomScrollView 理解为滑动容器。

在 CustomScrollView 中需要结合 Sliver 家族的组件来使用,包括 SliverToBoxAdapter、SliverPersistentHeader、SliverFixedExtentList、SliverList、SliverGrid、SliverAppBar、SliverPadding 等。

SliverAppBar 用来处理折叠标题,在 CustomScrollView 中配置也可实现 5.1.2 节中的折叠视图效果,SliverToBoxAdapter 用于实现非 Sliver 家族的组件也可在 CustomScrollView 中使用,宫格布局 SliverGrid 用来实现二维滑动视图,代码如下:

```
//代码清单 5-3-1 九宫格通过构造函数来创建
//lib/code5/example_505_CustomScrollView.dart
class Example505 extends StatefulWidget {

  …

  @override
  Widget build(BuildContext context) {
    return Scaffold(
      appBar: AppBar(
        title: Text("CustomScrollView"),
      ),
      body: CustomScrollView(
        //滑动控制器
        controller: _scrollController,
        slivers: [buildSliverGrid()],
```

```
      ),
    );
  }

  SliverGrid buildSliverGrid() {
    //使用构建方法来构建
    return SliverGrid(
      //用来配置每个子 Item 之间的关系
      gridDelegate: const SliverGridDelegateWithFixedCrossAxisCount(
        //Grid 按 2 列显示,也就是列数
        crossAxisCount: 2,
        //主方向每个 Item 之间的间隔
        mainAxisSpacing: 10.0,
        //次方向每个 Item 之间的间隔
        crossAxisSpacing: 10.0,
        //Item 的宽与高的比例
        childAspectRatio: 3.0,
      ),

      //用来配置每个子 Item 的具体构建
      delegate: SliverChildBuilderDelegate(
        //构建每个 Item 的具体显示 UI
        (BuildContext context, int index) {
          //创建子 Widget
          return Container(
            alignment: Alignment.center,
            //根据角标来动态计算生成不同的背景颜色
            color: Colors.cyan[100 * (index % 9)],
            child: new Text('grid item $ index'),
          );
        }, //Grid 的个数
        childCount: 20,
      ),
    );
  }
}
```

在上述代码中使用了 SliverGridDelegateWithFixedCrossAxisCount，这个 delegate 用来根据指定的每行显示多少列 Item，而依次换行显示，不同屏幕分辨率下的手机显示的列数是一样的。

还可以使用 SliverGridDelegateWithMaxCrossAxisExtent，这个 delegate 根据每个 Item 允许的最大宽度依次排列每个 Item，也就是不同屏幕分辨率下的手机显示的列数不一样，代码如下：

```
//代码清单 5-3-2 九宫格通过构造函数来创建
//lib/code5/example_505_CustomScrollView.dart
SliverGrid buildSliverGrid2() {
  //使用构建方法来构建
  return SliverGrid(
    //用来配置每个子 Item 之间的关系
    gridDelegate: const SliverGridDelegateWithMaxCrossAxisExtent(
      //主方向每个 Item 之间的间隔
      mainAxisSpacing: 10.0,
      //次方向每个 Item 之间的间隔
      crossAxisSpacing: 10.0,
      //Item 的宽与高的比例
      childAspectRatio: 3.0,
      //每个 Item 的最大宽度
      maxCrossAxisExtent: 200,
    ),
    delegate: SliverChildListDelegate([
      Container(
        color: Colors.redAccent,
        child: new Text('grid item'),
      ),
      Container(
        color: Colors.black,
        child: new Text('grid item'),
      )
    ]),
  );
}
```

在上述代码中分别使用了 SliverChildListDelegate 和 SliverChildBuilderDelegate，这两个 delegate 的区别如下：

（1）SliverChildListDelegate 用来构建少量的 Item 应用场景，在使用这个 delegate 时，会将用到的 Item 一次性构建出来。

（2）SliverChildBuilderDelegate 用来构建大量的 Item 应用场景，这个 delegate 不会将超出屏幕以外未显示的 Item 构建出来，会在滑动到时且需要显示时才会去构建。

通过 SliverGrid. count()来创建，会指定一行展示多少个 Item，实现的效果与使用 SliverGrid 构造函数中 SliverGridDelegateWithFixedCrossAxisCount 的 delegate 创建的效果一致。

通过 SliverGrid. extent()来创建，实现的效果与使用 SliverGrid 构造函数中 SliverGridDelegateWithMaxCrossAxisExtent 的 delegate 创建的效果一致，会指定每个 Item 允许展示的最大宽度来依次排列 Item。

SliverList 只有一个 delegate 属性，可以用 SliverChildListDelegate 或 SliverChildBuilderDelegate

这两个类实现，在上述也描述及对比过这两个 delegate 的区别，前者将会一次性全部渲染子组件，后者将会根据视窗渲染当前出现的元素，在实际开发中，SliverChildBuilderDelegate 使用得比较多，代码如下：

```
//代码清单 5-3-3 SliverList 列表
//lib/code5/example_505_CustomScrollView.dart
Widget buildSliverList() {
  return SliverList(
    delegate: SliverChildBuilderDelegate(
      //构建每个 Item 的具体显示 UI
      (BuildContext context, int index) {
        //创建子 Widget
        return new Container(
          height: 44,
          alignment: Alignment.center,
          //根据角标来动态计算生成不同的背景颜色
          color: Colors.cyan[100 * (index % 9)],
          child: new Text('grid item $ index'),
        );
      },
      //列表的条目个数
      childCount: 100,
    ),
  );
}
```

5.2 列表数据展示

ListView 是最常用的可滚动列表，GridView 用来构建二维网格列表，PageView 可用于 Widget 的整屏滑动切换。

5.2.1 ListView 用来构建常用的列表数据页面

ListView 有 4 种创建方式，描述如下：

(1) 默认构造函数(传入 List children)。

(2) 通过 ListView.builder 方式来创建，适用于有大量数据的情况。

(3) 通过 ListView.custom 方式来创建，提供了自定义子 Widget 的能力。

(4) 通过 ListView.separated 方式来创建，可以配置分割线，适用于具有固定数量列表项的 ListView。

通过 ListView 的构造函数来创建，适用于构建少量数据时的场景，如图 5-3 所示，代码如下：

```
ListView(
  //子 Item
  children: const [
    Text("测试数据 1"),
    Text("测试数据 2"),
    Text("测试数据 3"),
    Text("测试数据 4"),
  ],
)
```

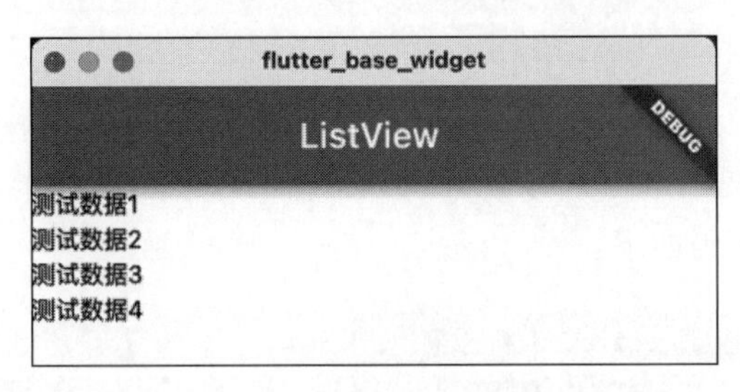

图 5-3　ListView 效果图

在实际项目业务开发中，以 ListView. builder 方式来创建使用得比较多，通常称为懒加载模式，适合列表项比较多的情况，因为只有当子组件真正显示时才会被创建，代码如下：

```
//代码清单 5-4-1 ListView 通过 builder 来构建
//lib/code5/example_510_ListView.dart
Widget buildListView1() {
  return ListView.builder(
    //列表子 Item 的个数
    itemCount: 10000,
    //每列表子 Item 的高度
    itemExtent: 100,
    //构建每个 ListView 的 Item
    itemBuilder: (BuildContext context, int index) {
      //子 Item 可单独封装成一个 StatefulWidget
      //也可以是一个 Widget
      return buildListViewItemWidget(index);
    },
  );
}
```

对于在 ListView 中构建的子 Item 的页面 UI，在实际很多应用场景中，代码量比较大，也要根据数据的不同来展示不同的效果，建议将子 Item 的构建放到单独的 StatelessWidget 或者 StatefulWidget 中，本节中显示的效果简单，所以构建了一个普通的 Widget，代码如下：

```
//代码清单 5-4-2 创建 ListView 使用的子布局
//lib/code5/example_510_ListView.dart
Widget buildListViewItemWidget(int index) {
  return Container(
    height: 84,//列表子 Item 的高度
    alignment: Alignment.center,  //内容居中
    //根据索引来动态计算生成不同的背景颜色
    color: Colors.cyan[100 * (index % 9)],
    child: Text('grid item $ index'),
  );
}
```

ListView.separated 可以在生成的列表项之间添加一个分割组件，它比 ListView.builder 多了一个 separatorBuilder 参数，该参数是一个分割组件生成器，常用于列表 Item 之间有分隔线的场景，代码如下：

```
//代码清单 5-4-3 通过 separated 来构建
//lib/code5/example_510_ListView.dart
Widget buildListView2() {
  return ListView.separated(
    //列表子 Item 的个数
    itemCount: 10000,
    //构建每个 ListView 的 Item
    itemBuilder: (BuildContext context, int index) {
      //ListView 的子 Item
      return buildListViewItemWidget(index);
    },
    //构建每个子 Item 之间的间隔 Widget
    separatorBuilder: (BuildContext context, int index) {
      //这里构建的是不同颜色的分隔线
      return Container(
        height: 4,
        //根据索引来动态计算生成不同的背景颜色
        color: Colors.cyan[100 * (index % 9)],
      );
    },
  );
}
```

ListView 的 custom()方法使用参数 childrenDelegate 来配置一个 SliverChildDelegate 代理构建子 Item，SliverChildDelegate 是抽象的，不可直接使用，一般在实际项目开发中使用它的两个子类 SliverChildListDelegate 和 SliverChildBuilderDelegate，SliverChildListDelegate 常用于构建少量数据 Item 的场景，它会一次性将所有的子 Item 绘制出来。

SliverChildBuilderDelegate 常用于构建大量的数据，采用懒加载的模式来加载数据与 ListView.builder 的加载原理一致。

一般在实际应用开发的特殊场景中，当上述 ListView 构建方式无法满足时，才使用 ListView.custom 方式，如这里需要获取页面上 ListView 屏幕中显示的第 1 个 Item 在 ListView 中对应的 postion 索引，代码如下：

```
//代码清单 5-4-4 通过 custom 来构建
//lib/code5/example_510_ListView.dart
buildListView6() {
  return ListView.custom(
    //缓存空间
    cacheExtent: 0.0,
    //自定义代理
    childrenDelegate: CustomScrollDelegate(
          (BuildContext context, int index) {
            //构建子 Item 显示布局
       return Container(
         height: 80,
         //根据索引来动态计算生成不同的背景颜色
         color: Colors.cyan[100 * (index % 9)],
       );
      },
      itemCount: 1000,//子 Item 的个数
      //滑动回调
      scrollCallBack: (int firstIndex, int lastIndex) {
        print("firstIndex $ firstIndex lastIndex $ lastIndex");
      },
    ),
  );
}
```

自定义滑动监听回调 CustomScrollDelegate，代码如下：

```
//代码清单 5-4-5 ListView 自定义滑动监听回调
//lib/code5/example_510_ListView.dart
class CustomScrollDelegate extends SliverChildBuilderDelegate {
  //定义滑动回调监听
  Function(int firstIndex, int lastIndex) scrollCallBack;
  //构造函数
  CustomScrollDelegate(builder,
      {required int itemCount, required this.scrollCallBack})
      : super(builder, childCount: itemCount);

  @override
  double? estimateMaxScrollOffset(int firstIndex, int lastIndex,
      double leadingScrollOffset, double trailingScrollOffset) {
      scrollCallBack(firstIndex, lastIndex);
```

```
        return super.estimateMaxScrollOffset(
            firstIndex, lastIndex, leadingScrollOffset, trailingScrollOffset);
    }
}
```

5.2.2 GridView 用来构建二维宫格页面

GridView 用来构建二维网格列表，GridView 创建方法有 5 种，如图 5-4 所示，描述如下：

（1）GridView 的构造函数方法，一次性构建所有的子条目，适用于少量数据。

（2）通过 GridView.builder 方式来构建，懒加载模式，适用于大量数据的情况。

（3）通过 GridView.count 方式来构建，适用于固定列的情况，适用于少量数据。

（4）通过 GridView.extent 方式来构建，适用于条目有最大宽度的限制情况，适用于少量数据。

（5）通过 GridView.custom 方式来构建，可配置子条目的排列规则，也可配置子条目的渲染加载模式。

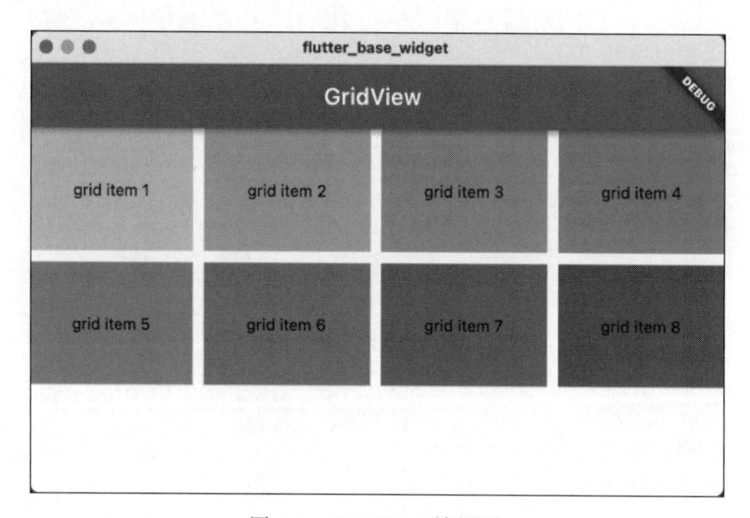

图 5-4　GridView 效果图

通过 GridView 构造函数、count 方法与 extent 方式来构建，都是一次性将所有的子 Item 构建出来，所以只适用于少量的数据，在实际业务开发中，如果数据少于一屏内容，则建议使用少数数据构建方式。

通过 GridView 的构造函数来构建，通过参数 children 来构建 GridView 中用到的所有的子条目，通过参数 gridDelegate 配置 SliverGridDelegate 来配置子条目的排列规则，可以使用 SliverGridDelegateWithFixedCrossAxisCount 和 SliverGridDelegateWithMaxCrossAxisExtent。

通过 SliverGridDelegateWithFixedCrossAxisCount 来构建一个横轴为固定数量的子条

目的 GridView，代码如下：

```
//代码清单 5-5-1 GridView 通过构造函数来创建
//1lib/code5/example_511_GridView.dart
Widget buildGridView1() {
  return GridView(
    //子 Item 排列规则
    gridDelegate: const SliverGridDelegateWithFixedCrossAxisCount(
        crossAxisCount: 4,        //横轴元素的个数
        mainAxisSpacing: 10.0,    //纵轴间距
        crossAxisSpacing: 10.0,   //横轴间距
        //子组件宽和高的比值
        childAspectRatio: 1.4),
    //GridView 中使用的子 Widegt
    children: [ … ],
  );
}
```

通过 SliverGridDelegateWithMaxCrossAxisExtent 来构建横轴 Item 数量不固定的 GridView，其水平方向 Item 的个数由 maxCrossAxisExtent 和屏幕的宽度及 padding 和 mainAxisSpacing 来共同决定，代码如下：

```
//代码清单 5-5-2 GridView 通过构造函数来创建
//1lib/code5/example_511_GridView.dart
Widget buildGridView2() {
  return GridView(
    //子 Item 排列规则
    gridDelegate: const SliverGridDelegateWithMaxCrossAxisExtent(
      maxCrossAxisExtent: 120,      //子 Item 的最大宽度
      mainAxisSpacing: 10.0,        //纵轴间距
      crossAxisSpacing: 10.0,       //横轴间距
      childAspectRatio: 1.4,        //子组件宽和高的比值
    ),
    //GridView 中使用的子 Widegt
    children: [ … ],
  );
}
```

GridView 的 count 用来构建每行有固定列数的宫格布局，参数 crossAxisCount 为必选参数，用来配置列数，与使用 GridView 通过 SliverGridDelegateWithFixedCrossAxisCount 方式来构建的效果一致，代码如下：

```
//代码清单 5-5-3 GridView 以 count 方式来创建,适用于少量数据
//1lib/code5/example_511_GridView.dart
```

```
Widget buildGridView3() {
  return GridView.count(
    crossAxisCount: 4,          //每行的列数
    mainAxisSpacing: 10.0,      //纵轴间距
    crossAxisSpacing: 10.0,     //横轴间距
    //所有的子条目
    children: [ … ],
  );
}
```

GridView 的 extent 用来构建列数不固定，限制每列的最大宽度或者高度的宫格布局，参数 maxCrossAxisExtent 为必选参数，用来配置每列允许的最大宽度或者高度，与使用 GridView 通过 SliverGridDelegateWithMaxCrossAxisExtent 方式来构建的效果一致，代码如下：

```
//代码清单 5-5-4 GridView 以 extent 方式来创建适用于少量数据
//llib/code5/example_511_GridView.dart
Widget buildGridView4() {
  return GridView.extent(
    maxCrossAxisExtent: 120,  //每列 Item 的最大宽度
    mainAxisSpacing: 10.0,    //纵轴间距
    crossAxisSpacing: 10.0,   //横轴间距
    //所有的子条目
    children: [ … ],
  );
}
```

GridView 的 builder 方式来构建，是通过懒加载模式来构建的，参数 gridDelegate 用来配置子 Item 的排列规则，与 GridView 的构造函数中 gridDelegate 使用一致，可分别使用 SliverGridDelegateWithFixedCrossAxisCount 构建固定列数的宫格和 SliverGridDelegateWithMaxCrossAxisExtent 构建不固定列数与固定 Item 最大宽度或者高度的宫格，代码如下：

```
//代码清单 5-5-5 GridView 以 builder 方式来创建
//懒加载模式适用于大量数据
//llib/code5/example_511_GridView.dart
Widget buildGridView5() {
  return GridView.builder(
    cacheExtent: 120,                      //缓存区域

    padding: const EdgeInsets.all(8),      //内边距
    itemCount: 100,                        //条目的个数
    //子 Item 排列规则
```

```
    gridDelegate: const SliverGridDelegateWithMaxCrossAxisExtent(
      maxCrossAxisExtent: 100,       //子 Item 的最大宽度
      mainAxisSpacing: 10.0,         //纵轴间距
      crossAxisSpacing: 10.0,        //横轴间距
      childAspectRatio: 1.4,         //子组件宽和高的比值
    ),
    //懒加载构建子条目
    itemBuilder: (BuildContext context, int index) {
      //index 子 Item 对应的索引
      return buildListViewItemWidget(index);
    },
  );
}
```

5.2.3 PageView 实现页面整屏切换

PageView 可用于 Widget 的整屏滑动切换,如常用的短视频 App 中的上下滑动切换的功能,也可用于横向页面的切换,如 App 第一次安装时的引导页面,也可用于开发轮播图功能,代码如下:

```
//代码清单 5-6-1 PageView 以 builder 方式来创建
//lib/code5/example_508_PageView.dart
//懒加载模式适用于大量数据
PageView buildPageView() {
  return PageView.builder(
    //当页面选中后回调此方法
    //参数[index]是当前滑动到的页面角标索引,从 0 开始
    onPageChanged: (int index) {
      print("当前的页面是 $ index");
      currentPage = index;
    },
    //当值为 false 时显示第 1 个页面,然后从左向右开始滑动
    //当值为 true 时显示最后一个页面,然后从右向左开始滑动
    reverse: false,
    //滑动到页面底部的回弹效果
    physics: const BouncingScrollPhysics(),
    scrollDirection: Axis.vertical,     //纵向滑动切换
    controller: pageController,         //页面控制器
    //所有的子 Widget
    itemBuilder: (BuildContext context, int index) {
      return SizedBox(
        width: MediaQuery.of(context).size.width,
        child: Image.asset(
          "assets/images/loginbg.png",
```

```
          ),
        );
      },
    );
  }
```

PageController 可以监听 PageView 的滑动监听，通过 PageView 的属性 controller 来绑定，代码如下：

```
//代码清单 5-6-2 PageView 控制器
//lib/code5/example_508_PageView.dart
//初始化控制器
late PageController pageController;
//PageView 当前显示页面索引
int currentPage = 0;

@override
void initState() {
  super.initState();
  //创建控制器的实例
  pageController = PageController(
    //用来配置 PageView 中默认显示的页面，从 0 开始
    initialPage: 0,
    //当为 true 时保持加载的每个页面的状态
    keepPage: true,
  );

  //PageView 设置滑动监听
  pageController.addListener(() {
    //PageView 滑动的距离
    double offset = pageController.offset;
    //当前显示的页面的索引
    double? page = pageController.page;
    print("pageView 滑动的距离 $offset 索引 $page");
  });
}
```

通过控制器 PageController 可以将 PageView 主动滑动到指定的位置，核心代码如下：

```
//代码清单 5-6-3 控制器常用方法
//lib/code5/example_508_PageView.dart
void pageViewController() {
  //以动画的方式滚动到指定的页面
  pageController.animateToPage(
    0,//子 Widget 的索引
```

```
      curve: Curves.ease,//动画曲线
      //滚动时间
      duration: const Duration(milliseconds: 200),
    );

    //以动画的方式滚动到指定的位置
    pageController.animateTo(
      100,
      //动画曲线
      curve: Curves.ease,
      //滚动时间
      duration: Duration(milliseconds: 200),
    );

    //无动画切换到指定的页面
    pageController.jumpToPage(0);
    //无动画切换到指定的位置
    pageController.jumpTo(100);
  }
```

5.3 滑动视图的应用

RefreshIndicator可结合滑动视图实现下拉刷新效果，ListView、GridView、PageView在加载多数据时，时常会用到分页加载功能，本节通过滚动监听，可以实现当滑动列表视图时动态隐藏操作按钮功能。

5.3.1 ListView下刷新与分页加载

RefreshIndicator是Material风格的滑动刷新Widget，可在ListView与GridView的外层直接嵌套使用，如图5-5所示。

图5-5 RefreshIndicator下拉刷新效果图

代码如下：

```
//代码清单 5-7-1 RefreshIndicator 基本使用
//lib/code5/example_513_ListView_RefreshIndicator.dart
buildRefreshIndicator() {
  return RefreshIndicator(
      color: Colors.blue,                    //圆圈进度颜色
      displacement: 44.0,                    //下拉停止的距离
      backgroundColor: Colors.grey[200],  //背景颜色
      //下拉刷新的回调
      onRefresh: () async {
        //模拟网络请求
        await Future.delayed(Duration(milliseconds: 2000));
        //清空数据业务操作

        //结束刷新
        return Future.value(true);
      },
      child: buildListView(),//ListView
  );
}
```

手动触发 RefreshIndicator 刷新，需要定义 GlobalKey，类型为 RefreshIndicatorState，然后通过 RefreshIndicator 的 key 绑定，代码如下：

```
final GlobalKey<RefreshIndicatorState>
      _indicatorKey = GlobalKey<RefreshIndicatorState>();

_refresh() {
  _indicatorKey.currentState?.show();
}
```

对于上滑加载更多数据，在这里通过 ListView 的滑动控制器来监听 ListView 当前滑动的距离，当滑动的距离超出总长的 2/3 时，静默加载更多数据，代码如下：

```
//代码清单 5-7-2 ListView 加载更多数据
//lib/code5/example_513_ListView_RefreshIndicator.dart
//ListView 中使用的滑动控制器
final ScrollController _scrollController = ScrollController();
//是否在加载更多
bool isLoadingMore = false;
//当前加载的页数
int pageIndex = 1;

@override
```

```
void initState() {
  super.initState();

  //添加滑动监听,在这里实现的是预一屏加载
  //也就是当数据滑动查看还只剩下一屏显示时
  //如果用户还在滑动就触发加载更多数据功能
  //在网络正常的情况下就可达成静默加载效果
  _scrollController.addListener(() {
    //获取滑动的距离
    double offset = _scrollController.offset;
    //ListView 可滑动的最大距离
    double maxOffset = _scrollController.position.maxScrollExtent;
    //当前视图的一屏高度
    double height = MediaQuery.of(context).size.height;
    if (offset >= maxOffset * 2/3 && !isLoadingMore) {
      print("上拉加载更多数据");
      //更新标识,防止重复调用,加载完成后更新标识
      isLoadingMore = true;
      //当前数据页数更新
      pageIndex++;
      //加载更多数据方法
      loadMoreData();
    }
  });
}
```

5.3.2 苹果风格下拉刷新

一个苹果风格的下拉刷新效果,在 Flutter 中需要结合滑动视图 CustomScrollView 来实现,如图 5-6 所示。

图 5-6 CupertinoSliverRefreshControl 下拉刷新效果

代码如下：

```
//代码清单 5-8 ListView 加载更多数据
//lib/code4/example_403_progress_page.dart
buildCustomScrollView() {
  return CustomScrollView(
    slivers: <Widget>[
      //下拉刷新组件
CupertinoSliverRefreshControl(
        //下拉刷新回调
        onRefresh: () async {
          //模拟网络请求
          await Future.delayed(const Duration(milliseconds: 5000));

        },
      ),
      //列表
      SliverList(
        delegate: SliverChildBuilderDelegate((content, index) {
          return ListTile(
            title: Text('测试数据 $ index'),
          );
        }, childCount: 100),
      )
    ],
  );
}
```

5.3.3 PageView 实现轮播图特效

1min

通过 PageView 实现水平轮播图效果，如图 5-7 所示。读者可观看视频讲解【5.3.3 PageView 实现轮播图特效】。

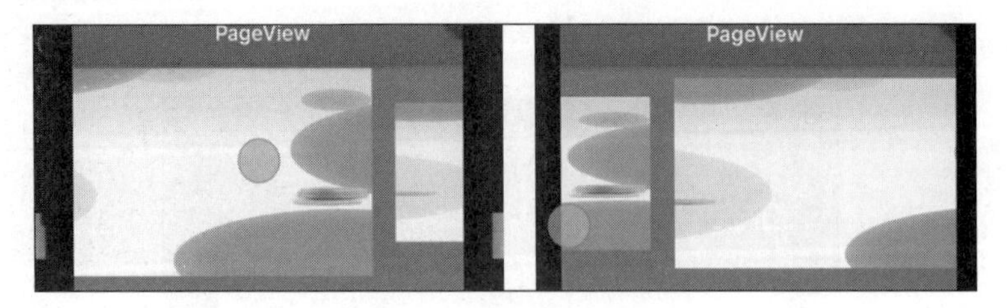

图 5-7 水平轮播图

代码如下：

```
//代码清单 5-9-1 PageView 构建
//lib/code5/example_509_PageView.dart
```

```
buildPageView() {
  return SizedBox(
    height: 200,
    child: PageView.custom(
      controller: _pageController,  //控制器
      //子 Item 的构建器,当前显示的和即将显示的子 Item 都会回调
      childrenDelegate: SliverChildBuilderDelegate(
        (BuildContext context, int index) {
          //子 Item 变换操作,移出和移进都会回调
          return buildTransform(index);
        },
        //普通数据集合
        childCount: _imageList.length,
      ),
    ),
  );
}
```

需要用到 PageView 滑动控制器来监听 PageView 的滑动,并获取当前页面中显示的 Page 的索引,代码如下:

```
//代码清单 5-9-2 PageView 控制器
//lib/code5/example_509_PageView.dart
//初始化控制器
late PageController _pageController;
//PageView 当前显示页面的索引
double? _currentPage = 0;
@override
void initState() {
  super.initState();

  //创建控制器的实例
  _pageController = PageController(
    //用来配置 PageView 中默认显示的页面,从 0 开始
    initialPage: 0,
    //当为 true 时保持加载的每个页面的状态
    keepPage: true,
  );
  //PageView 设置滑动监听
  _pageController.addListener(() {
    //PageView 滑动的距离
    setState(() {
      _currentPage = _pageController.page;
    });
  });
}
```

结合 Transform，实现 PageView 展示切换特效，代码如下：

```
//代码清单 5-9-3 Transform 构建
//lib/code5/example_509_PageView.dart
buildTransform(int index) {
  if (_currentPage != null) {
    //计算
    if (index == _currentPage!.floor()) {
      //出去的 Item
      return Transform(
          alignment: Alignment.center,
          transform: Matrix4.identity()
            ..rotateX(_currentPage! - index)
            ..scale(0.98, 0.98),
          child: buildItem(index));
    } else if (index == _currentPage!.floor() + 1) {
      //进来的 Item
      return Transform(
          alignment: Alignment.center,
          transform: Matrix4.identity()
            ..rotateX(_currentPage! - index)
            ..scale(0.9, 0.9),
          child: buildItem(index));
    } else {
      print("当前显示 $index");
      return buildItem(index);
    }
  } else {
    return buildItem(index);   //buildItem 用来构建具体的 UI 视图
  }
}
```

5.3.4 NestedScrollView 下拉刷新失效问题

在使用 NestedScrollView 结合 RefreshIndicator 实现下拉刷新功能时，会出现无法触发刷新问题，解决方法如下：

```
//代码清单 5-10 NestedScrollView 下拉刷新
//lib/code5/example_506_NestScrollView.dart
buildRefreshIndicator(){
  return RefreshIndicator(
    //可滚动组件在滚动时会发送 ScrollNotification 类型的通知
    notificationPredicate: (ScrollNotification notifation) {
      //该属性包含当前 ViewPort 及滚动位置等信息
```

```
        ScrollMetrics scrollMetrics = notifation.metrics;
        if (scrollMetrics.minScrollExtent == 0) {
          return true;
        } else {
          return false;
        }
      },
      //下拉刷新回调方法
      onRefresh: () async {
        //模拟网络刷新,等待 2s
        await Future.delayed(Duration(milliseconds: 2000));
      },
      //NestedScrollView
      child: buildNestedScrollView(),
    );
  }
```

5.3.5 滚动监听 NotificationListener

当滑动组件开始滑动时,实际上手指接触屏幕那一刻就会触发 ScrollStartNotification 通知消息,滑动组件结束滑动时会触发 ScrollEndNotification,滑动中会有 ScrollUpdateNotification,滑动越界时会触发 OverscrollNotification 通知,在 Flutter 中,可通过 NotificationListener 来捕捉这些滑动通知,代码如下:

```
//代码清单 5-11 滚动监听者
//lib/code5/example_503_NotificationListener.dart
buildNotificationListener() {
  return NotificationListener<ScrollNotification>(
    onNotification: (ScrollNotification notification) {
      //滚动信息封装对象
      ScrollMetrics metrics = notification.metrics;
      //可滑动的最大距离
      double max = metrics.maxScrollExtent;
      double extentBefore = metrics.extentBefore;
      //距离底部边距
      double extentAfter = metrics.extentAfter;
      if (_scrollController.hasClients) {
        //根据滑动组件绑定的控制器来判断监听的是哪个组件触发的滑动
        double maxScrollExtent = _scrollController.position.maxScrollExtent;
        if (max == maxScrollExtent) {
          print("netsScrollView 滑动 extentAfter $extentAfter");
        }
      }
```

```
      return false;
    },
    child: buildScrollView,
  );
}
```

对于 NotificationListener 的 onNotification()方法可以理解为事件分发，在这里如果返回值为 true，则代表消费事件，滑动通知不会再向上发送，也就是上级的 NotificationListener 无法捕捉到滑动通知，反之为 false 时，就会一级一级地向上传递。

5.3.6 ListView 实现自动滚动标签效果

3min

如视频播放时，在视频播放小窗口下会有播放集的横向列表显示，单击对应的小集，标签会自动向前或者向后滚动，效果如图 5-8 所示。读者可观看视频讲解【5.3.6 ListView 实现自动滚动标签效果】。

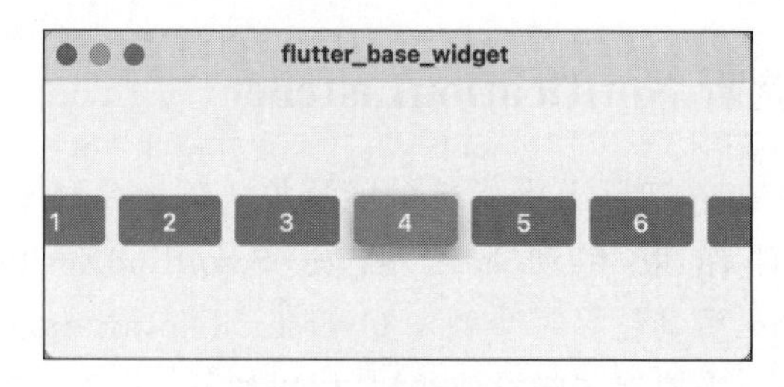

图 5-8 水平滚动标签

在页面中，组合页面可能非常复杂，本节使用 StreamBuilder 实现局部刷新功能，代码如下：

```
//代码清单 5-12 横向滚动标签
class _DemoListViewFlagPageState extends State<DemoListViewFlagPage> {

  @override
  Widget build(BuildContext context) {
    return Scaffold(
      //层叠布局
      body: Center(
        child: Stack(children: [
          //页面中的其他布局

          //横向滚动的标签
          SizedBox(
            height: 44,
            child: buildStreamBuilder(),
```

```
          )
        ]),
      ),
    );
  }
  //以多订阅流的方式来创建
  final StreamController<int> _streamController
                               = StreamController.broadcast();
  StreamBuilder<dynamic> buildStreamBuilder() {
    return StreamBuilder<int>(
      stream: _streamController.stream,
      builder: (BuildContext context, AsyncSnapshot<int> snapshot) =>
          buildListView(),
    );
  }

  @override
  void dispose() {
    _streamController.close();
    super.dispose();
  }
}
```

横向滚动的标签通过 ListView 来构建，需要获取当前屏幕上显示的第 1 个 Item 与最后一个 Item 的位置，所以使用 5.2.1 节代码清单 5-4-5 中自定义的代理 CustomScrollDelegate，代码如下：

```
//代码清单 5-12-1 横向 ListView 构建
  //滑动控制器
  final ScrollController _scrollController = ScrollController();
  int _firstIndex = 0;                //屏幕上列表显示第 1 个标签位置
  int _lastIndex = 0;                 //屏幕上列表显示最后一个标签位置
  int _currentIndex = 0;              //当前选中的标签

  buildListView() {
    return ListView.custom(
      padding: const EdgeInsets.only(left: 5, right: 5),
      controller: _scrollController,
      cacheExtent: 0.0,               //缓存空间
      scrollDirection: Axis.horizontal, //横向滑动
      //自定义代理
      childrenDelegate: CustomScrollDelegate(
        (BuildContext context, int index) {
          //ListView 的子条目
          return Row(
```

```
                mainAxisSize: MainAxisSize.min,
                children: [buildItemContainer(index)],
              );
            },
            //条目的个数
            itemCount: 1000,
            //滑动回调
            scrollCallBack: (int firstIndex, int lastIndex) {
              _firstIndex = firstIndex;
              _lastIndex = lastIndex;
            },
          ),
        );
      }
```

然后在子 Item 布局构建中，通过 Container 来限制标签的大小与外边距，通过 ElevatedButton 实现单击事件，代码如下：

```
//代码清单 5-12-2 标签构建
Container buildItemContainer(int index) {
  return Container(
      margin: EdgeInsets.only(left: 4, right: 4),
      width: 60,
      height: 28,
      child: ElevatedButton(
          style: ButtonStyle(
            elevation: MaterialStateProperty.all(
              index == _currentIndex ? 10 : 0,
            ),
            backgroundColor: MaterialStateProperty.resolveWith(
              (states) {
                if (index == _currentIndex) {
                  return Colors.orange;
                }
                //默认状态使用灰色
                return Colors.grey;
              },
            ),
          ),
          onPressed: () => clickAction(index),
          child: Text("$index")));
}
```

单击按钮时，调用 ListView 绑定的滑动控制器 scrollController 实现标签列表的自动滚动功能，代码如下：

```
//代码清单 5-12-3 标签单击事件自动滚动
  clickAction(int index) {
    _currentIndex = index;
    //计算当前屏幕中间显示的 Item 索引
    double mid = (_firstIndex + _lastIndex) / 2;
    //获取当前 ListView 滑动的距离
    double offset = _scrollController.offset;
    if (index > mid) {
      //向左
      _scrollController.animateTo(offset + 100,
          duration: Duration(milliseconds: 400), curve: Curves.easeInSine);
_streamController.add(_currentIndex);
    return;
    }
    //当前显示的是第 1 个 Item
    if (_firstIndex == 0) {
      //完全显示的第 1 个不滑动
      if (offset == 0) {
_streamController.add(_currentIndex);
        return;
      }
      //未完全显示的第 1 个滑动到顶部
      _scrollController.animateTo(0,
          duration: Duration(milliseconds: 400), curve: Curves.easeInSine);
_streamController.add(_currentIndex);
      return;
    }
    //向右
    _scrollController.animateTo(offset - 100,
        duration: Duration(milliseconds: 400), curve: Curves.easeInSine);
    //刷新页面显示
    _streamController.add(_currentIndex);
  }
```

5.4 小结

本章概述了 Flutter 项目中用来处理滑动视图的系列 Widget，App 运行在多种手机上，需要多种机型与屏幕尺寸适配，使用滑动视图是一个不错的选择，CustomScrollView 与 NestedScrollView 用来结合 Sliver 家族的 Widget 实现酷炫的滑动折叠图特效，ListView 用来处理列表视图，大部分 App 有这样的应用场景，GridView 用来处理宫格排版类，PageView 的合理应用，可以实现各种轮播图效果及页面的横向或者纵向的整屏内容切换。

核心功能篇

第6章 动画专题

CHAPTER 6

精心设计的动画会让用户界面感觉更直观、流畅，能改善用户体验，本章将概述 Flutter 的基本动画（透明度动画、缩放动画、平移动画、旋转动画）、页面交互动画、绘图动画及 Transform 矩阵变换等。

6.1 基本动画

基本动画包括透明度动画、缩放动画、平移动画、旋转动画，每种基本动画，在 Flutter 中均提供了多种实现方式，在本章中只概述最常用的方式。

6.1.1 透明度渐变动画

通过 AnimatedOpacity 组件可实现基本的透明度变化动画，读者可观看视频讲解【6.1.1 透明度渐变动画 AnimatedOpacity】。适用于简单过渡业务场景，如当其初始透明度为 1.0，然后将透明度修改为 0.0，再刷新页面 Widget 的状态（setState 方法），此时 AnimatedOpacity 的透明度会在配置的时间内慢慢地从 1.0 过渡到 0.0，呈现出来的就是一个透明渐变的效果，代码如下：

0.5min

```
//代码清单 6-1
//lib/code6/example_601_AnimatedOpacity.dart
//
//当前页面显示组件的透明度
double _opacityLevel = 1.0;

//构建透明动画组件[AnimatedOpacity]
AnimatedOpacity buildAnimatedOpacity() {
  return AnimatedOpacity(
    //透明度
    opacity: _opacityLevel,
```

```
      //过渡时间
      duration: const Duration(milliseconds: 2000),
      //动画插值器
      curve: Curves.linear,
      //动画过渡完毕的回调
      onEnd: () {},
      //子 Widget
      child: Container(
        height: 100,width: 100,color: Colors.red,
      ),
    );
  }
```

1min

FadeTransition 组件可以用来实现对子 Widget 的透明度变换，读者可观看视频讲解【6.1.1 透明度渐变动画 FadeTransition】，代码如下：

```
//代码清单 6-2-1 构建渐变动画[FadeTransition] 基本使用
//lib/code6/example_602_FadeTransition.dart
Widget buildFadeTransition() {
  //透明度渐变动画组件
  return FadeTransition(
    //过渡值
    opacity: _animationController,
    //将要执行动画的子 view
    child: Container(
      width: 200,
      height: 200,
      color: Colors.red,
    ),
  );
}
```

在这里使用了 AnimationController 动画控制器，AnimationController 默认值的范围是 0.0～1.0，AnimationController 用来对动画进行操作及监听，其创建方法如下：

```
//代码清单 6-2-2 构建渐变动画[FadeTransition] 基本使用
//lib/code6/example_602_FadeTransition.dart
class _Example602State extends State<Example602>
    with SingleTickerProviderStateMixin {
  //动画控制器
  late AnimationController _animationController;
  @override
  void initState() {
    super.initState();
```

```
    //创建动画控制器
    _animationController = AnimationController(
      //绑定 Ticker
      vsync: this,
      //正向执行的执行时间
      duration: const Duration(milliseconds: 3000),
      //反向执行的执行时间
      reverseDuration: const Duration(milliseconds: 3000),
      //执行起始值,默认为 0.0,需要比 upperBound 小
      lowerBound: 0.0,
      //执行终点值,默认为 1.0
      upperBound: 1.0,
    );

    //动画执行过程中的实时回调
    _animationController.addListener(() {});

    //添加动画控制器的动画状态监听
    _animationController.addStatusListener((status) {
      switch (status) {
        case AnimationStatus.dismissed:
          //动画从 controller.reverse()反向执行,结束时会回调此方法
          break;
        case AnimationStatus.forward:
          //执行 controller.forward()时会回调此状态
          break;
        case AnimationStatus.reverse:
          //执行 controller.reverse()时会回调此状态

          break;
        case AnimationStatus.completed:
          //动画从 controller.forward() 正向执行,结束时会回调此方法
          break;
      }
    });
  }

  …

}
```

FadeTransition 需要通过其绑定的 AnimationController 来控制动画的开始与结束,代码如下:

```
//通过控制器反向执行动画,值由 1.0 变为 0.0
_animationController.reverse();


//通过控制器正向执行动画,值由 0.0 变为 1.0
_animationController.forward(from: 0);
```

在这里对应的 State 也绑定了一个 SingleTickerProviderStateMixin,在页面 Widget 中如果只有一个 AnimationController,则绑定 SingleTickerProviderStateMixin 就可以了,如果是有多个 AnimationController,就需要绑定 TickerProviderStateMixin。

SingleTickerProviderStateMixin 与 TickerProviderStateMixin 都实现于 TickerProvider, TickerProvider 用来发送 Ticker 对象,Ticker 对象主要的作用是获取 Widget 每一帧刷新的通知,相当于一个帧定时器,在执行 start 之后会一直在指定时间内执行回调。

Flutter 应用在启动时都会绑定一个 SchedulerBinding,通过 SchedulerBinding 可以给每一次屏幕刷新添加回调,而 Ticker 就是通过 SchedulerBinding 来添加屏幕刷新回调的,每当屏幕刷新时都会调用 TickerCallback 回调。

使用 Ticker 来驱动动画会有效避免资源浪费,例如锁屏时避免执行动画,因为 Flutter 中屏幕刷新时会通知绑定的 SchedulerBinding,而 Ticker 是受 SchedulerBinding 驱动的,由于锁屏后屏幕会停止刷新,所以 Ticker 就不会再被触发了。

6.1.2 缩放动画

组件 ScaleTransition 可用来构建缩放动画效果,通过参数 alignment 来配置缩放中心,通过 scale 来配置缩放动画规则(Animation<double>类型),此处实现的是等比缩放,读者可观看视频讲解【6.1.2 缩放动画】,代码如下:

0.5min

```
//代码清单 6-3 [ScaleTransition]
//lib/code6/example_607_ScaleTransition.dart
class _Example607State extends State<Example607>
    with SingleTickerProviderStateMixin {
  //动画控制器
  late AnimationController _animationController;
  @override
  void initState() {
    super.initState();
    //创建动画控制器
    _animationController = new AnimationController(
      //绑定 Ticker
      vsync: this,
      //正向执行的执行时间
      duration: Duration(milliseconds: 1000),
```

```
    );
  }

  @override
  Widget build(BuildContext context) {…}

//构建缩放动画[ScaleTransition]
  Widget buildScaleTransition() {
    //实现的是等比缩放
    return ScaleTransition(
      //配置缩放中心
      alignment: Alignment.center,
      //过渡
      scale: _animationController,
      //将要执行动画的子 view
      child: Container(
        width: 200,
        height: 200,
        color: Colors.red,
      ),
    );
  }
}
```

6.1.3 平移动画

SlideTransition 组件可实现对子 Widget 的平移变换，SlideTransition 的 position 属性是一个 Animation < Offset >类型，用来动态设置平移位置，所以在这里构建了一个 Animation < Offset >来配合 AnimationController 综合使用，读者可观看视频讲解【6.1.3 平移动画】，代码如下：

0.5min

```
//代码清单 6-4 [SlideTransition]
//lib/code6/example_608_SlideTransition.dart
class _Example608State extends State<Example608>
    with SingleTickerProviderStateMixin {
  //动画控制器
  late AnimationController _animationController;

  late Animation<Offset> _animation;

  @override
  void initState() {
    super.initState();
```

```
    //创建动画控制器
    _animationController = AnimationController(
      vsync: this,
      duration: const Duration(milliseconds: 1000),
    );

    //通过 animate 方法将 Animation 与动画控制器 AnimationController 结合在一起
    _animation = Tween(
      //begin: Offset.zero, end: Offset(1, 0)
      //以左上角为参考点,相对于左上角坐标 x 轴方向向右平
      //平移执行动画的 view 的 1 倍宽度,y 轴方向不动,也就是水平向右平移
      begin: const Offset(-1, 0),
      //end: Offset.zero, end: Offset(1, 1)
      //以左上角为参考点,相对于左上角坐标 x 轴方向向右
      //平移执行动画的 Widget 的 1 倍宽度
      //y 轴方向向下平移执行动画 view 的 1 倍的高度
      //也就是向右下角平移
      end: const Offset(0, 0),
    ).animate(_animationController);
  }

  //代码清单[SlideTransition]
  //lib/code/code6/example_608_ScaleTransition.dart
  @override
  Widget build(BuildContext context) { … }

  //构建平移变换
  Widget buildScalContainer() {
    return SlideTransition(
      //设置平移
      position: _animation,
      //执行缩放变换的子 Widget
      child: Container(
        height: 100,
        width: MediaQuery.of(context).size.width,
        color: Colors.blue,
      ),
    );
  }
}
```

6.1.4 旋转动画

在 Flutter 中使用 RotationTransition 组件实现旋转效果,代码如下:

```
//代码清单 6-5 [RotationTransition] 旋转动画
//lib/code6/example_609_RotationTransition.dart
class _Example609State extends State<Example609>
    with SingleTickerProviderStateMixin {
  //动画控制器
  late AnimationController _animationController;

  @override
  void initState() {
    super.initState();
    //创建动画控制器
    _animationController = AnimationController(
      vsync: this,
      //正向执行的执行时间
      duration: const Duration(milliseconds: 3000),
    );
  }

  @override
  Widget build(BuildContext context) { … }

  //构建旋转变换
  Widget buildRotationTransition() {
    return RotationTransition(
      //设置旋转中心
      alignment: Alignment.center,
      //设置旋转的倍率,实际旋转的角度为 value * 2 * pi
      //如 _animationController 的值是 0.0 - 1.0
      //旋转的角度就是从 0 到 2×pi,即 360°
      //顺时针
      turns: _animationController,
      //执行缩放变换的子 Widget
      child: Container(
        height: 100,
        width: 300,
        color: Colors.blue,
      ),
    );
  }
}
```

6.2 高级动画应用提升用户视觉体验

Tween 可以实现诸多数据类型的变换,所以在 Flutter 中定义了比较多 Tween 的子类,如表 6-1 所示。

表 6-1　Tween 动画概述

类　别	描　述
ColorTween	Color 类型的动画
BoxConstraintsTween	针对于 ConstrainedBox 组件来使用
DecorationTween	容器 Container 设置装饰 Decoration 动画变化
EdgeInsetsTween	一般用于 Container 容器中进行内边距 padding 或者外边距 margin
BorderRadiusTween	用来设置边框圆角的动画变换
BorderTween	用来设置边框的动画变换
Matrix4Tween	通过动画的方式实现矩阵变换
TextStyleTween	文本样式的动态过渡
TweenSequence	串行动画，按照一定的序列来组合动画
ConstantTween	常量值动画，常结合 TweenSequence 实现保持一定时间内的值不变
SizeTween	size 类型的动画
RectTween	矩形变化动画
StepTween	TweenAnimationBuilder 适用于 Tween 系列动画的构建者，可显著提高动画效率

6.2.1　ColorTween 颜色动画过渡

当 Tween 指定 Color 类型时，可使用 ColorTween 实现颜色的动画切换，本节实现的颜色过渡动画如图 6-1 所示。

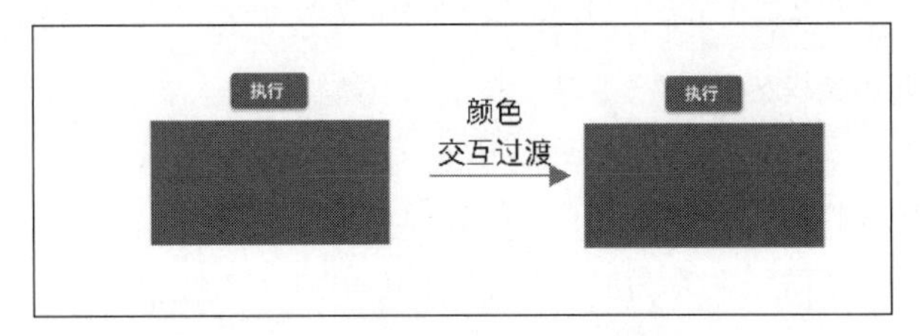

图 6-1　ColorTween 动画变换

首先创建动画控制器，代码如下：

```
//代码清单 6-6-1 [ColorTween] 的基本使用
//lib/code/code6/example_611_ColorTween.dart
class _Example611State extends State<Example611>
    with SingleTickerProviderStateMixin {
  //动画控制器
  late AnimationController _animationController;
  late Animation<Color?> _animation;

  @override
```

```
  void initState() {
    super.initState();
    //创建动画控制器
    _animationController = AnimationController(vsync: this,
        duration: const Duration(seconds: 1));
    //添加动画监听实时刷新布局
    _animationController.addListener(() {
      setState(() {});
    });
    //创建一个 Tween,值从 Colors.red 到 Colors.blue
    _animation = ColorTween(begin: Colors.red, end: Colors.blue)
        .animate(_animationController);
  }

…

}
```

然后在页面中,通过 Column 构建按钮与布局容器,单击按钮时触发动画效果,代码如下:

```
//代码清单 6-6-2 [ColorTween] 的基本使用
//lib/code/code6/example_611_ColorTween.dart
@override
Widget build(BuildContext context) {
  return Scaffold(
    backgroundColor: Colors.white,
    body: Center(
      child: Column(
        mainAxisAlignment: MainAxisAlignment.center,
        children: <Widget>[
          //单击一个按钮
          ElevatedButton(
            child: const Text("执行 "),
            onPressed: () {
              //当前动画的执行进度
              double progress = _animationController.value;
              if (progress == 1.0) {
                //反向执行
                _animationController.reverse();
              } else {
                //通过控制器正向执行动画
                _animationController.forward(from: 0);
              }
            },
```

```
          ),
          Container(
            margin: const EdgeInsets.symmetric(vertical: 10.0),
            //背景颜色从红色变为蓝色
            color: _animation.value,
            width: 200, height: 100,
          )
        ],
      ),
    ),
  );
}
```

6.2.2 跳动动画效果

在 Flutter 开发中，通过 Curve 曲线来描述动画的过程，可以是线性的 Curves. linear，也可以是非线性的 ono-linear，因此，整个动画过程可以是匀速的、加速的、先加速后减速的等。

在 Flutter 中，通过 CurvedAnimation 来描述非线性执行的动画曲线，通过 curve 来绑定不同的曲线类型，实现一个文本跳动的动画效果，代码如下：

```
//代码清单 6-7
//lib/code6/example_619_CurvedAnimation.dart
class _Example620State extends State<Example620>
    with SingleTickerProviderStateMixin {
  //动画控制器
  late AnimationController _animationController;
  //文本样式动画
  late Animation<TextStyle> _animation;
  @override
  void initState() {
    super.initState();
    //创建动画控制器
    _animationController = AnimationController(
        duration: const Duration(milliseconds: 2000), vsync: this);
    //创建曲线动画
    Animation<double> _curvedAnimation =
        CurvedAnimation(parent: _animationController, curve: Curves.bounceIn);
    //创建一个 Tween
    _animation = TextStyleTween(
      begin: const TextStyle(
          color: Colors.blue, fontSize: 14, fontWeight: FontWeight.normal),
      end: const TextStyle(
          color: Colors.deepPurple, fontSize: 24,
```

```
                    fontWeight: FontWeight.bold),
        ).animate(_curvedAnimation);
//-----------------------------------------------------------
        //添加动画状态监听
        _animation.addStatusListener((status) {
          if (status == AnimationStatus.completed) {
            //正向执行完毕后立刻反向执行(倒回去)
            _animationController.reverse();
          } else if (status == AnimationStatus.dismissed) {
            //反向执行完毕后立刻正向执行
            _animationController.forward();
          }
        });

        //添加监听
        _animation.addListener(() {});
        //正向执行
        _animationController.forward();
      }

      @override
      void dispose() {
        //销毁
        _animationController.dispose();
        super.dispose();
      }

      @override
      Widget build(BuildContext context) {
        return Scaffold(
          body: Center(
            child: Text(
              "这是一行 Flutter 代码",
              style: _animation.value,
            ),
          ),
        );
      }
}
```

6.2.3 Hero 屏幕共享元素动画

在 Flutter 中，通过 Hero 组件来组合不同路由页面过渡元素，在页面一中，通过 Hero 绑定需要过渡的元素，如一个按钮或者一组图片，读者可观看视频讲解【6.2.3 Hero 屏幕共

享元素动画】,代码如下:

```
//代码清单 6-8-1 Hero 动画
//lib/code6/example_622_Hero.dart
class Example622 extends StatelessWidget {
  @override
  Widget build(BuildContext context) {
    return Scaffold(
      appBar: AppBar(
        title: Text("Hero 动画"),
      ),
      body: Container(
        padding: EdgeInsets.all(10),
        child: Column(
          children: [
            //左上角的一个按钮
            Hero(
              //Hero 动画标签
              tag: "tag1",
              child: ElevatedButton(
                //+ 号图标
                child: Icon(Icons.add),
                onPressed: () {
                  //打开新的页面
                  openPage(context,"tag1");
                },
              ),
            )
          ],
        ),
      ),
    );
  }
```

然后在页面二中,通过 Hero 组件来绑定过渡交互的组件,需要保证两个页面中的 Hero 对应的 tag 一致,代码如下:

```
//代码清单 6-8-2 目标页面
//lib/code6/example_622_Hero.dart
class ItemDetailsPage extends StatelessWidget {
  String heroTag;
  ItemDetailsPage(this.heroTag, {Key? key}) : super(key: key);
  @override
  Widget build(BuildContext context) {
    return Scaffold(
```

```
        appBar: AppBar(
          title: const Text("页面二"),
        ),
        body: Center(
          child: Hero(
            tag: heroTag,
            child: ElevatedButton(
              child: const Icon(Icons.add),
              onPressed: () {
                Navigator.of(context).pop();
              },
            ),
          ),
        ),
      );
    }
  }
```

在页面一中，通过 Navigator 的 push 方式打开页面二，会有明显的页面切换效果，并且动画速度很快，用户视觉感不是很好，所以可以结合自定义路由动画，并且配置路由时间来控制 Hero 动画时间，达成一个舒适的视觉体验，代码如下：

```
//代码清单 6-8-3 自定义路由
//lib/code6/example_622_Hero.dart
//以透明渐变动画方式打开新的页面
void openPage2(BuildContext context, String heroTag) {
  //以动态方式打开
  Navigator.of(context).push(
    PageRouteBuilder(
      pageBuilder: (BuildContext context, Animation<double> animation,
          Animation<double> secondaryAnimation) {
        //目标页面，即页面二
        return ItemDetailsPage(heroTag);
      },
      //透明方式
      opaque: false, //动画时间
      transitionDuration: const Duration(milliseconds: 800),
      //过渡动画
      transitionsBuilder: (
        BuildContext context,
        Animation<double> animation,
        Animation<double> secondaryAnimation,
        Widget child,
      ) {
```

```
          //渐变过渡动画
          return FadeTransition(
            //透明度为 0.0~1.0
            opacity: Tween(begin: 0.0, end: 1.0).animate(
              CurvedAnimation(
                parent: animation,
                //动画曲线规则,这里使用的是先快后慢
                curve: Curves.fastOutSlowIn,
              ),
            ),
            child: child,
          );
        },
      ),
    );
}
```

6.2.4 Path 绘图高级动画

在 Android、iOS、JS、Flutter 中都可以使用 Canvas 画布结合 Path 路径来绘制任意想要的自定义图形,再结合动画,就可以实现更加酷炫的效果。

在 Flutter 中,通过 PathMetric 来度量完整的 Path 路径,PathMetric 可以获取完整 Path 路径下的任意一截数据,形成新的 Path 路径。

从一个完整路径的开始位置,在一定时间内,执行一定的更新次数,每次绘制只度量完整 Path 的一小段,就可以形成绘制 Path 的画线动画效果,如图 6-2 所示,绘制了一个矩形来形成画线动画。

在这里是通过动画控制器 AnimationController 实现在 1400ms 内从 0.0 到 1.0 的匀速控制,代码如下:

```
//代码清单 6-9-1 Path 动画
//以动态画线动画方式绘制矩形
//lib/code6/example_623_Path.dart
class Example623 extends StatefulWidget {
  @override
  _PageState createState() => _PageState();
}

class _PageState extends State with TickerProviderStateMixin {

  //创建动画控制器
  Late AnimationController animationController;

  @override
  void initState() {
```

```
    super.initState();
    //创建动画控制器 0.0 ～ 1.0,执行时间 1400 毫秒
    animationController = new AnimationController(
      vsync: this,
      duration: Duration(milliseconds: 1400),
    );

    //实时刷新
    animationController.addListener(() {
      setState(() {});
    });

    //动画状态监听
    animationController.addStatusListener((status) {
      //反复执行
      if (status == AnimationStatus.completed) {
        animationController.reset();
        animationController.forward();
      }
    });
  }

  @override
  void dispose() {
    animationController.dispose();
    super.dispose();
  }

  …
}
```

然后页面显示的 UI 布局包括两方面,代码如下:

```
//代码清单 6-9-2
  //lib/code6/example_623_Path.dart
  @override
  Widget build(BuildContext context) {
    return Scaffold(
      backgroundColor: Colors.grey,
      appBar: AppBar(
        title: Text("Path 动画 "),
      ),
      //线性排列
      body: Column(
        children: [
```

```
            //第一部分画布
            Container(
              width: MediaQuery.of(context).size.width,
              height: 200,
              color: Color(0xfffbfbfb),
              child: CustomPaint(
                //画布
                painter: PathAnimationPainter(animationController.value),
              ),
            ),
            //第二部分按钮区域
            buildContainer()
          ],
        ),
      );
    }
```

第二部分就是通过线性布局 Row 在水平方向排列两个按钮，代码如下：

```
//代码清单 6-9-3
//lib/code6/example_623_Path.dart
Container buildContainer() {
  return Container(
    child: Row(
      //子 Widget 居中
      mainAxisAlignment: MainAxisAlignment.center,
      children: [
        ElevatedButton(
          child: Text("开始"),
          onPressed: () {
            //正向执行动画
            animationController.forward();
          },
        ),
        SizedBox(
          width: 20,
        ),
        ElevatedButton(
          child: Text("停止"),
          onPressed: () {
            //正向执行动画
            animationController.reset();
          },
        ),
      ],
    ),
  );
}
```

第一部分中用到的画布 PathAnimationPainter 是一个自定义的 CustomPainter，真正的绘制操作就是在这里完成的，代码如下：

```
//代码清单 6-9-4
//lib/code6/example_623_Path.dart
class PathAnimationPainter extends CustomPainter {
  //[定义画笔]
  Paint _paint = Paint()
    ..strokeCap = StrokeCap.round       //画笔笔触类型
    ..isAntiAlias = true                //是否启动抗锯齿
    ..style = PaintingStyle.stroke      //绘画风格,默认为填充
    ..strokeWidth = 5.0;                //画笔的宽度
  //当前绘制的进度
double _progress;
  //构建函数
  PathAnimationPainter(this._progress);

  //绘制操作
  @override
  void paint(Canvas canvas, Size size) {
    canvasFunction(canvas);
  }

  //实时刷新
  @override
  bool shouldRepaint(CustomPainter oldDelegate) {
    return true;
  }
  ...
}
```

canvasFunction 中封装的是这里实现的画线动画的核心，代码如下：

```
//代码清单 6-9-5
//lib/code6/example_623_Path.dart
void canvasFunction(Canvas canvas) {
  //创建一个路径
  Path startPath = new Path();
  //向路径中添加一个矩形
  startPath.addRect(
      Rect.fromCenter(center: Offset(100, 100), width: 150, height: 100));

  //测量路径获取这个路径中所有的组合单元
  //将每个单元信息封装到[PathMetric]中
  PathMetrics pathMetrics = startPath.computeMetrics();
```

```
    //遍历 Path 中的每个单元信息
    pathMetrics.forEach((PathMetric element) {
      //路径长度
      double length = element.length;
      //是否闭合
      bool isColosed = element.isClosed;
      //角标索引
      int index = element.contourIndex;

      print("测量当前单元的长度为 $length 闭合 $isColosed 角标索引 $index");
    });

    //获取第 1 个单元
    PathMetric pathMetric = startPath.computeMetrics().first;

    //测量并裁剪路径
    Path extractPath = pathMetric.extractPath(
        //参数一开始测量的路径长度位置
        //参数二结束测量的路径长度位置
        0.0,
        pathMetric.length * _progress);

  _paint.color = const Color(0xFFE8E8E8E8);
    //绘制原路径充当背景
    canvas.drawPath(startPath, _paint);

    _paint.color = Colors.blue;
    //绘制测量裁剪后的路径
    canvas.drawPath(extractPath, _paint);
  }
```

运行效果如图 6-2 所示。

图 6-2　Path 画线动画效果图

6.2.5 AnimatedSwitcher 实现页面局部动画切换

AnimatedSwitcher 组件用来组合动画对相邻或者有关联的 Widget 进行切换过渡，如在实际应用开发中，如果页面进入后有加载数据，则需要显示加载中，如果加载完成后有数据，则构建数据页面，当无数据时构建无数据页面，这中间过程就涉及三个页面的切换，通过 AnimatedSwitcher 组件可实现这三个组件的动画切换，如这里实现的是透明渐变动画切换，代码如下：

```
//代码清单 6 - 10 AnimatedSwitcher 的基本使用
//llib/code6/example_629_AnimatedSwitcher.dart
AnimatedSwitcher buildAnimatedSwitcher() {
  return AnimatedSwitcher(
    //动画执行切换时间
    duration: const Duration(milliseconds: 1000),
    //动画构建器构建指定动画类型
    //每次修改都会回调两次,移进时回调一次,移出时回调一次
    transitionBuilder: (Widget child, Animation < double > animation) {
      //构建切换使用动画
      return FadeTransition(
        child: child,
        //透明度从 0.0 到 1.0
        opacity: Tween(begin: 0.0, end: 1.0).animate(
          CurvedAnimation(
            parent: animation,
            curve: Curves.linear,
          ),
        ),
      );
    },
    //执行动画的子 Widget
    //只有子 Widget 被切换时才会触发动画
    child: buildBodyWidget(),
  );
}

Widget buildBodyWidget() {
  //业务代码逻辑构建不同的 Widget

}
```

6.2.6 左右抖动动画效果

TweenSequence 在 Flutter 中用来将一系列的动画组合到一起，按顺序执行，可定义为

串行动画，如这里通过 TweenSequence 组合多组 Tween < double >变化值而实现左右抖动动画效果，代码如下：

```
//代码清单 6 - 11
//lib/code6/example_620_TweenSequence.dart
class _Example620State extends State<Example620>
    with SingleTickerProviderStateMixin {
  //动画控制器
  late AnimationController _animationController;
  //文本样式动画
  late Animation<double> _animation;

  @override
  void initState() {
    super.initState();
    //创建动画控制器
    _animationController = AnimationController(
        duration: const Duration(milliseconds: 200), vsync: this);
    //创建串行动画
    //使用 TweenSequence 进行多组补间动画
    _animation = TweenSequence<double>([
      //TweenSequenceItem 来组合其他的 Tween
      TweenSequenceItem<double>(tween: Tween(begin: 0, end: 20), weight: 1),
      TweenSequenceItem<double>(tween: Tween(begin: 20, end: 0), weight: 2),
      TweenSequenceItem<double>(tween: Tween(begin: 0, end: -20), weight: 3),
      TweenSequenceItem<double>(tween: Tween(begin: -20, end: 0), weight: 4),
    ]).animate(_animationController);

    //-----------------------------------------------------------------
    //添加动画状态监听
    _animation.addStatusListener((status) {
      if (status == AnimationStatus.completed) {
        //正向执行完毕后立刻反向执行(倒回去)
        _animationController.reverse();
      } else if (status == AnimationStatus.dismissed) {
        //反向执行完毕后立刻正向执行
        _animationController.forward();
      }
    });

    //添加监听
    _animation.addListener(() {
      setState(() {
        print('${_animationController.value} - ${_animation.value}');
      });
    });
```

```
    //正向执行
    _animationController.forward();
  }

  @override
  void dispose() {
    //销毁
    _animationController.dispose();
    super.dispose();
  }

  @override
  Widget build(BuildContext context) {
    return Scaffold(
      appBar: AppBar(
        title: const Text("Tween"),
      ),
      //执行平移,快速地左右平移而形成左右抖动的动画效果
      body: Transform.translate(
        offset: Offset(_animation.value, 0),
        child: Container(
          width: 200,
          height: 200,
          color: Colors.blue,
        ),
      ),
    );
  }
}
```

如在用户输入框中,当用户没有输入内容而单击"提交"按钮时,可通过抖动动画效果来提醒用户输入内容,读者可以基于 TweenSequence 原理实现,抖动效果笔者已封装为依赖库,读者可查看 2.5.6 节的 shake_animation_widget。

```
dependencies:
#效果图查看 pub 仓库说明文档
shake_animation_widget: ^3.0.1
```

6.3 AnimatedWidget 应用分析

AnimatedWidget 主要通过 Listenable 来监听子 Widget 动画,AnimatedWidget 已封装好 setState()状态更新模块,无须单独维护一个 State 状态来保存动画,所以在使用 xxxTransition 系列的动画组件时,不需要再次调用 setState()就可以实现动画更新。SlideTransition、

ScaleTransition、RotationTransition、SizeTransition、PositionedTransition、RelativePositionedTransition、DecoratedBoxTransition、AlignTransition、DefaultTextStyleTransition、AnimatedBuilder、AnimatedModalBarrier 均是 AnimatedWidget 的子类。

6.3.1 实现单击高亮自定义按钮

自定义按钮，实现的效果是按下时边框加粗高亮显示，抬起时恢复默认方式，代码如下：

```
//代码清单 6-12-1
//lib/code6/example_629_AnimatedSwitcher.dart
class ButtonContainer extends AnimatedWidget {
  GestureTapCallback? onTapUp;
  GestureTapDownCallback? onTapDown;
  String buttonText;

  ButtonContainer({
    Key? key,
    this.onTapUp,
    this.onTapDown,
    required this.buttonText,
    required AnimationController controller,
  }) : super(key: key, listenable: controller);

  Animation<double> get width => listenable as Animation<double>;

  @override
  Widget build(BuildContext context) {
    return GestureDetector(
      onTap: onTapUp,
      onTapDown: onTapDown,
      child: Container(
        child: Text(buttonText),
        padding: EdgeInsets.only(left: 10, right: 10, top: 4, bottom: 4),
        decoration: BoxDecoration(
            borderRadius: BorderRadius.all(Radius.circular(8)),
            border: Border.all(color: Colors.red, width: 2 * width.value)),
      ),
    );
  }
}
```

然后在代码中使用：

```
//代码清单 6-12-2
//lib/code6/example_629_AnimatedSwitcher.dart
```

```
class Example612 extends StatefulWidget {
  @override
  _Example612State createState() => _Example612State();
}

class _Example612State extends State<Example612>
                        with TickerProviderStateMixin {

  //创建动画控制器
  late final AnimationController _controller = AnimationController(
    duration: const Duration(milliseconds: 400),
    vsync: this,
  );

  @override
  void dispose() {
    _controller.dispose();
    super.dispose();
  }

  @override
  Widget build(BuildContext context) {
    return Scaffold(
      backgroundColor: Colors.white,
      body: Center(
        child: Column(
          mainAxisAlignment: MainAxisAlignment.center,
          children: <Widget>[

            ButtonContainer(
              buttonText: '测试一下',
              controller: _controller,
              onTapDown: (TapDownDetails details) {
                _controller.reset();
                _controller.forward();
              },
              onTapUp: () {
                _controller.reverse();
              },
            ),
          ],
        ),
      ),
    );
  }
}
```

AnimatedWidget 可以结合 Transform 变换实现更为复杂的动画效果，如这里定义一个向上滑动渐变显示的电话小按钮，代码如下：

```
//代码清单 6-12-3 平移显示
//lib/code6/example_629_AnimatedSwitcher.dart
class PhoneButtonWidget extends AnimatedWidget {
  const PhoneButtonWidget({
    Key? key,
    required AnimationController controller,
  }) : super(key: key, listenable: controller);

  Animation<double> get _progress => listenable as Animation<double>;

  @override
  Widget build(BuildContext context) {
    return Transform.translate(
        //竖直方向的位移
        offset: Offset(00, 100 * _progress.value),
        child: Opacity(
            opacity: _progress.value,
            child: Container(
                width: 100.0,
                height: 100.0,
                color: Colors.orange,
                child: Center(
                  child: GestureDetector(
                    onTap: () {},
                    child: Icon(Icons.phone, color: Colors.white),
                  ),
                ))));
  }
}
```

6.3.2 AnimatedBuilder 应用实践

AnimatedBuilder 继承于 AnimatedWidget，是一个优化实现，如图 6-3 所示。将动画定义与子 Widget 构建分离出来，代码如下：

```
//代码清单 6-13-1
//lib/code6/example_626_AnimatedBuilder.dart
AnimatedBuilder buildAnimatedBuilder() {
  return AnimatedBuilder(
    animation: _controller,
    child: Container(
```

```
      width: 200.0,
      height: 200.0,
      color: Colors.orange,
      child: const Center(
        child: Text('测试一下!'),
      ),
    ),
    builder: (BuildContext context, Widget? child) {
      return Transform.rotate(
        angle: _controller.value * 2.0 * pi,
        child: child,
      );
    },
  );
}
```

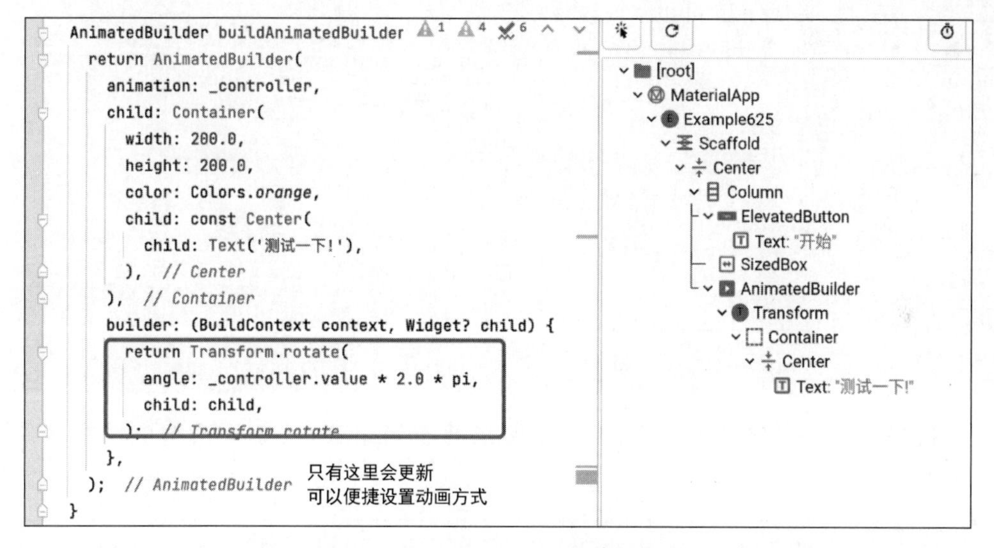

图 6-3　AnimatedBuilder 构建效果图

6.3.3　AnimatedModalBuilder 应用实践

AnimatedModalBarrier 继承于 AnimatedWidget，它的工作原理是防止用户与自身的 Widget 交互，例如在显示一个提示消息时，可以禁止用户单击按钮，代码如下：

```
//代码清单 6-14-1
//lib/code6/example_627_AnimatedModalBarrier.dart
class _AnimatedModalBarrierAppState extends State<Example627>
    with SingleTickerProviderStateMixin {
```

```
  bool _isLoading = false;
  late Widget _animatedModalBarrier;
  late AnimationController _animationController;
  late Animation<Color?> _colorTweenAnimation;

  @override
  void initState() {
    //构建动画控制器
    _animationController = AnimationController(
      vsync: this,
      duration: const Duration(seconds: 3),
    );
    //构建 Animation Color 类型
    _colorTweenAnimation = ColorTween(
      begin: Color.fromARGB(100, 255, 255, 255),
      end: Color.fromARGB(100, 127, 127, 127),
    ).animate(_animationController);

    //创建 AnimatedModalBarrier
    _animatedModalBarrier = AnimatedModalBarrier(
      color: _colorTweenAnimation,
      dismissible: true,
    );
    super.initState();
  }
}
```

按钮与 AnimatedModalBarrier 以 Stack 层叠布局组合，按钮无法响应单击事件，如图 6-4 所示。

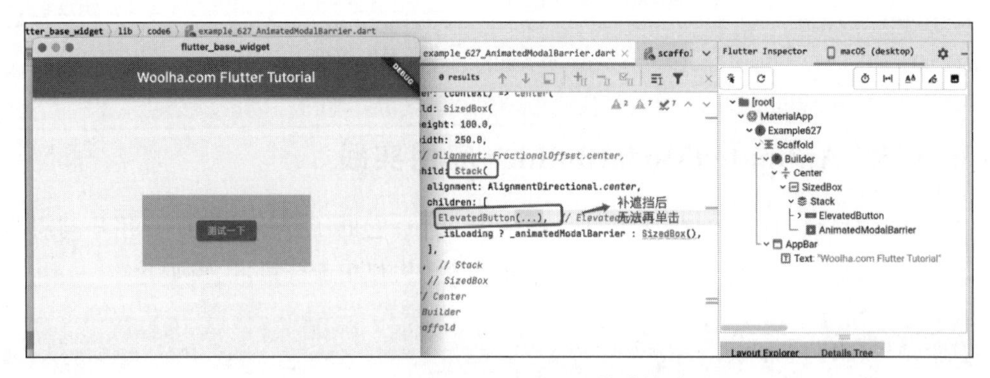

图 6-4　AnimatedModalBarrier 构建效果图

代码如下：

```
//代码清单 6-14-2
@override
Widget build(BuildContext context) {
  return Scaffold(
    body: Builder(
      builder: (context) => Center(
        child: SizedBox(
          height: 100.0,
          width: 250.0,
          //alignment: FractionalOffset.center,
          child: Stack(
            alignment: AlignmentDirectional.center,
            children: [
              ElevatedButton(
                child: Text('测试一下'),
                onPressed: () {
                  setState(() {
                    _isLoading = true;
                  });
                  _animationController.reset();
                  _animationController.forward();
                  //页面底部的弹框效果
                  ScaffoldMessenger.of(context).showSnackBar(
                    const SnackBar(
                      content: Text('按钮被单击了'),
                    ),
                  );
                  Future.delayed(const Duration(seconds: 5), () {
                    setState(() {
                      //_isLoading = false;
                    });
                  });
                },
              ),
              _isLoading ? _animatedModalBarrier : SizedBox(),
            ],
          ),
        ),
      ),
    ),
  );
}
```

6.3.4 弹簧动画应用实践

通常 UI 布局在整合业务功能时，如果页面多一些微动画，则可以使应用程序充满生机，如页面上的任何小 Widget 可以拖动一下，有微回弹效果，如弹簧上或是在重力作用下坠落(弹簧动画使用 SpringSimulation 实现),如本实例实现的是拖动一个小 Widget,然后以回弹效果移动回去,实现步骤如下:

(1) 设置动画控制器。

(2) 使用手势移动 Widget。

(3) 手指抬起时,计算速度以模拟弹簧运动。

第一步,定义动画控制器,代码如下:

```
//lib/code13/code1301.dart
class DraggableWidget extends StatefulWidget {
  const DraggableWidget({required this.child, Key? key}) : super(key: key);
  final Widget child;
  @override
  _DraggableWidgetState createState() => _DraggableWidgetState();
}

class _DraggableWidgetState extends State<DraggableWidget>
    with SingleTickerProviderStateMixin {
  late AnimationController _controller;

  @override
  void initState() {
    super.initState();
    _controller = AnimationController(vsync: this);
  }


  @override
  void dispose() {
    _controller.dispose();
    super.dispose();
  }
}
```

第二步,设置 GestureDetector 手势监听,在手指移动时,构建 Alignment,通过 Alignment 来控制对应的 Widget 移动,代码如下:

```
Alignment _dragAlignment = Alignment.center;

@override
Widget build(BuildContext context) {
  final size = MediaQuery.of(context).size;
  return GestureDetector(
    onPanDown: (details) {
      _controller.stop();  //手指按下时停止动画
```

```
      },
      onPanUpdate: (details) {
        setState(() {
          _dragAlignment += Alignment(
            details.delta.dx / (size.width / 2),
            details.delta.dy / (size.height / 2),
          );
        });
      },
      onPanEnd: (DragEndDetails details) {
        //自动将拖动的 Widget 移动回去
        _runAnimation(details.velocity.pixelsPerSecond, size);
      },
      child: Align(
        alignment: _dragAlignment,
        child: Card(
          child: widget.child,
        ),
      ),
    );
  }
```

第三步，使用 Animation<Alignment>将拖动的 Widget 通过动画回弹的方式自动移动到原来中心位置，手指抬起时，DragEndDetails 可以获取手指离开屏幕的位置与速度，代码如下：

```
Alignment _dragAlignment = Alignment.center;
late Animation<Alignment> _animation;

@override
  void initState() {
    super.initState();

    _controller.addListener(() {
      setState(() {
        _dragAlignment = _animation.value;
      });
    });
  }


  @override
  void dispose() {
    _controller.dispose();
    super.dispose();
```

```
    }

    void _runAnimation(Offset pixelsPerSecond, Size size) {
      _animation = _controller.drive(
        AlignmentTween(
          begin: _dragAlignment,
          end: Alignment.center,
        ),
      );
      final unitsPerSecondX = pixelsPerSecond.dx / size.width;
      final unitsPerSecondY = pixelsPerSecond.dy / size.height;
      final unitsPerSecond = Offset(unitsPerSecondX, unitsPerSecondY);
      //计算距离
      final unitVelocity = unitsPerSecond.distance;
      //弹簧动画效果,弹簧系数
      const spring = SpringDescription(
        mass: 30,//质量弹簧的质量,越小就越轻,弹性越强,越大越笨重,弹性越小
        stiffness: 1,//硬度
        damping: 1,//阻尼系数,阻尼系数越大,弹性越小
      );
      //弹簧动画效果
      final simulation = SpringSimulation(spring, 0, 1, -unitVelocity);
      //关联执行动画
      _controller.animateWith(simulation);
    }
```

6.4 小结

在实际项目开发中,笔者建议,对于一些简单的过渡微动画效果,可以使用 Animated 系列的组件,对于有交互的动画效果,可通过动画控制器来结合 Transition 系列的组件实现,对于变化情况复杂点的动画,如抖动、颜色过渡等动画效果可以使用 Tween 系列实现。

对于其他与用户操作或者数据访问结合度高的动画,可通过定时器或者动画控制器,在一定时间内动态修改 Widget 的位置、大小,从而体现为动画效果。

第7章 状态管理专题

CHAPTER 7

在 Flutter 中，更新页面数据显示，可以简单粗暴地执行 setState 方法，这种适用于少量代码的简单页面或者封装的 Widget，InheritedWidget 可实现数据的跨组件传输，Flutter 内置，可以基于 InheritedWidget 达到逻辑和视图解耦的效果。

模拟实际业务的一个场景，如图 7-1 所示，在页面 A 中更新数据，分别在不同的组件 B、C 中同步更新数据显示，也就是跨组件同步数据。

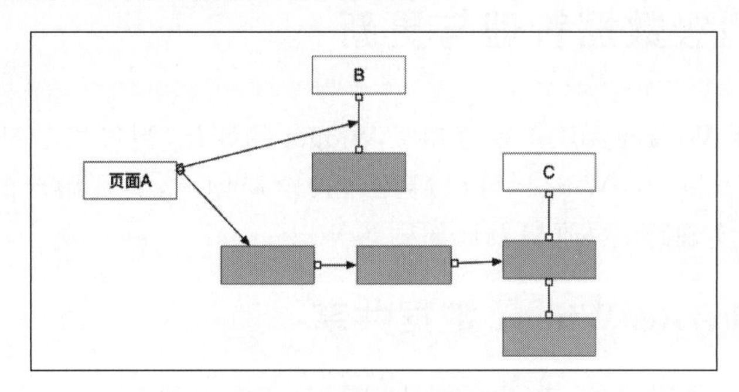

图 7-1　业务模型分析图

Provider 是谷歌官方仓库下的一种状态管理框架，本章讲解 Provider 的基本使用，提供视频讲解，视频目录如表 7-1 所示。

表 7-1　Provider 讲解视频目录

序　　号	描　　述
7-1 案例效果预览	讲解案例最终使用效果
7-2 Provider 基本使用三步曲	讲解 Provider 入门使用
7-3 Provider 获取数据的 3 种方式	讲解使用 Provider 主动获取数据的方式
7-4 Provider 通信原理简析	分析 Provider 状态管理、数据通信原理
7-5 TabBar 顶部分类标签	商品分类列表案例，顶部分类标签 UI 构建
7-6 分类页面 TabBarView	商品分类对应的分类商品列表页面 UI 构建
7-7 ListView 数据动态	商品列表，ListView 的使用
7-8 单击添加购物车	添加购物车，同步数据
7-9 购物车全局刷新问题	修改同步显示购物车数据问题

Provider
讲解视频

GetX 是 Flutter 的一个轻量级高性能的状态管理框架,同时提供了智能的依赖注入和便捷的路由管理,本章中会讲解 GetX 的基本使用、状态管理及依赖注入,也提供了视频教程,如表 7-2 所示。

表 7-2 GetX 讲解视频目录

序　　号	描　　述
7-10 GetX 的路由管理	讲解使用 GetX 的路由功能
7-11 GetX 基本使用三步曲	讲解 GetX 的基本使用
7-12 GetX 状态管理的局部更新	讲解使用 GetBuilder 实现局部更新功能
7-13 GetX 响应式编程	概述 GetX 的响应编程,以及基本使用案例
7-14 GetX 的工作流搜索框的实时搜索解决方案	概述 GetX 的工作流机制,实现一个搜索框搜索的案例
7-15 GetX 商品列表页面 UI 构建	商品分类对应的分类商品列表页面 UI 构建
7-16 GetX 实现商品列表页面动态数据	商品列表,ListView 的使用
7-17 GetX 商品列表添加购物车功能	添加购物车,同步数据

GetX
讲解视频

7.1 入门级数据管理与更新

通常情况下,子 Widget 无法单独感知父 Widget 的变化,当父状态变化时,可通过其重建所有子 Widget; InheritedWidget 可以避免这种全局创建,实现局部的子 Widget 更新; ValueNotifier 可以实现跨组件通信及局部刷新 Widget。

7.1.1 InheritedWidget 数据共享

InheritedWidget 提供了在 Widget 树中从上到下共享数据的方式,例如在应用的根 Widget 中通过 InheritedWidget 共享了一个数据,便可以在任意子 Widget 中获取该共享的数据,如图 7-1 所示的业务数据模型。本节通过 InheritedWidget 原理实现数据的共享,如图 7-2 所示,是本节构建的页面模型。

首先创建自定义的 InheritedWidget 来提供数据共享模型,代码如下:

```
//代码清单 7-1-1
//lib/code7/code701_main_page.dart
@immutable
class CountInherited extends InheritedWidget {
  //共享数据
  final int data;
  const CountInherited({Key? key, this.data = 0, required Widget child})
      : super(key: key, child: child);
```

```
  //定义一种方法,方便子树中的 Widget 获取这个 Widget,从而获得共享数据
  static CountInherited? of(BuildContext context) {
    return context.dependOnInheritedWidgetOfExactType<CountInherited>(
        aspect: CountInherited);
  }

  @override
  bool updateShouldNotify(CountInherited oldWidget) {
    return true;
  }
}
```

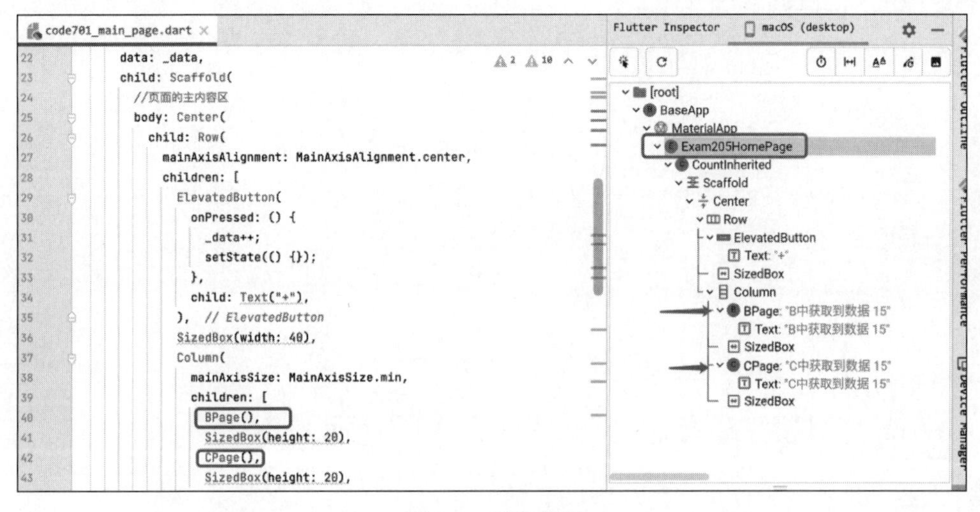

图 7-2　页面模型

然后在主页面中通过自定义的 CountInherited 来构建页面根视图,并设置初始的数据模型,如这里实现的是单击按钮修改 data 的值,代码如下:

```
//代码清单 7-1-2
//lib/code7/code701_main_page.dart
class _Exam205HomePageState extends State<Exam205HomePage> {
  int _data = 0;
  @override
  Widget build(BuildContext context) {
    //Scaffold 用来搭建页面的主体结构
    return CountInherited(
      data: _data,
      child: Scaffold(
        //页面的主内容区
        body: Center(
```

```
            child: Row(
              mainAxisAlignment: MainAxisAlignment.center,
              children: [

                ElevatedButton(
                  onPressed: () {
                    _data++;
                    setState(() {});
                  },
                  child: const Text(" + "),
                ),

                Column(
                  mainAxisSize: MainAxisSize.min,
                  children: [

                    BPage(),

                    CPage(),

                  ],
                )
              ],
            ),
          ),
        ),
      );
  }
}
```

最后在子组件 Bpage 与 Cpage 中获取共享数据，代码如下：

```
//代码清单 7-1-3
//lib/code7/code701_main_page.dart
class BPage extends StatelessWidget {
  @override
  Widget build(BuildContext context) {
    //获取实例
    CountInherited? countInherited = CountInherited.of(context);
    if (countInherited != null) {
      //获取数据
      return Text("B 中获取数据 ${countInherited.data}");
    }
    return Text("B 中未获取数据");
  }
}
```

7.1.2 ValueNotifier 单数据模型通信

在 Flutter 中使用 setState(){}刷新页面时，会导致整个界面重新构建，在实际需求中想更新的 Widget 只有某一个或几个，所以整个界面的重绘，会造成不必要的性能损耗，此时可以实现跨组件通信及局部刷新 Widget。

使用 ValueNotifier，第一步是创建实例对象，如在这里通过 ValueNotifier 更新的数据类型为 String 类型，定义一个全局 ValueNotifier，可以定义在一个的单独 dart 文件中，代码如下：

第一步，定义 ValueNotifier，这里传递的数据类型为 String

```
ValueNotifier<String> testValueNotifier = ValueNotifier<String>('');
```

第二步，在需要同步更新数据的位置使用 ValueListenableBuilder 来订阅第一步创建的 ValueNotifier 实例，代码如下：

```
//代码清单 7-2
//第二步在需要同步数据的地方使用
//lib/code7/code701_main_page.dart
class BPage extends StatelessWidget {
  const BPage({Key? key}) : super(key: key);
  @override
  Widget build(BuildContext context) {
    return ValueListenableBuilder(
      //数据发生变化时回调
      builder: (context, value, child) {
        return Text("B未获取数据 $value");
      },
      //监听的数据
      valueListenable: testValueNotifier,
    );
  }
}
```

第三步，操作数据更新，如单击按钮后通过 ValueNotifier 来发送更新的数据，代码如下：

```
//第三步是数据变化后赋值更新
testValueNotifier.value = '----';
```

当 ValueNotifier 中绑定的 value 值发生变化后，对应的 ValueListenableBuilder 中会收到回调自动刷新布局。

7.1.3 ValueNotifier 自定义模型通信

在实际应用开发中，业务中用到的数据通常比较复杂。

第一步，自定义一个用户数据模型，代码如下：

```
//本节源码存放在 flutter_base_widget 项目的
//lib/code7/code703_ValueNotifier_page.dart 文件中

//代码清单 7-2-1 在实际应用开发中变量可能足够多
class UserInfo {
  String? name;
  int? age;
  UserInfo({this.name, this.age});
}
```

在业务中，可能有很多组件需要同步更新用户的信息，有时会更新用户信息中的某一个值，而不需要更新所有的信息。

第二步，定义一个 ValueNotifier，代码如下：

```
//代码清单 7-2-2
//UserInfo 为数据类型
class UserNotifier extends ValueNotifier<UserInfo> {
  UserNotifier(UserInfo userInfo) : super(userInfo);
  //自定义更新值方法
  void setAge(int age) {
    value.age = age;
    //通知更新
    notifyListeners();
  }
}
```

第三步，创建 UserNotifier 实例，通过 UserNotifier 来更新数据与绑定监听者，代码如下：

```
//第三步，创建的实例参数为用户信息的初始信息
UserNotifier userNotifier
              = UserNotifier(UserInfo(name: "张三", age: 0));
```

第四步，通过 ValueListenableBuilder 订阅 UserNotifier 实现数据同步的功能，ValueListenableBuilder 的 child 属性可以用来构建不需要刷新的子视图，其构建的 Widget 在 builder 回调中，代码如下：

```
//代码清单 7-2-3
//第四步,设置监听
class BPage extends StatelessWidget {
  const BPage({Key? key}) : super(key: key);
  @override
  Widget build(BuildContext context) {
    return ValueListenableBuilder(
      //数据发生变化时回调
      builder: (context, UserInfo value, Widget? child) {
        return Column(
          mainAxisSize: MainAxisSize.min,
          children: [
            Text("B未获取数据 ${value.age}"),
            child ?? const SizedBox(),
          ],
        );
      },
      //监听的数据
      valueListenable: userNotifier,
      child: const Text(
        '子视图中的内容',
        style: TextStyle(color: Colors.red),
      ),
    );
  }
}
```

最后一步,更新用户的数据,代码如下:

```
//赋值更新
userNotifier.setAge(22);
```

7.2 Stream 流通信

Stream 是一个抽象类,用来表示一个序列的异步数据,Stream 流分类如下:

(1) 单订阅流(Single Subscription),这种流最多只能有一个监听器(listener)。

(2) 多订阅流(Broadcast),这种流可以有多个监听器(listener)。

Stream 可以接收任何类型的数据,Stream 有同步流和异步流之分,它们的区别在于同步流会在执行 add、addError 或 close 方法会立即发送事件,而异步流总是在事件队列中的代码执行完成后再发送事件。

7.2.1 多订阅流实现多组件同步数据

如本章开头图 7-1 所示的数据模型，在页面 A 中更新数据，在组件 B、C 中同步更新数据，可通过多订阅流结合 StreamBuilder 同步更新数据，第一步是创建多订阅流控制器，代码如下：

```
//本节源码存放在 flutter_base_widget 项目的
//lib/code7/code705_Stream_page.dart 文件中
```

第一步，创建多订阅流控制器，代码如下：

```
StreamController<String> streamController = StreamController.broadcast();
```

第二步，在需要同步更新的 Widget 中通过 StreamBuilder 来绑定 Stream，代码如下：

```
//代码清单 7-3
class BPage extends StatelessWidget {
  const BPage({Key? key}) : super(key: key);
  @override
  Widget build(BuildContext context) {
    return StreamBuilder<String>(
      //绑定 stream
      stream: streamController.stream,
      //默认的数据初始
      initialData: "00:00:00",
      //构建绑定数据的 UI
      builder: (BuildContext context, AsyncSnapshot<String> snapshot) {
        //snapshot.data 是传递的数据对象
        return Text(
          'B 中更新数据 ${snapshot.data} ',
          style: const TextStyle(fontSize: 22, color: Colors.blue),
        );
      },
    );
  }
}
```

第三步，当数据发生变化时，通过第一步创建的 StreamController 实例发送数据，如图 7-3 所示，单击按钮发送数据，在组件 B、C 中同步更新数据，代码如下：

```
//StreamController 发送数据
streamController.add("5");
```

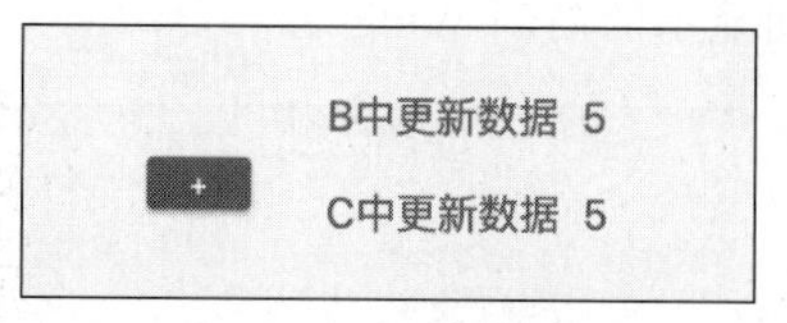

图 7-3 跨组件更新数据

可以通过 StreamController 为 Stream 添加事件监听，当有数据变化时，监听接收到回调，这方面可用作 Flutter 开发中的消息通道，代码如下：

```
//代码清单 7-4
class BPageState extends State {
  //消息订阅对象
  StreamSubscription? _streamSubscription;
  @override
  void initState() {
    super.initState();
    //监听二
    _streamSubscription = streamController.stream.listen((event) {
      print("页面 B 接收到数据 $event");
    });
  }

…

  @override
  void dispose() {
    super.dispose();
    if(_streamSubscription!= null) {
      //取消监听
      _streamSubscription?.cancel();
    }
  }

}
```

通过 StreamController.broadcast()方式创建的是多订阅流，单订阅流与多订阅流的区别是，单订阅流只能添加一个监听，单订阅流的创建方式如下：

```
StreamController<String> streamController2 = StreamController();
```

7.2.2 单订阅流实现计时功能

本节实现的计时功能如图 7-4 所示，在页面显示中，计时功能是实时刷新的，所以可以通过 StreamBuilder 刷新时间显示的局部 Widget。

图 7-4 计时功能

在本节中，通过 Timer 实现计时功能，在页面销毁时，为避免内存泄漏，需要在页面销毁的方法中关闭计时器，同时也关闭订阅流，代码如下：

```
//代码清单 7-5-1
//本节源码在 lib/code7/code706_StreamTimer_page.dart 文件中
class _TestABPageState extends State<Exam205HomePage> {

  //使用单订阅流即可
  final StreamController<String> _streamController = StreamController();
  //计时器
  Timer ?_timer;

  @override
  void initState() {
    super.initState();
    //间隔 1s 执行时间
    _timer = Timer.periodic(const Duration(milliseconds: 1000), (timer) {
      //获取当前的时间
      DateTime dateTime = DateTime.now();
      //格式化时间
      String formatTime = DateFormat("HH:mm:ss").format(dateTime);
      //流数据更新
      _streamController.add(formatTime);
    });

  }

  @override
  void dispose() {
    super.dispose();
    //关闭
    _streamController.close();
    if(_timer!= null&&_timer!.isActive) {
      //取消计时器
      _timer!.cancel();
      _timer = null;
    }
  }

}
```

在这里通过 DateFormat 来格式化时间显示，需要导入 intl.dart 文件，代码如下：

```
import 'package:intl/intl.dart';
```

然后显示时间计时的是 StreamBuilder 包裹的一个 Text 文本，代码如下：

```
//代码清单 7-5-2
@override
Widget build(BuildContext context) {
  //页面主体脚手架
  return Scaffold(
    body:Center(child: buildStreamBuilder(),),
  );
}

//监听 Stream，每次值改变的时候，更新 Text 中的内容
StreamBuilder<String> buildStreamBuilder() {
  return StreamBuilder<String>(
    //绑定 stream
    stream: _streamController.stream,
    //默认的数据
    initialData: "00:00:00",
    //构建绑定数据的 UI
    builder: (BuildContext context, AsyncSnapshot<String> snapshot) {
      //snapshot.data 是传递的数据对象
      return Text(
        '当前时间 ${snapshot.data} ',
        style: const TextStyle(fontSize: 22, color: Colors.blue),
      );
    },
  );
}
```

7.3 Provider 状态管理

Provider 是一个组件，也是一种编程思想，在使用 Provider 时首先要添加依赖，先到 pub 仓库中查看最新版本：

```
//pub 仓库地址
https://pub.flutter-io.cn/
```

然后复制最新的版本，到项目的配置文件 pubspec.yaml 中添加依赖，本书使用的版本信息如图 7-5 所示。

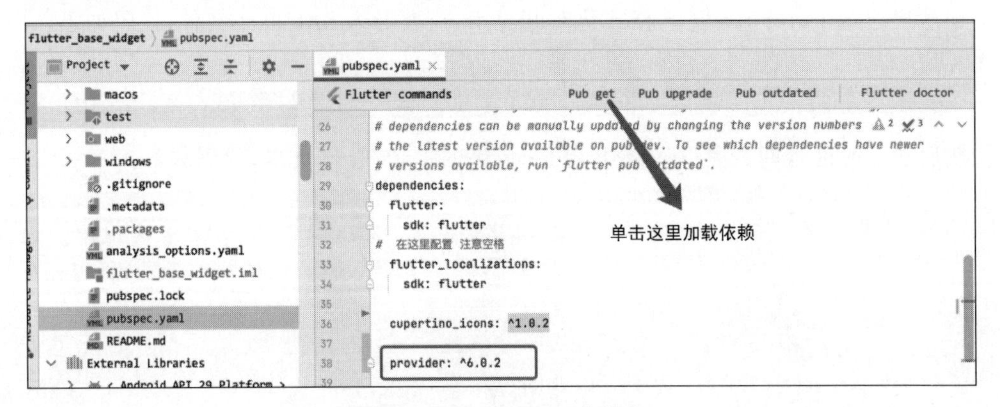

图 7-5　添加 Provider 依赖库

Provider 库中常用的组件有 ChangeNotifierProvider、MultiProvider、Provider、Consumer、Consumer2 等。

7.3.1 Provider 单数据模型通信

Provider 可实现父子组件、跨组件通信，如图 7-1 所示的数据模型。本节通过 Provider 结合 ChangeNotifierProvider、Consumer 实现一个计时显示功能。

第一步，创建数据监听模型，代码如下：

```
//代码清单 7-6
//lib/code7/provider/demo1/time_counter_model.dart
import 'package:flutter/cupertino.dart';
import 'package:intl/intl.dart';
class TimeCounterModel with ChangeNotifier {

  String _formatTime = "00:00:00";

  void getCurrentTime() {
    //获取当前的时间
    DateTime dateTime = DateTime.now();
    //格式化时间 import 'package:intl/intl.dart';
    //需要添加 intl 依赖
    _formatTime = DateFormat("HH:mm:ss").format(dateTime);
    notifyListeners();                  //2
  }

  get formatTime => _formatTime;        //3
}
```

第二步，通过 ChangeNotifierProvider 绑定上述创建的 TimeCounterModel，代码如下：

```
//代码清单 7-7-1
//lib/code7/provider/demo1/consumer_page.dart
void main() => runApp(
  ChangeNotifierProvider(
    create: (BuildContext context) {
      return TimeCounterModel();
    },
    child: const MaterialApp(
      //默认的首页面
      home: ConsumerTimePage(),
    ),
  ),
);
```

在这里直接将 ChangeNotifierProvider 应用在根视图，读者可根据实际业务版本，将其应用在对应代码模块父布局处，ConsumerTimePage 是自定义的一个首页面，在其中通过 Timer 提供了计时功能，代码如下：

```
//代码清单 7-7-2
class ConsumerTimePage extends StatefulWidget {
  const ConsumerTimePage({Key? key}) : super(key: key);

  @override
  State<StatefulWidget> createState() {
    return _TestABPageState();
  }
}
class _TestABPageState extends State {
  //计时器
  Timer? _timer;
  @override
  void initState() {
    super.initState();
    //间隔 1s 执行时间
    _timer = Timer.periodic(const Duration(milliseconds: 1000), (timer) {
      //发送事件
      Provider.of<TimeCounterModel>(context,listen: false).getCurrentTime(); //2
    });
  }
  @override
  void dispose() {
    super.dispose();
    //取消计时器
    _timer!.cancel();
```

```
  }
  @override
  Widget build(BuildContext context) {
    //页面主体脚手架
    return const Scaffold(
      body: Center(child: BPage(),),
    );
  }
}
```

第三步，在页面中加载使用 BPage，此处的消费者就是需要同步更新数据的 Widget，代码如下：

```
//代码清单 7-7-2
class BPage extends StatelessWidget{
  const BPage({Key? key}) : super(key: key);

  @override
  Widget build(BuildContext context) {
    return Consumer<TimeCounterModel>(
      //参数 value 是绑定的事件结果 TimeCounterModel
      builder: (BuildContext context, value, Widget ?child) {
        return Container(
          //外边距
          margin: EdgeInsets.only(left: 12, top: 12),
          child: Text(
            '${value.formatTime}',
            style: TextStyle(fontSize: 22.0, color: Colors.red),
          ),
        );
      },
    );
  }
}
```

7.3.2 Provider 多数据模型通信

当有多个数据模型时，可通过 MultiProvider 来组合多个 ChangeNotifierProvider，在 7.3.1 节中定义 TimeCounterModel 来处理时间数据，本节中再定义 RandomNumberModel 来处理随机数据模型，代码如下：

```
//代码清单 7-8-1
//lib/code7/provider/demo1/provider_mult.dart
```

```
class RandomNumberModel with ChangeNotifier {
  int _randomNumber = 0;
  //指定数据
  void testNumber(int number) {
    _randomNumber = number;
    notifyListeners(); //2
  }
  //随机数据
  void testRandom() {
    _randomNumber = Random().nextInt(100);
    notifyListeners();
  }
  get randomNumber => _randomNumber; //3
}
```

然后通过 MultiProvider 来组合这两个数据模型，代码如下：

```
//代码清单 7-8-2
//lib/code7/provider/demo1/provider_mult.dart
void main() => runApp(
      const MaterialApp(
        //默认的首页面
        home: TestProviderMulPage(),
      ),
    );

class TestProviderMulPage extends StatelessWidget {
  const TestProviderMulPage({Key? key}) : super(key: key);
  @override
  Widget build(BuildContext context) {
    //组合多个 Provider
    return MultiProvider(
      providers: [
        //计时器
        ChangeNotifierProvider(
          create: (BuildContext context) {
            return TimeCounterModel();
          },
        ),

        //随机数据
        ChangeNotifierProvider(
          create: (BuildContext context) {
            return RandomNumberModel();
          },
```

```
      )
    ],
    child: const TestConsumerTimePage(),
  );
 }
}
```

TestConsumerTimePage 是显示的首页面，在其中定义一个计时器，间隔 1s 时间来模拟更新数据，代码如下：

```
//代码清单 7-8-3
class TestConsumerTimePage extends StatefulWidget {
  const TestConsumerTimePage({Key? key}) : super(key: key);
  @override
  State<StatefulWidget> createState() {
    return _TestABPageState();
  }
}

class _TestABPageState extends State {
  //计时器
  Timer? _timer;
  @override
  void initState() {
    super.initState();
    //间隔 1s 执行时间
    _timer = Timer.periodic(Duration(milliseconds: 1000), (timer) {
      //发送事件
      Provider.of<TimeCounterModel>(context, listen: false).getCurrentTime();
      Provider.of<RandomNumberModel>(context, listen: false).testRandom(); //2
    });
  }

  @override
  Widget build(BuildContext context) {
    //页面主体脚手架
    return const Scaffold(
      body: CPage(),
    );
  }

  @override
  void dispose() {
    super.dispose();
    //取消计时器
```

```
    if(_timer!= null) {
      _timer!.cancel();
    }
  }
}
```

Consumer 用来接收单数据，Consumer2 可以同时接收双数据监听，依次还有 Consumer3、Consumer4、Consumer5、Consumer6，在本节中，使用 Consumer2 处理两个数据模型的同时更新监听，其他多数据模型监听与此类似，代码如下：

```
//代码清单 7-8-4
class CPage extends StatelessWidget{
  const CPage({Key? key}) : super(key: key);
  @override
  Widget build(BuildContext context) {
    //通过 Consumer2 同时监听处理两个结果
    return Consumer2<TimeCounterModel, RandomNumberModel>(
      //参数 value 为 TimeCounterModel 类型
      //参数 value2 为 RandomNumberModel 类型
      builder: (BuildContext context, value, value2, Widget? child) {
        return Container(
          margin: EdgeInsets.only(left: 12, top: 12),
          child: Text(
            '当前时间 ${value.formatTime} 随机数 ${value2.randomNumber}',
            style: TextStyle(fontSize: 22.0, color: Colors.red),
          ),
        );
      },
    );
  }
}
```

7.4 GetX 状态管理

GetX 是一个超轻量的 Flutter 状态管理框架，可通过 GetX 实现状态管理、智能地依赖注入及路由管理，在使用 GetX 时，首先需要在项目的配置文件 pubspec.yaml 中添加依赖，代码如下：

```
#https://pub.flutter-io.cn/ 中搜索 get
get: ^4.6.1
```

使用 GetX 的第一步，在应用根视图使用 GetMaterialApp，在这可以很方便地使用

GetX 的 routes、snackbars、dialog、bottomsheets 等,7.4 节的源码在本书配套源码 flutter_getx_shop 项目中,程序入口代码如下:

```
//代码清单 7-9
//程序入口
void main() {
  runApp(RootApp());
}

class RootApp extends StatelessWidget {
  @override
  Widget build(BuildContext context) {
    //使用 GetX 的第一步
    return GetMaterialApp(
      theme: ThemeData(
        primarySwatch: Colors.blue,
      ),
      //静态路径
      routes: {
        "/testa": (context) => TestAPage(),
      },
      home: MyHomePage(),   //默认显示的首页页面
    );
  }
}
```

7.4.1 GetX 路由管理功能

在前面章节中,通过 Navigator 实现页面的跳转,GetX 提供了路由管理功能,可以实现 Navigator 的一系列功能,如基本的路由功能——打开新页面,代码如下:

```
void fun11() async {
  //动态路由的方式
  dynamic result = await Navigator.of(context).push(
    new MaterialPageRoute(
      builder: (BuildContext context) {
        return new TestAPage();
      },
    ),
  );
  //获取上一个页面的数据
}
```

GetX 打开新页面,代码如下:

```
//GetX 以动态的方式获取数据
void fun5() async {
  //以动态的方式获取 A 页面的返回值
  var value = await Get.to(new TestAPage());
  print("A 页面的返回值 $value");
}
```

在 TestAPage 中，退出页面时将回传数据，Navigator 实现方式的代码如下：

```
Navigator.of(context).pop("A 页面回传的数据");
```

GetX 实现方式的代码如下：

```
//退出页面
//Get.back();

//退出页面返回数据
Get.back(result: "GetX 关闭的页面");
```

在业务开发中，通常需要打开一个新页面，然后关闭当前面，代码如下：

```
void fun6() {
  //GetX 替换当前的页面
  Get.off(new TestAPage());
  //Navigator 方式
  Navigator.of(context).pushReplacement(
    new MaterialPageRoute(
      builder: (BuildContext context) {
        return new TestAPage();
      },
    ),
  );
}
```

一种应用场景是当用户单击退出登录时需要关闭所有的页面，然后打开一个新的页面，代码如下：

```
void fun7() {
  //GetX 打开新的页面并关闭之前所有的页面
  Get.offAll(new TestAPage());

  //Navigator 打开新的页面并关闭之前所有的页面
  Navigator.of(context).pushAndRemoveUntil(
    new MaterialPageRoute(
      builder: (BuildContext context) {
```

```
        return new TestAPage();
      },
    ),
    (Route route) => false,
  );
}
```

7.4.2 GetX 局部数据更新

GetxController 结合 GetBuilder 实现局部数据更新，第一步创建自定义的 GetxController，在其中实现业务代码，如基本的网络请求、数据处理等，在这里定义两个用来将数据相加的方法来模拟实际中的业务处理，代码如下：

```
//代码清单 7-10
//第一步,定义 Controller
class CountController extends GetxController{

  int _count = 0;
  int _count2 = 0;

  int get count => _count;
  int get count2 => _count2;

  //根据 ID 更新
  void addCount2(){
    _count2++;
    //更新指定位置(GetBuilder),count2 是自定义的 ID,可以随意定义
    update(["count2"]);
  }

  void add(){
    _count++;
    //更新所有绑定此 Controller 的 GetBuilder
    update();
  }
}
```

第二步，在页面 UI 中在需要同步更新数据的位置使用 GetBuilder 来构建，如图 7-6 所示。

图 7-6 GetBuilder 同步数据

在这里构建的文本用来同步数据，在单击按钮时，通过 Get.find 方式获取对应的 Controller，用来调用对应的方法，代码如下：

```
//代码清单 7-10
//lib/page/count/CountMainPage.dart
buildColumn2() {
  return Column(
    mainAxisSize: MainAxisSize.min,
    children: [
      SizedBox(height: 30),
      //第二步,使用 GetBuilder 获取数据
      GetBuilder<CountController>(
        init: CountController(),
        builder: (CountController controller) {
          return Text("当前显示计数 ${controller.count}");
        },
      ),
      SizedBox(height: 10),
      ElevatedButton(
        onPressed: () {
          //第三步,使用 find ,调用方法通信
          Get.find<CountController>().add();
        },
        child: Text("添加"),
      ),
    ],
  );
}
```

在一个页面或者一个 Widgets 树结构中，可能会有多个 GetBuilder 绑定了同一个 GetxController，可以通过 GetBuilder 的 id 方法设置一个唯一的 ID，在自定义控制器中，如这里的 CountController 的 addCount2()方法中，只更新对应的 ID 的 GetBuilder，代码如下：

```
//代码清单 7-11
//lib/page/count/CountMainPage.dart
//第二步,使用 GetBuilder 获取数据
GetBuilder<CountController>(
    init: CountController(),
    id: "count2",
    builder: (CountController controller) {
      return Text("当前显示计数 ${controller.count2}");
    }),


ElevatedButton(
```

```
  onPressed: () {
    //第三步,使用 find 调用方法通信
    Get.find<CountController>().addCount2();
  },
  child: Text("添加"),
),
```

7.4.3 GetX 依赖注入

依赖注入是将一个类的实例注入另一个类的技术,如在实际业务开发中,当进入一个新页面 A 时,需要调用 GetxController 的方法来加载处理数据,然后在这个页面 A 的子组件 B 和 C 中同步更新数据,在页面 A 中处理数据,需要用到 GetxController 的实例对象,这里就用到了 GetX 的依赖注入功能,首先是对应的 GetxController,代码如下:

```
//代码清单 7-12
//第一步定义 Controller
class TestController extends GetxController{

  int _count = 0;
  int get count => _count;


  //模拟业务处理数据
  void add(){
    _count++;
    //更新所有绑定此 Controller 的 GetBuilder
    update();
  }

  void requestNetData() async{
    //模拟网络请求
    await Future.delayed(Duration(milliseconds: 1400),(){});
    _count = 100;
    //数据请求处理完成后更新页面内容并显示
    update();
  }
}
```

然后在用到异步加载数据的页面中注入 TestController 对象,代码如下:

```
//代码清单 7-13-1
//lib/page/count/CountTestAPage.dart
class CountTestAPage extends StatefulWidget {
```

```
  @override
  _CountTestAPageState createState() => _CountTestAPageState();
}
class _CountTestAPageState extends State<CountTestAPage> {
  //注入 GetxController 对象
  TestController _controller = Get.put(TestController());
  @override
  void initState() {
    super.initState();
    //异步请求网络数据
    _controller.requestNetData();
  }

  @override
  Widget build(BuildContext context) {
    return Scaffold(
      body: Container(
        child: Column(
          children: [
            BPage(),
            SizedBox(
              height: 40,
            ),
            CPage(),
          ],
        ),
      ),
    );
  }
}
```

CountTestAPage 中构建的 Widget 效果与结构如图 7-7 所示，在 CountTestAPage 中，通过 TestController 调用异步方法去加载数据，加载完成后，更新组件 Bpage 与组件 CPage 中的值并显示。

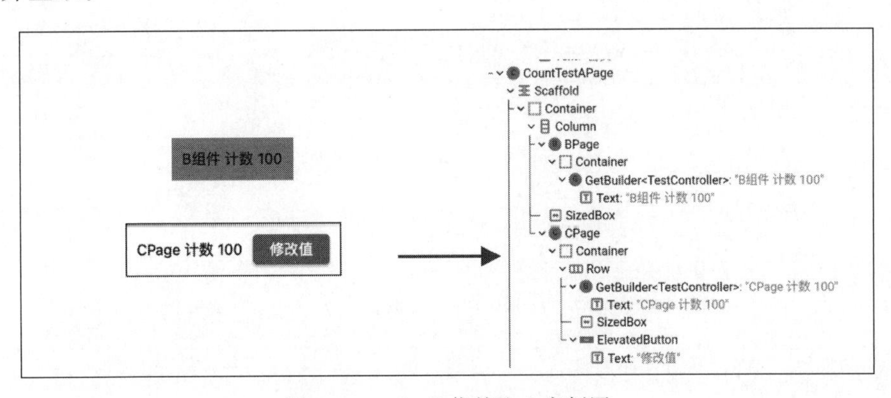

图 7-7　GetX 依赖注入案例图

组件 BPage 中的代码如下：

```
//代码清单 7-13-2
class BPage extends StatelessWidget {
  @override
  Widget build(BuildContext context) {
    return Container(
      padding: EdgeInsets.all(10),
      color: Colors.grey,
      child: GetBuilder<TestController>(
        builder: (TestController controller) {
          return Text("B组件计数 ${controller.count}");
        },
      ),
    );
  }
}
```

在组件 CPage 中，可以单击按钮调用 TestController 中的方法，更新值，在 BPage 中也可以同步显示，代码如下：

```
//代码清单 7-13-3
class CPage extends StatelessWidget {
  @override
  Widget build(BuildContext context) {
    return Container(
      padding: EdgeInsets.all(10),
      decoration: BoxDecoration(border: Border.all()),      //边框
      child: Row(//水平排列文本与按钮
        mainAxisSize: MainAxisSize.min,                     //包裹子 Widget
        children: [
          //数据监听
          GetBuilder<TestController>(
            builder: (TestController controller) {
              return Text("CPage 计数 ${controller.count}");
            },
          ),
          SizedBox(width: 10,),                             //间隔
          ElevatedButton(
            onPressed: () {
              //操作数据
              Get.find<CountController>().add();
            },
            child: Text("修改值"),
          ),
```

```
        ],
      ),
    );
  }
}
```

可通过 Get 的 isRegistered 方法来校验对应的 GetxController 是否注入成功，只有注入成功的控制器才可以通过 find 方法加载出来，否则会抛相应的异常，代码如下：

```
  //判断是否已成功注入依赖
  bool isRegister = Get.isRegistered<CountController>();
  if(isRegister) {
    //操作数据
CountController _controller = Get.find<CountController>();
//再调用对应的方法
_controller.add();
  }
```

Get.put()方法可以实现类的注入，默认情况下 Get.put()的实例的生命周期和 put 所在的 Widget 生命周期绑定，如果在全局 main 方法里使用 put，则这个实例就一直存在。如果在一个 Widget 里使用 put，则这个 Widget 从内存中被删除，这个实例也会被销毁，如果期望对应的控制器一直存在，则可将 Get.put()的参数 permanent 的值设置为 true，代码如下：

```
//注入 GetxController 对象
TestController _controller = Get.put(TestController(),permanent: true);
```

也可以通过 Get.lazyPut 方法实现依赖注入，懒加载一个依赖，只有在使用时才会被实例化，适用于不确定是否会被使用的依赖或者计算代价高昂的依赖。

7.4.4 GetXBinding 自动管理内存

在前面实现的 GetxController 依赖注入，为了生命周期和使用的 Widget 绑定，需要在 Widget 里注入和使用，并没有完全解耦，可以结合 Bindings 类实现自动注入，本节使用 7.4.2 节中创建的 CountController 和 7.4.3 节中创建的 TestController。

首先创建一个类并实现 Binding，代码如下：

```
//代码清单 7-14
//lib/page/count/binding/TestBinding.dart
class TestBinding extends Bindings {
  @override
  void dependencies() {
```

```
    //7.4.3 节中已创建
    Get.lazyPut<TestController>(() => TestController());
    //7.4.2 节中已创建
    Get.lazyPut<CountController>(() => CountController());
  }
}
```

然后在打开对应的页面里，需要配置这个自定义的 Bindings，代码如下：

```
//以静态的方式配置
GetMaterialApp(
  getPages: [
    GetPage(
      name: "/test",
      page: () => TestBindingPage(),
      binding: TestBinding(),
    )
  ],
  ...
)

//以动态的方式
Get.to(() => TestBindingPage(), binding: TestBinding());
```

TestBindingPage 是这里定义的普通页面，在这个页面中可能会有多个小组件，在本实例中，为测试效果，只用到 BPage 组件，在 Bpage 中定义按钮与 Text 文本，来测试数据修改与同步显示，代码如下：

```
//代码清单 7-15
//lib/page/count/binding/TestBindingPage.dart
class TestBindingPage extends StatelessWidget {
  @override
  Widget build(BuildContext context) {
    return Scaffold(
      body: BPage(),
    );
  }
}

class BPage extends GetView<TestController> {
  @override
  Widget build(BuildContext context) {
    return Container(
      padding: EdgeInsets.all(10),
```

```
        decoration: BoxDecoration(border: Border.all()),   //边框
        child: Row(
          //水平排列文本与按钮
          mainAxisSize: MainAxisSize.min,                   //包裹子 Widget
          children: [
            //数据监听
            GetBuilder<TestController>(
              builder: (TestController controller) {
                return Text("CPage 计数 ${controller.count}");
              },
            ),
            SizedBox(width: 10),                            //间隔
            ElevatedButton(
              onPressed: () {
                controller.add();
              },
              child: Text("修改值"),
            ),
          ],
        ),
      );
    }
  }
```

可以使用 BindingsBuilder 来简单地实现一个 Bindings 实例，在其中绑定对判断的控制器，代码如下：

```
GetPage(
  name: '/details',
  page: () => TestBindingPage(),
  binding: BindingsBuilder(
    () => {
      Get.lazyPut<TestController>(
        () => TestController(),
      )
    },
  ),
)
//以动态路由方式
Get.to(
  () => TestBindingPage(),
  binding: BindingsBuilder(
    () => {
      Get.lazyPut<TestController>(
        () => TestController(),
```

```
      )
    },
  ),
);
```

7.4.5 Obx 响应式编程

响应式编程需要结合 Rx 类型数据，创建 GetxController，代码如下：

```
//代码清单 7-16
class TestObsController extends GetxController {
  //声明为被观察者
  RxInt _count = 0.obs;
  RxInt get getCount => _count;
  //操作方法
  void addCount() {
    _count++;
  }
}
```

然后在页面中在需要同步更新显示数据的位置使用 Obx 组件进行监听，代码如下：

```
//代码清单 7-17
//GetX 响应编程
class TestObsPage extends StatelessWidget {
  @override
  Widget build(BuildContext context) {
    //注入控制器
    TestObsController _controllr = Get.put(TestObsController());
    return Scaffold(
      body: Center(
        child: Obx(() {
          return Text("Obx 当前 count 的值为 ${_controllr.getCount}");
        }),
      ),
      //单击按钮修改值
      floatingActionButton: FloatingActionButton(
        child: Icon(Icons.add),
        onPressed: () {
          //修改值
          _controllr.addCount();
        },
      ),
```

```
    );
  }
}
```

7.5 小结

在 Flutter 中,状态管理是一个比较深刻的主题,也经历了比较长的变革,本章中所概述的 Stream 与 Provider、GetX 可以相互结合使用,以及合理搭配使用。

第 8 章 绘图专题

CHAPTER 8

在 Flutter 中，绘图需要用到 CustomPaint 和 CustomPainter。CustomPainter 可理解为画板，承载画布，CustomPaint 理解为画布，承载绘制的具体内容。

画布(Canvas)顾名思义就是用来画图形，如绘制点、线、路径、矩形、圆形及添加图像等。

画笔(Paint)用来决定在画布上绘制图形的颜色、粗细、是否抗锯齿、笔触形状及作画风格等。

8.1 绘制基本图形

基本图片包括点、线、矩形(正方形、长方形)、弧、椭圆等，绘制功能需要结合 CustomPaint 和 CustomPainter 来实现。

8.1.1 绘图基础知识概述

在屏幕中绘制一条直线，如图 8-1 所示。

图 8-1 绘图基本入门

代码如下：

```
//代码清单 8-1-1
//lib/code8/example_801_baseUse.dart
```

```
class Example801 extends StatelessWidget {
  const Example801({Key? key}) : super(key: key);

  @override
  Widget build(BuildContext context) {
    return Scaffold(
      body: Container(
        width: MediaQuery.of(context).size.width,
        color: Colors.white,
        height: 200,
        //创建画板
        child: CustomPaint(
          //定义画板的大小
          size: Size(300, 300),
          //配置画布
          painter: LinePainter(),
        ),
      ),
    );
  }
}
```

对于 CustomPaint，参数 painter 与参数 foregroundPainter 都可配置 CustomPainter，child 配置的是 Widget，意味着在这里可渲染任意的子 Widget，这三个参数配置的结果都是在画板上显示出图形，所以必然存在着层次。

参数 painter 是绘制在 backgroud 层，也就是最底层，child 是在 backgroud 之上绘制，foregroundPainter 是在 child 之上绘制。

实现绘制功能的 LinePainter 是一个自定义的 CustomPainter，CustomPainer 被定义为抽象类型，在使用时需要自定义实现其子类，需要重写 paint()和 shouldRepaint()这两种方法，一个是绘制流程，另一个是在刷新布局时配置是否需要重绘。在 paint()方法中的 size 参数是 CustomPaint 中定义的 size 属性，它包含了基本的画布大小信息，代码如下：

```
//代码清单 8-1-2
//lib/code8/example_801_baseUse.dart
//自定义绘图者
class LinePainter extends CustomPainter {
  //[定义画笔]
  final Paint _paint = Paint()
    ..color = Colors.blue          //画笔颜色
    ..strokeWidth = 4;             //画笔宽度

  //绘制功能主要在这里进行
  @override
```

```
  void paint(Canvas canvas, Size size) {
    //绘制一条直线
    canvas.drawLine(const Offset(20, 20), const Offset(100, 20), _paint);
  }

  //是否重新绘制
  @override
  bool shouldRepaint(CustomPainter oldDelegate) {
    return false;
  }
}
```

如果 CustomPaint 有子节点，官方建议将子节点包裹在 RepaintBoundary 组件中，这样会在绘制时创建一个新的图层 Layer，然后将其子 Widget 在新创建的 Layer 上绘制，而 painter 与 foregroundPainter 还在原来的图层 Layer 上绘制，RepaintBoundary 会隔离其子节点和 CustomPaint 本身的绘制边界，也就是说 RepaintBoundary 包裹的子 Widget 的绘制将独立于父 CustomPaint 的绘制，这样就避免了子节点不必要的重绘而达到提高性能的目的，RepaintBoundary 的基本使用代码如下：

```
//代码清单 8-2
//lib/code8/example_802_baseUse.dart
class Example801 extends StatelessWidget {
  @override
  Widget build(BuildContext context) {
    return Scaffold(
      body: Container(
        width: MediaQuery.of(context).size.width,
        color: Colors.white,
        height: 200,
        //创建画板
        child: CustomPaint(
          //指定画布大小
          size: const Size(400, 400),
          //与代码清单 8-1-1 中的代码一致
          painter: LinePainter(),
          //child 不为空时使用 RepaintBoundary 通过独立图层的方式
          //分离与父 Widget 绘制图层，避免不必要的重复绘制
          //在更多业务场景中如自定义的 StatefulWidget
          child: const RepaintBoundary(child: Text("测试数据")),
        ),
      ),
    );
  }
}
```

画笔 Panit 用来决定画布上绘制图形的颜色、粗细、是否抗锯齿、笔触形状及作画风格等，常用属性配置如下：

```
//[定义画笔]
final Paint _paint = Paint()
  //画笔颜色
  ..color = Colors.blue
  //画笔笔触类型,将在本书 8.2 节中概述
  ..strokeCap = StrokeCap.round
  //拐角类型,将在本书 8.2 节中概述
  ..strokeJoin = StrokeJoin.round
  //是否启动抗锯齿
  ..isAntiAlias = true
  //颜色混合模式
  ..blendMode = BlendMode.exclusion
  //绘画风格,默认为填充
  ..style = PaintingStyle.fill
  //颜色渲染模式,一般通过矩阵效果来改变,但是 Flutter 中只能使用颜色混合模式
  ..colorFilter = ColorFilter.mode(Colors.blueAccent, BlendMode.exclusion)
  //模糊遮罩效果,Flutter 中只有这个
  ..maskFilter = MaskFilter.blur(BlurStyle.inner, 3.0)
  //颜色渲染模式的质量
  ..filterQuality = FilterQuality.high
  //画笔的宽度
  ..strokeWidth = 15.0;
```

在自定义的 CustomPainter 中，真正的绘制则是在 paint()方法中，通过 canvas 和 Paint 实现，可先定义 Paint 画笔，然后结合 canvas.drawXXX()方法来绘制各种图形，一些常用的 canvas 绘制方法如下：

```
//移动到指定点,当前绘制的开始
moveTo()
//将当前点绘制到设置的新起点,通常用来绘制直线
lineTo():
//通常用于闭合绘制的路径
close()
//绘制点
drawPoints(PointMode pointMode, List<Offset> points, Paint paint)
//绘制线条
drawLine(Offset p1, Offset p2, Paint paint)
//绘制弧线
drawArc(Rect rect, double startAngle, double sweepAngle, bool useCenter, Paint paint)
//绘制图片
drawImage(Image image, Offset p, Paint paint)
```

```
//绘制圆
drawCircle(Offset c, double radius, Paint paint)
//绘制椭圆
drawOval(Rect rect, Paint paint)
//绘制文字
drawParagraph(Paragraph paragraph, Offset offset)
//绘制路径
drawPath(Path path, Paint paint)
//绘制 Rect
drawRect(Rect rect, Paint paint)
//绘制阴影
drawShadow(Path path, Color color, double elevation, bool transparentOccluder)
```

从另一个角度来讲，canvas 画布的方法大致可以分为以下三类：

（1）drawXXX 等是一系列绘制基本图形及各种自定义图形与曲线相关的方法。

（2）scale、rotate、clipXXX 等是对画布进行变换操作的方法。

（3）save、restore 等是与层的保存和回滚相关的方法。

默认情况下，对于画布 canvas 的左上角是原点(0,0)，基于左上角往右为 x 轴正方向，往下为 y 轴正方向，反之为负。

8.1.2 绘制点与线

在绘图中，可以认为点是最小的单位，由点构成线，由线构成面，由面构成空间，通过 canvas 的 drawPoints 来绘制点，其函数源代码如下：

```
void drawPoints(PointMode pointMode, List points, Paint paint)
```

参数一为 PointMode 枚举类型，可取值为 points(点)、lines(线，隔点连接)和 polygon(线，相邻连接)，参数二为所要绘制的点的集合，参数三为所使用的画笔。

如图 8-2 所示，绘制 5 个点，代码如下：

```
//代码清单 8-3 绘制点 drawPoints
//lib/code8/example_803_Point.dart
class PointPainter extends CustomPainter {
  //[定义画笔]
  final Paint _paint = Paint()
    //画笔颜色
    ..color = Colors.blue
    //画笔笔触类型
    ..strokeCap = StrokeCap.round
    //是否启动抗锯齿
    ..isAntiAlias = true
    //绘画风格，默认为填充
```

```
    ..style = PaintingStyle.fill
    //画笔的宽度
    ..strokeWidth = 20.0;

  @override
  void paint(Canvas canvas, Size size) {
    //绘制点
    canvas.drawPoints(
        PointMode.lines,
        [
          const Offset(0.0, 0.0),
          const Offset(60.0, 10.0),
          const Offset(50.0, 50.0),
          const Offset(90.0, 90.0),
          const Offset(190.0, 60.0),
        ],
        _paint);
  }

  @override
  bool shouldRepaint(CustomPainter oldDelegate) {
    return true;
  }
```

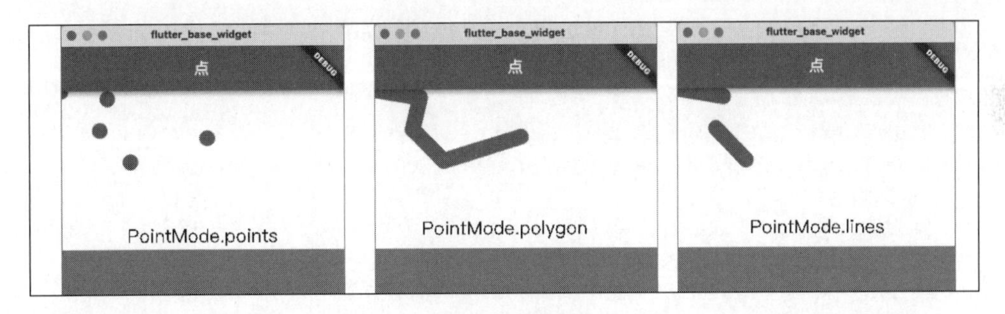

图 8-2 绘制点

当绘制模式为 PointMode. lines 时，会发现第 1 个点与第 2 个点连接后形成线段，第 3 个点与第 4 个点连接后形成线段，以此类推，所以可以得出的结论就是 lines 通过点来绘制线段。

当绘制模式为 PointMode. polygon 时，会发现所有的点会依次连接，形成折线，也就是相邻的两个点会连接。

Paint 画笔的 strokeCap 属性用来配置在绘制图形过程中拐角的类型，对于 StrokeCap. round 类型，在绘制图形拐角处使用圆形过渡，当使用 StrokeCap. butt 时在绘制图形拐角处不使用过渡方式，即直来直去，所以会形成直角。

两点确定一条直线，通过 canvas 画布提供的 drawLine()方法可以直接绘制直线，如图 8-3 所示，绘制直线需要用到两个点，核心代码如下：

```
@override
void paint(Canvas canvas, Size size) {
  //定义点
  Offset p1 = Offset(20,40);
  Offset p2 = Offset(160,40);
  //绘制线
  canvas.drawLine(p1, p2, _paint);
}
```

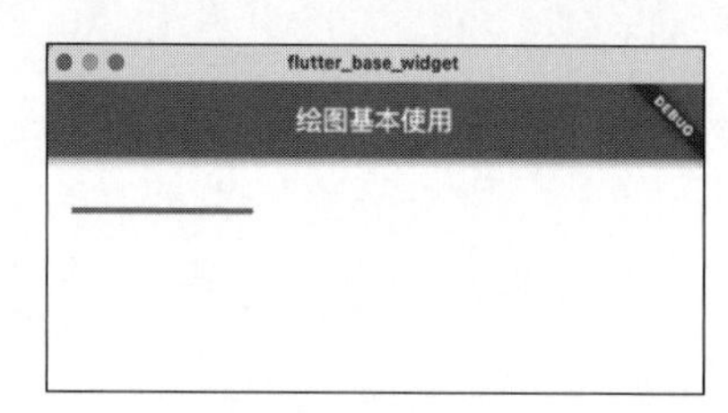

图 8-3　绘制线

8.1.3　绘制矩形与圆角矩形

通过 canvas 画布的 drawRect()方法可以绘制矩形，如图 8-4 所示，代码如下：

```
//代码清单 8 - 4 矩形
//lib/code8/example_805_Rect.dart
class RectPainter extends CustomPainter {
  //[定义画笔]
  final Paint _paint = Paint()
    ..color = Colors.blueAccent             //画笔颜色
    ..strokeCap = StrokeCap.round           //画笔笔触类型
    ..isAntiAlias = true                    //是否启动抗锯齿
    ..style = PaintingStyle.fill            //绘画风格,默认为填充
    ..strokeWidth = 2.0;                    //画笔的宽度

  @override
  void paint(Canvas canvas, Size size) {
    //创建一个矩形
    Rect rect = buildRect1();
    //绘制矩形
    canvas.drawRect(rect, _paint);
  }
```

```
  //创建矩形方式一
  Rect buildRect1() {
    //根据以屏幕左上角为坐标系圆点,分别设置上、下、左、右四个方向的距离
    //left, top, right, bottom
    return const Rect.fromLTRB(20, 40, 150, 100);
  }
}
```

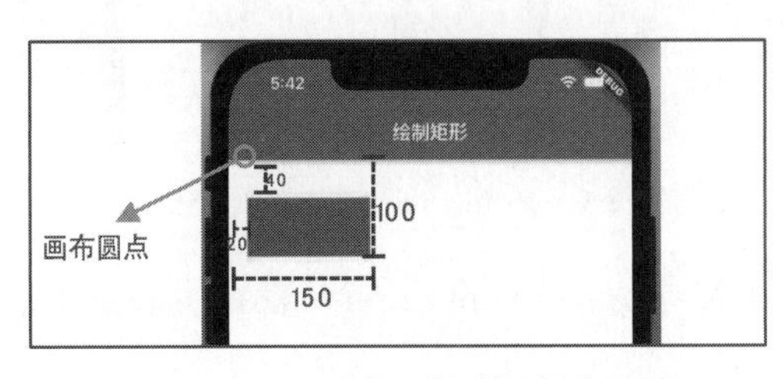

图 8-4 fromLTRB 创建矩形分析图

绘制矩形的核心内容主要是创建 Rect 矩形,通过 Rect 的静态方法 fromLTWH()根据设置左上角的点与矩形的宽和高来绘制,坐标分析如图 8-5 所示,这样创建出来的矩形实际上 width 为 150,height100,代码如下:

```
//创建矩形方式二
  Rect buildRect2() {
    //根据设置左上角的点与矩形宽和高来绘制;
    //left, top, width, height
    return const Rect.fromLTWH(20, 40, 150, 100);
  }
```

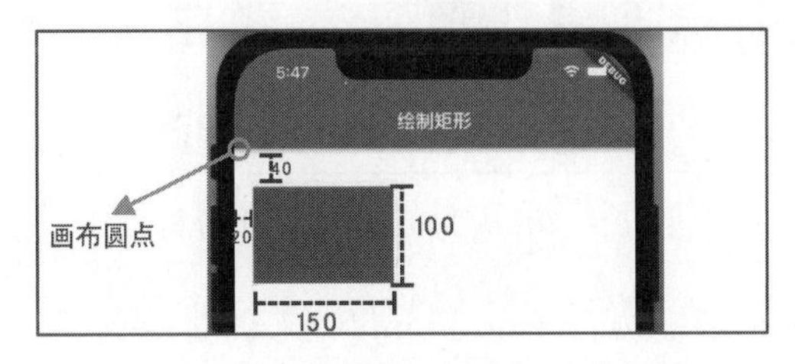

图 8-5 fromLTWH 创建矩形分析图

方式三是通过 Rect 的静态方法 fromCircle()根据圆形来绘制正方形,如图 8-6 所示,这里参考的圆形为所要绘制的正方形的内切圆,代码如下:

```
//创建矩形方式三
Rect buildRect3() {
  //根据圆形绘制正方形
  //参数一 center Offset 类型,参考圆的圆心
```

```
    //参数二 radius 以 center 为圆心,以 radius 为半径
    return const Rect.fromCircle(center: Offset(100, 100), radius: 50);
  }
```

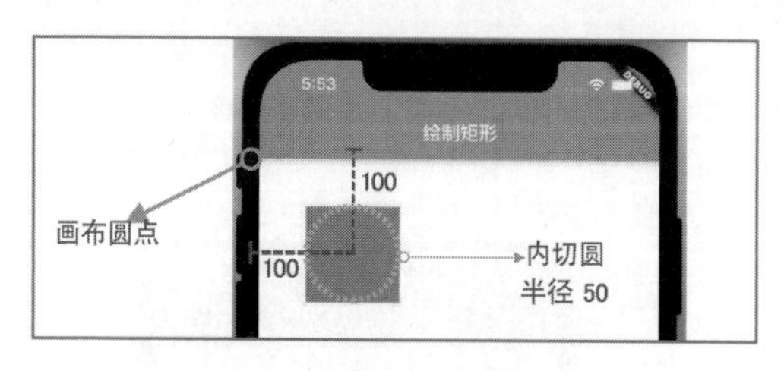

图 8-6 fromCircle 创建矩形分析图

方式四是通过 Rect 的静态方法 fromCenter()根据中心点来绘制矩形,如图 8-7 所示,代码如下:

```
//创建矩形方式四
Rect buildRect4() {
  //根据中心点绘制正方形
  //参数一 center Offset 类型,参考圆的圆心
  //参数二 width
  //参数三 height
  return Rect.fromCenter(center: Offset(100, 100), width: 100, height: 100);
}
```

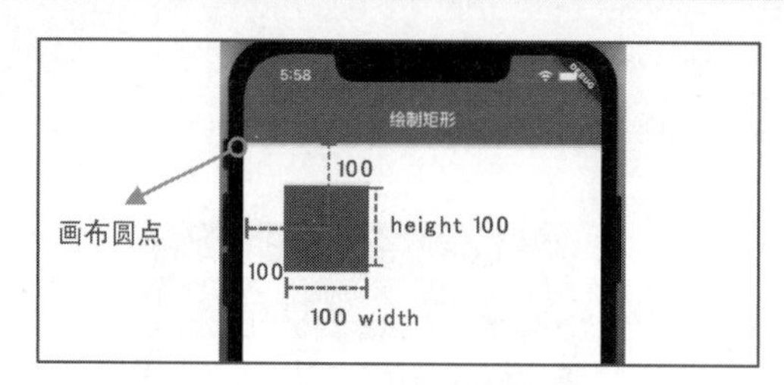

图 8-7 fromCenter 创建矩形分析图

方式五是通过 Rect 的静态方法 fromPoints()来创建矩形,fromPoints 需要两个点,即矩形的左上角的点与右下角的点,如图 8-8 所示,代码如下:

```
//创建矩形方式五
Rect buildRect5() {
  //根据对角点来绘制矩形
  //参数一矩形的左上角的点
  //参数二矩形右下角的点
  return Rect.fromPoints(Offset(60, 50), Offset(200, 100));
}
```

图 8-8 Points 创建矩形分析图

通过 canvas 的 drawRRect()方法来绘制圆角矩形，如图 8-9 所示，代码如下：

```
//代码清单 8-5 绘制圆角矩形
//lib/code8/example_807_RRect.dart
class RRectPainter extends CustomPainter {
  //[定义画笔]
  final Paint _paint = Paint()
    //画笔颜色
    ..color = Colors.red
    //画笔的宽度
    ..style = PaintingStyle.stroke
    ..strokeWidth = 3.0;

  @override
  void paint(Canvas canvas, Size size) {
    //创建圆角矩形
    RRect rect = buildRect3();
    //绘制
    canvas.drawRRect(rect, _paint);
  }

//创建圆角矩形方式一
  RRect buildRect1() {
    //以画板左上角为坐标原点,分别设置上、下、左、右四个方向的距离来创建矩形
    //left, top, right, bottom
    //最后两个参数用于设置圆角的大小
    return const RRect.fromLTRBXY(20, 40, 250, 200, 60, 40);
  }
}
```

图 8-9 圆角矩形

创建圆角矩形的方法 RRect.fromLTRBXY 的最后两个参数为 radiusX 与 radiusY，用来设置圆角的两个半径，圆角半径分析图如图 8-10 所示。

通过 RRect 的 fromLTRBAndCorners 方法来创建可配置矩形的四个角，每个圆的弧度不同，如图 8-11 所示。

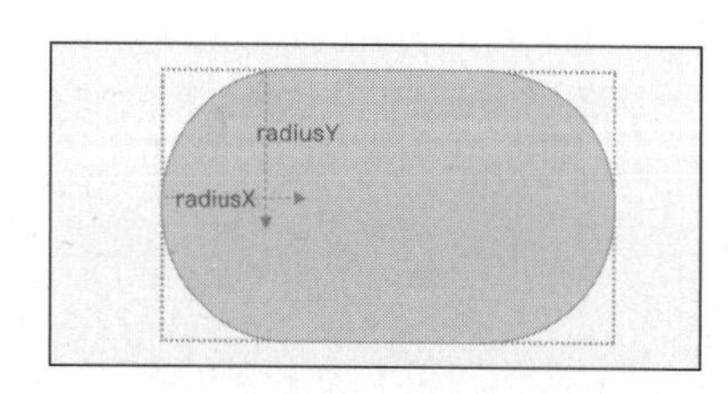

图 8-10　圆角半径分析图

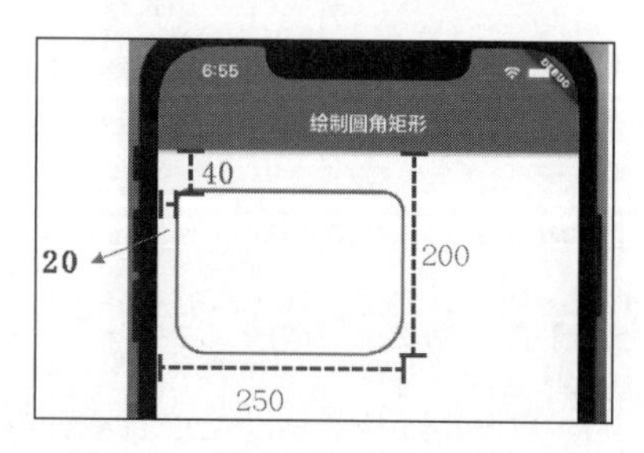

图 8-11　圆角不同的矩形效果图

代码如下：

```
//创建圆角矩形方式二
RRect buildRect3() {
  //以画板左上角为坐标原点,分别设置上、下、左、右四个方向的距离来创建矩形
  //前四个参数分别为 left、top、right、bottom,用来创建矩形
  return RRect.fromLTRBAndCorners(
    20, 40, 250, 200,
    //左上角的圆角
    topLeft: Radius.circular(10),
    //右上角的圆角
    topRight: Radius.circular(20),
    //左下角的圆角
    bottomLeft: Radius.circular(30),
    //右下角的圆角
    bottomRight: Radius.circular(40),
  );
}
```

8.1.4　绘制圆与椭圆

通过 canvas 的 drawCircle 可实现绘制实心圆或者圆环，代码如下：

```
//代码清单 8-6 绘制圆形
//lib/code8/example_808_Circl.dart
class CirclePainter extends CustomPainter {
  //[定义画笔]
  final Paint _paint = Paint()
  ..color = Colors.blue
```

```
    ..style = PaintingStyle.stroke
    ..strokeWidth = 4.0;

  @override
  void paint(Canvas canvas, Size size) {
    //参数一圆心的位置
    //参数二半径大小
    canvas.drawCircle(const Offset(100,100), 40, _paint);

  }
  @override
  bool shouldRepaint(CustomPainter oldDelegate) {
    return true;
  }
}
```

将画笔 Paint 的 style 配置为 PaintingStyle.fill 模式，绘制的是实心圆，PaintingStyle.stroke 模式用于绘制圆环。

绘制椭圆时，首先需要定义一个矩形(Rect)，然后绘制这个矩形的内切圆，如果这个矩形正好是个正方形，则绘制出来的椭圆也是个圆形，通过 canvas 的 drawOval 方法来绘制一个椭圆，如图 8-12 所示。

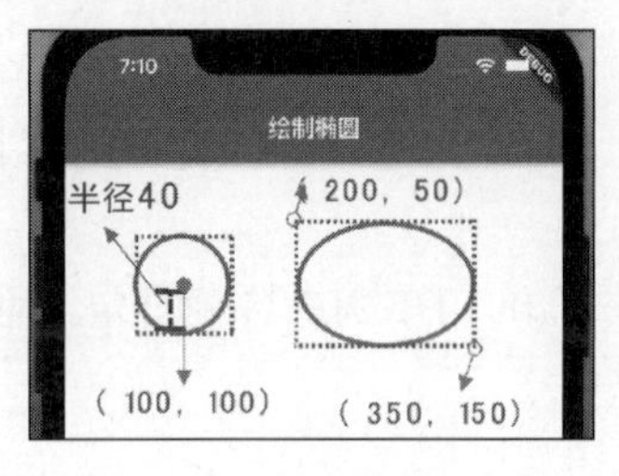

图 8-12　绘制椭圆效果图

核心绘制代码如下：

```
//代码清单 8-7 绘制椭圆
//lib/code8/example_809_Oval.dart
class OvalPainter extends CustomPainter {
  //[定义画笔]
  final Paint _paint = Paint()
    ..color = Colors.blue
    ..style = PaintingStyle.stroke
    ..strokeWidth = 4.0;

  @override
  void paint(Canvas canvas, Size size) {
    //用 Rect 构建一条边长为 50,中心点坐标为(100,100)的矩形
    Rect rect = Rect.fromCircle(
      center: const Offset(100.0, 100.0),
      radius: 40.0,
    );
    //绘制椭圆
    canvas.drawOval(rect, _paint);
```

```
    //使用两个对角点来创建 width 为 150,height 为 100 的矩形
    Rect rect2 = Rect.fromPoints(
      const Offset(200, 50),
      const Offset(350, 150),
    );
    //绘制椭圆
    canvas.drawOval(rect2, _paint);
  }

  @override
  bool shouldRepaint(CustomPainter oldDelegate) {
    return true;
  }
}
```

8.2 Path 自定义图形

Path 用于构建各种自定义曲线、图形等，如表 8-1 所示，此表列出了一些 Path 的相关操作方法。

表 8-1　Path 常用方法概述

类　　别	描　　述
moveTo	将路径起始点移动到指定的位置，即画笔的起点，相对于画布的原点位置，当不设置时，默认从画布原点的位置开始
relativeMoveTo	这种方法与 moveTo 方法的功能一致，不同的是相对于当前的位置移动到新的位置
lineTo	从当前的位置通过直线的方式连接到新的位置
relativeLineTo	相对于当前的位置连接到新的位置
add **	添加其他图形，如 addArc，在路径中是添加圆弧
arcTo	二阶贝塞尔曲线
conicTo	三阶贝塞尔曲线
close	关闭路径，连接路径的起始点
reset	重置路径，即清空路线，恢复到默认状态

8.2.1 Path 构建基本图形

Path 的 moveTo()方法是将画笔移动到指定的位置，默认情况下画笔的位置在原点（画布的左上角），lineTo 是从当前画笔的位置将直线画到指定的点，再结合 colose 方法可实现直线、折线、三角形、矩形、平行四边形、梯形及其他不规则图形的绘制，如图 8-13 所示。

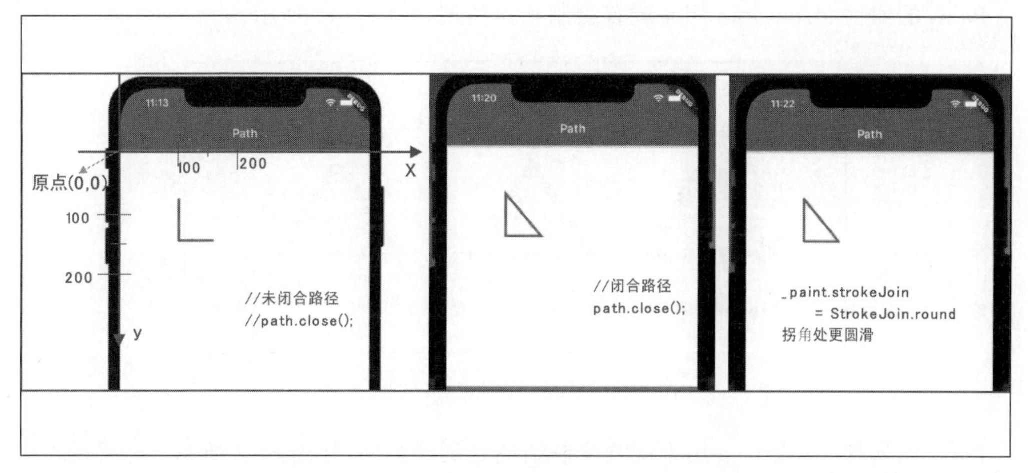

图 8-13 Path 通过线构建基本图形效果图

代码如下：

```
//代码清单 8-8-1 Path 路径相关操作
//lib/code8/example_810_Path.dart
//构建 Path, Path 的基本使用
void buildPath1(Canvas canvas, Size size) {
  //创建 Path
  Path path = Path();
  //将画笔移动到起点
  path.moveTo(100, 80);
  //将直线画到点
  path.lineTo(100, 150);
  //继续画
  path.lineTo(160, 150);
  //闭合路径
  path.close();
  //绘制 Path
  canvas.drawPath(path, _paint);
}
```

本节用到的画笔 Paint 的构建如下：

```
final Paint _paint = Paint()
  ..color = Colors.blue
  //画笔笔触类型
  ..strokeCap = StrokeCap.round
  ..strokeJoin = StrokeJoin.round
  //是否启动抗锯齿
  ..isAntiAlias = true
  ..style = PaintingStyle.stroke
  ..strokeWidth = 4.0;
```

Paint 的属性 strokeJoin 用来配置绘制拐角类型，如图 8-14 所示。

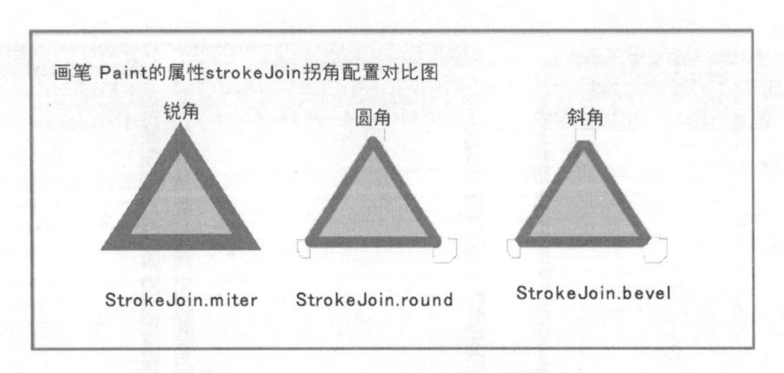

图 8-14　Paint strokeJoin 效果图

Paint 的属性 strokeCap 用来配置绘制结尾处延伸类型，如图 8-15 所示。

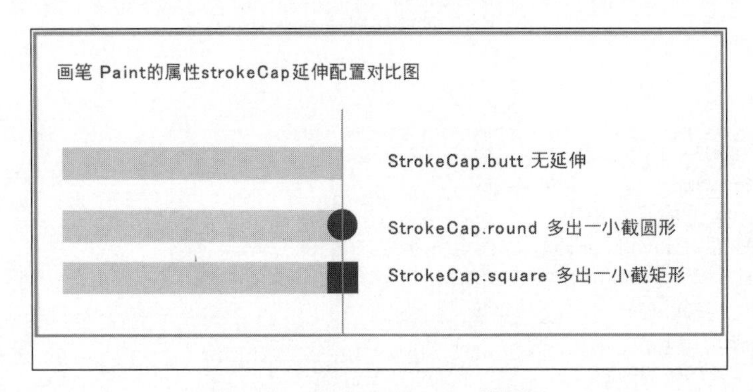

图 8-15　Paint strokeCap 效果图

Path 的 relativeLineTo 方法用于构建相对于当前点的坐标再递增所定义数值的坐标，如当前画笔的位置为(50,50)，再定义数值 relativeLineTo(100,10)，实际上是从点(50,50)将直线绘制到目标点(50＋100,50＋10)。

Path 可通过 addXXX 系列方法将基本图形添加至 Path 路径中，addRect 表示添加矩形、addOval 表示添加椭圆、addPolygon 表示添加多边形、addRRect 表示添加圆角矩形、addPath 表示添加另一个路径，本实例通过 addArc 方法来添加一段弧(其他方法类似)，如图 8-16 所示。

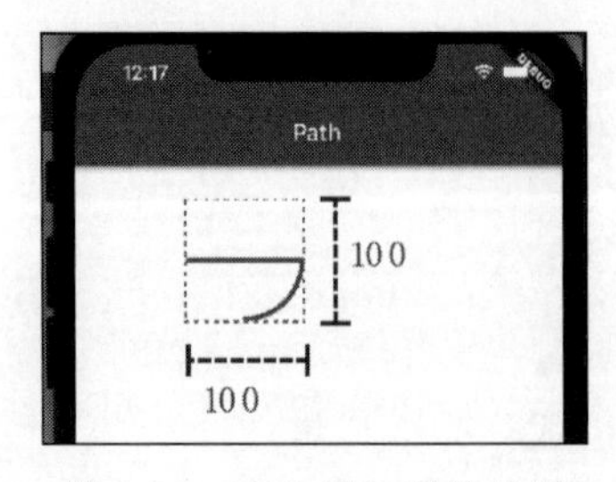

图 8-16　Path 添加弧效果图

代码如下：

```
//代码清单 8-8-2
//lib/code8/example_810_Path.dart
void buildPath4(Canvas canvas, Size size) {
  //创建 Path
  Path path = new Path();
  //将画笔移动到起点
  path.moveTo(100, 80);
  //将直线画到点
  path.lineTo(200, 80);

  //添加一段弧

  Rect rect = Rect.fromCenter(
    center: const Offset(150, 80),
    width: 100,
    height: 100,
  );
  //参数一为绘制弧所参考的外切矩形
  //参数二为起始位置
  //参数三为结束位置,2 * pi 表示 360°
  path.addArc(rect, 0, 0.5 * pi);
  //绘制 Path
  canvas.drawPath(path, _paint);
}
```

8.2.2 二阶贝塞尔曲线绘制弧线

两个点构成一条基本的直线，可称为一阶贝塞尔曲线，当这两个点之间的线段被其他的点牵引后，形成曲线，如图 8-17 所示，称为二阶贝塞尔曲线，这个牵引点称为控制点。

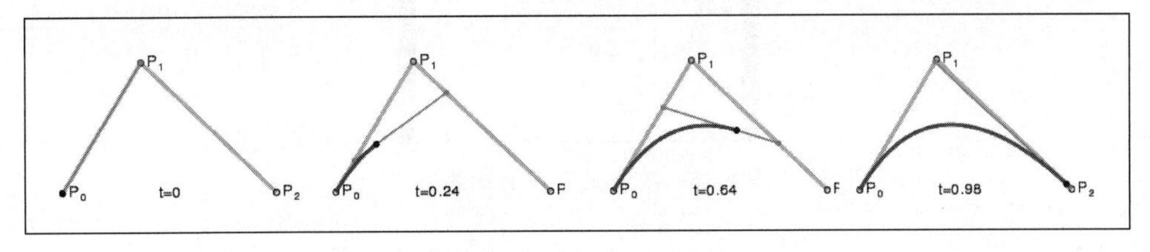

图 8-17 二阶贝塞尔曲线

在 Flutter 中，通过路径 Path 来构建贝塞尔曲线，其中方法 quadraticBezierTo 用来构建二阶贝塞尔曲线，代码如下：

```
//代码清单 8-8-3 Path 二阶贝塞尔曲线
//lib/code8/example_810_Path.dart
void buildPath5(Canvas canvas, Size size) {
  //定义起点
  Offset startPoint = const Offset(50, 50);
  //定义控制点
  Offset controllPoint = const Offset(100, 130);
  //定义终点
  Offset endPoint = const Offset(250, 50);

  //创建 Path
  Path path = Path();
  //将画笔移动到起点
  path.moveTo(startPoint.dx, startPoint.dy);
  //画二阶贝塞尔曲线

  //path.quadraticBezierTo(x1, y1, x2, y2)
  //参数(x1, y1) 为控制点
  //参数 x2, y2 为目标点
  path.quadraticBezierTo(
      controllPoint.dx, controllPoint.dy, endPoint.dx, endPoint.dy);

  //绘制 Path
  canvas.drawPath(path, _paint);

}
```

运行结果如图 8-18 所示。

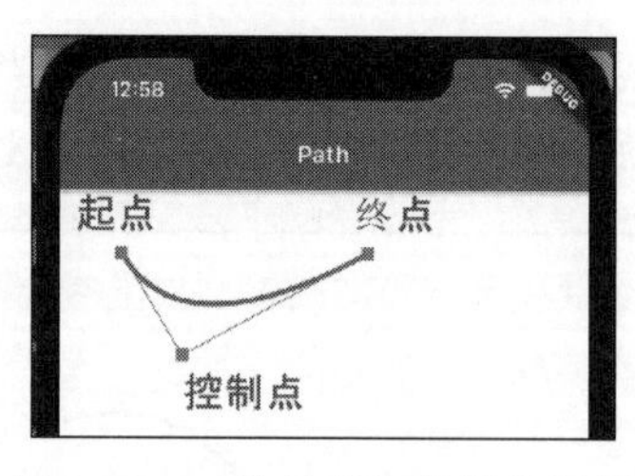

图 8-18　二阶贝塞尔曲线效果图

8.2.3　三阶贝塞尔曲线绘制弧线

当有两个点牵引时，相当于用两个控制点来控制曲线，如图 8-19 所示，称为三阶贝塞尔曲线。

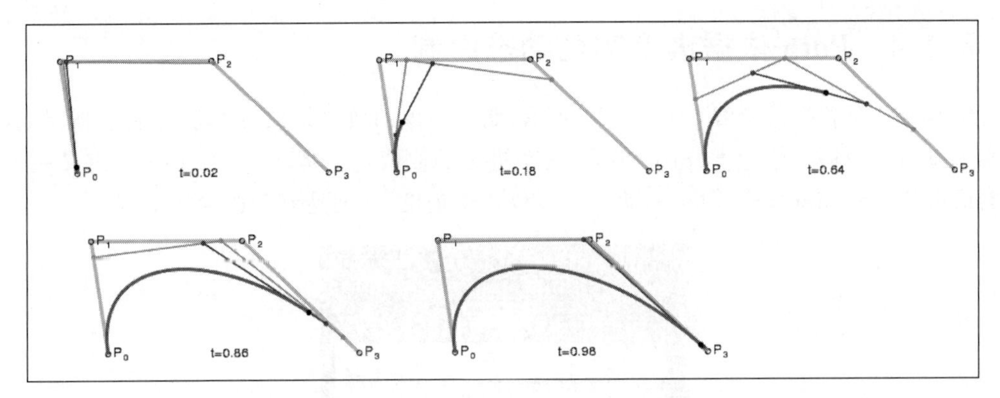

图 8-19 三阶贝塞尔曲线

可通过 Path 的 cubicTo 方法来构建，代码如下：

```
//代码清单 8-8-4 Path 三阶贝塞尔曲线
  //lib/code8/example_810_Path.dart
  void buildPath6(Canvas canvas, Size size) {
    //定义起点
    Offset startPoint = const Offset(50, 50);
    //定义控制点
    Offset controllPoint1 = const Offset(100, 150);
    Offset controllPoint2 = const Offset(200, 130);
    //定义终点
    Offset endPoint = const Offset(250, 50);

    //创建 Path
    Path path = Path();
    //将画笔移动到起点
    path.moveTo(startPoint.dx, startPoint.dy);
    //画三阶贝塞尔曲线
    path.cubicTo(
        //控制点 1
        controllPoint1.dx, controllPoint1.dy,
        //控制点 2
        controllPoint2.dx, controllPoint2.dy,
        //终点
        endPoint.dx, endPoint.dy),
    //绘制 Path
    canvas.drawPath(path, _paint);

  }
```

8.2.4 Path 依据进度实现动态绘制

绘制的一条普通的直线，如图 8-20 所示，期望实现的效果是从左侧动态地绘制完成，一个实现思路是使用一个进度(progress)，这个进度的取值范围是 0.0～1.0，然后定义一个计时器或者 AnimationController 来动态地修改这个进度，从而达到动画绘制效果。

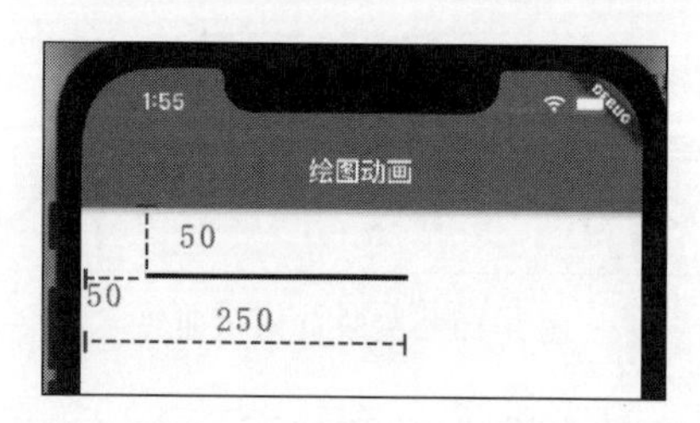

图 8-20 绘制直线效果图

绘制直线的代码如下：

```
//代码清单 8-9-1 绘制直线
//lib/code8/example_816_Line_Animation.dart
class ProgressLinePainter extends CustomPainter {
  //进度 0.0 ～ 1.0
  final double progress;
  ProgressLinePainter({required this.progress});
  //画笔
  final Paint _paint = Paint()
    ..color = Colors.blue
    ..strokeWidth = 4.0
    ..style = PaintingStyle.stroke
    ..strokeJoin = StrokeJoin.round;
  //路径
  Path path = Path();
  @override
  void paint(Canvas canvas, Size size) {
    //重置路径
    path.reset();
    canvas.save();                     //用来保存当前的画布状态
    path.moveTo(50, 50);
    //动态修改
    path.lineTo(
      50 + 150 * progress,
      50,
    );
    canvas.drawPath(path, _paint);  //绘制路径
```

```
    canvas.restore();            //重置画布状态
  }

  @override
  bool shouldRepaint(ProgressLinePainter oldDelegate) {
    return true;
  }
}
```

动画控制器的代码如下：

```
//代码清单 8-9-2 动画控制器
//lib/code8/example_816_Line_Animation.dart
class _ExampleState extends State with SingleTickerProviderStateMixin {
  //动画控制器
  late AnimationController _animationController;

  @override
  void initState() {
    super.initState();
    //创建
    _animationController = AnimationController(
        vsync: this,
        lowerBound: 0.0,
        upperBound: 1.0,
        duration: const Duration(
          milliseconds: 1600,
        ));
    //添加监听
    _animationController.addListener(() {
      setState(() {});
    });
    //添加状态监听
    _animationController.addStatusListener((status) {
      if (status == AnimationStatus.completed) {
        _animationController.reset();
        _animationController.forward();
      }
    });
  }

  @override
```

```
  Widget build(BuildContext context) {
    return Scaffold(
      body: Column(
        children: [

          Container(
            child: CustomPaint(
              size: const Size(400, 150),
              painter:ProgressLinePainter(
                                    progress: _animationController.value),

            ),
          ),
          Row(
            mainAxisAlignment: MainAxisAlignment.center,
            children: [
              ElevatedButton(
                child: const Text("开始"),
                onPressed: () {
                  _animationController.reset();
                  _animationController.forward();
                },
              ),
              const SizedBox(width: 12),
              ElevatedButton(
                child: const Text("结束"),
                onPressed: () {
                  _animationController.stop();
                },
              )
            ],
          )
        ],
      ),
    );
  }
}
```

8.2.5 Path 结合 PathMetric 实现动态绘制

在这里绘制的是直线，直线进度可以很好地计算出来，对于复杂点的图形，可通过 Path 来构建，然后结合 PathMetric 来测量 Path 的进度，从而实现动态绘制的功能，在 8.2.4 节的基础上修改，动态绘制直线的方法修改后的代码如下：

```
//代码清单 8-9-3
//lib/code8/example_816_Line_Animation.dart
class LinePathMetricPainter extends CustomPainter {
  //进度 0.0 ~ 1.0
  final double progress;

  LinePathMetricPainter({required this.progress});

  //画笔
  final Paint _paint = Paint();
  //路径
  Path path = Path();

  @override
  void paint(Canvas canvas, Size size) {
    //重置路径
    path.reset();
    canvas.save(); //用来保存当前的画布状态
    path.moveTo(50, 50);
    path.lineTo(150, 50);
    //核心功能实现
    //测量 Path
    PathMetrics pathMetrics = path.computeMetrics();
    //获取第一节信息
    PathMetric pathMetric = pathMetrics.first;

    //测量并裁剪 Path
    Path extrPath = pathMetric.extractPath(0, pathMetric.length * progress);
    //绘制
    canvas.drawPath(extrPath, _paint);

    canvas.restore();
  }

  @override
  bool shouldRepaint(LinePathMetricPainter oldDelegate) {
    return true;
  }
}
```

PathMetric 专门用来度量 Path 路径，例如由 Path. lineTo、Path. moveTo 和另一个 Path. lineTo 组成的路径将包含两个 PathMetric，因此由两个 PathMetric 度量对象表示，如图 8-21 所示，使用 Path 来构建一条直线与一个矩形，单击“开始”按钮，实现动态绘制。

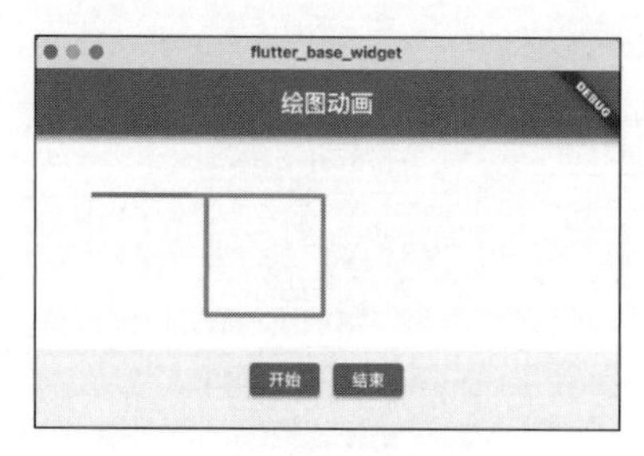

图 8-21 Path 组合绘制效果图

代码如下：

```
//代码清单 8-9-4
//lib/code8/example_816_Line_Animation.dart
class LineAndArcPainter extends CustomPainter {
  //进度 0.0 ~ 2.0
  final double progress;

  LineAndArcPainter({required this.progress});

  final Paint _paint = Paint()
    ..color = Colors.blue
    ..strokeWidth = 4.0
    ..style = PaintingStyle.stroke
    ..strokeJoin = StrokeJoin.round;
  Path path = Path();

  @override
  void paint(Canvas canvas, Size size) {
    //重置路径
    path.reset();
    canvas.save();
    //画一条直线,这条直线的长度为 100
    //也就是这条直线对应的 Path 长度为 400
    path.moveTo(50, 50);
    path.lineTo(150, 50);

    //添加一个矩形,宽和高都为 100,即周长为 400
    //也就是矩形对应的 Path 长度为 400
    path.addRect(const Rect.fromLTWH(150, 50, 100, 100));

    //获取 Path 中所有的组成单元(在这里应该是两个)
    PathMetrics computeMetrics = path.computeMetrics();
    int number = computeMetrics.length;
```

```
    print("path number $ number");

    //获取直线相关的 PathMetric
    PathMetric linePathMetric = path.computeMetrics().first;
    //直线的长度
    double linePathLength = linePathMetric.length;
    print("直线的长度为 $ linePathLength");

    //获取矩形相关的 PathMetric
    PathMetric rectPathMetric = path.computeMetrics().elementAt(1);
    //矩形的长度
    double rectPathLength = rectPathMetric.length;
    print("矩形的长度为 $ rectPathLength");

    //绘制代码

    canvas.restore();
  }

  @override
  bool shouldRepaint(covariant CustomPainter oldDelegate) {
    return true;
  }
}
```

运行程序,在 Android Studio 日志控制台输出测量的信息如图 8-22 所示。

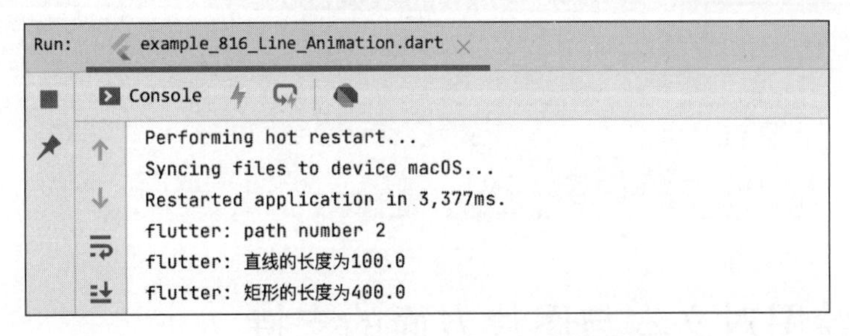

图 8-22 Path PathMetric 运行调试图

有了测量信息后,就可以根据长度来重新构建 Path,在本代码块中,需要实时改变进度的 Progress 值的范围为 0.0～2.0,代码如下:

```
// 0.0 ~ 1.0 内绘制直线
// 1.0 ~ 2.0 内绘制矩形
double lineExtracProgress = 0;
double rectExtracProgress = 0;
```

```
if (progress > 1.0) {
  //根据进度绘制矩形的长度
  rectExtracProgress = rectPathMetric.length * (progress - 1.0);
  //绘制直线的全部
  lineExtracProgress = linePathMetric.length;
} else {
  //动态绘制直线
  lineExtracProgress = linePathMetric.length * progress;
}

//构建指定长度的 Path
//测量并裁剪路径
Path lineExtractPath = linePathMetric.extractPath(
    //参数一,开始测量的路径长度位置
    //参数二,结束测量的路径长度位置
    0.0,
    lineExtracProgress);

Path rectExtractPath = rectPathMetric.extractPath(
    //参数一,开始测量的路径长度位置
    //参数二,结束测量的路径长度位置
    0.0,
    rectExtracProgress);

//合并这两个 Path
lineExtractPath.addPath(rectExtractPath, Offset.zero);


//绘制新组合创建的 Path
_paint.color = Colors.red;
canvas.drawPath(lineExtractPath, _paint);
```

8.3 绘图对文本与图片方面的支持

在绘制特殊形状时，时常会用到显示文本与背景图片的功能，所以本节讲解绘制文本与图片的功能。

8.3.1 绘制文本段落

在 Flutter 中，canvas 通过方法 drawParagraph 绘制文本段落，在绘制文本时，需要用到 ParagraphBuilder，在本节中，构建 TextPainter 的代码如下：

```
//代码清单 8-10 绘制文本段落
//lib/code8/example_811_Text.dart

import 'dart:ui' as ui;

class TextPainter extends CustomPainter {
  @override
  void paint(Canvas canvas, Size size) {
    //绘制文本代码
    …
  }

  @override
  bool shouldRepaint(CustomPainter oldDelegate) {
    return true;
  }
}
```

绘制一文本需要以下四步操作。

第一步，使用给定的样式信息生成包含文本的段落，通过配置的 ParagraphStyle 设置段落的对齐、截断和省略功能，然后使用 ParagraphBuilder 的构造函数来构建，代码如下：

```
//代码清单 8-10-1
//新建一个段落建造器，然后将文字基本信息填入
ui.ParagraphBuilder paragraphBuilder = ui.ParagraphBuilder(
  ui.ParagraphStyle(
    //文字方向为从左向右
    textDirection: TextDirection.ltr,
    //最大行数
    maxLines: 2,
    //文本居中
    textAlign: TextAlign.center,
    //粗体
    fontWeight: FontWeight.w400,
    //文字样式，FontStyle.italic 为斜体
    fontStyle: FontStyle.normal,
    //文本大小
    fontSize: 24.0,
    //配置超出范围时文本段落结尾时显示的内容
    ellipsis: "…",
    //设置行间距，实际的行高为 height * fontSize
    height: 1.2,
    textHeightBehavior: const TextHeightBehavior(
        //是否对段落中第一行的上行应用 TextStyle.height 修饰符
        //默认值为 true，TextStyle.height 修饰符将应用于第一行的上升
```

```
        //如果为 false,将使用字体的默认提升
        //并且 TextStyle.height 对第一行的提升没有影响
        applyHeightToFirstAscent: true,

        //是否将 TextStyle.height 修饰符应用于段落中最后一行的下行
        //默认为 true,将对最后一行的下行应用 TextStyle.height 修饰符
        //当为 false 时,将使用字体的默认下降
        //并且 TextStyle.height 对最后一行的下降没有影响
        applyHeightToLastDescent: true),
  ),
);
```

第二步,调用 pushStyle、addText 和 pop 的组合来向对象添加文本样式,代码如下:

```
//代码清单 8-10-2 第二步设置文字的样式
//这里配置的部分属性(如 fontSize)会覆盖 ParagraphStyle 中配置的 fontSize
paragraphBuilder.pushStyle(ui.TextStyle(
    color: Colors.red,
    fontSize: 20,
    height: 1,
    fontWeight: FontWeight.w500));

String text = "执剑天涯,从你的点滴积累开始,所及之处,必精益求精,即是折腾每一天";
//绑定要绘制的文本
paragraphBuilder.addText(text);
```

第三步,调用 build 以获得构造的 Paragraph 对象。

第四步,使用 Canvas.drawParagraph 绘制它,代码如下:

```
//代码清单 8-10-3 第三步设置文本的宽度约束
//参数为允许文本绘制的最大宽度
ui.ParagraphConstraints pc = const ui.ParagraphConstraints(width: 300);

//这里需要先调用 layout,将宽度约束填入,否则无法绘制
ui.Paragraph paragraph = paragraphBuilder.build()..layout(pc);

//第四步,绘制
//偏移量在这里指的是文字左上角的位置
canvas.drawParagraph(paragraph, const Offset(40, 40));
```

8.3.2 绘制图片

画布 canvas 通过方法 drawImage(常用)来绘制图片,使用 dart.ui 类中的 ui.Image,绘制图片的代码如下:

```
//代码清单 8-11 绘制图片
//lib/code8/example_812_Image.dart

import 'dart:ui' as ui;

class ImagePainter extends CustomPainter {
  final ui.Image _image;
  ImagePainter(this._image) : super();
  //定义画笔
  final Paint _paint = Paint()
    ..color = Colors.blue
    ..style = PaintingStyle.fill
    ..isAntiAlias = true
    ..strokeCap = StrokeCap.butt
    ..strokeWidth = 30.0;

  @override
  void paint(Canvas canvas, Size size) {
    //绘制图片
    canvas.drawImage(_image, const Offset(0, 0), _paint);
  }

  @override
  bool shouldRepaint(CustomPainter oldDelegate) {
    return true;
  }
}
```

绘制图片前，需要将图片加载到内存，如在这里加载的是资源目录下的图片，加载图片是一个耗时过程，所以放在异步线程中，代码如下：

```
//代码清单 8-11-1 加载图片
//lib/code8/example_812_Image.dart
class UIImageState extends State<Example812> {
  @override
  void initState() {
    super.initState();
    //加载本地资源目录图片
    _getAssetImage();
  }
  //获取本地图片
  Future<ui.Image>? _getAssetImage() async {
    ui.Image imageFrame = await getAssetImage(
      'assets/images/loginbg.png',
      width: 200,
```

```
      height: 200,
    );
    return imageFrame;
  }

  //获取资源目录图片，返回 ui.Image
  Future<ui.Image> getAssetImage(String asset, {width, height}) async {
    ByteData data = await rootBundle.load(asset);
    ui.Codec codec = await ui.instantiateImageCodec(
      data.buffer.asUint8List(),
      targetWidth: width,
      targetHeight: height,
    );
    ui.FrameInfo fi = await codec.getNextFrame();
    return fi.image;
  }
```

上述加载图片完成后，返回的是一个 Future，在 Flutter 中可以通过 FutureBuilder 来绑定 Future 任务，如在这里，当图片加载完成时，通过回调 FutureBuilder 来绘制并显示图片，代码如下：

```
//代码清单 8-11-2 显示图片
//lib/code8/example_812_Image.dart
@override
Widget build(BuildContext context) {
  return Scaffold(
    body: Center(
      child: buildFutureBuilder(),
    ),
  );
}

Widget buildFutureBuilder() {
  return FutureBuilder<ui.Image>(
    future: _getAssetImage(),
    builder: (context, snapshot) {
      if (snapshot.data == null) {
        //未加载完成时显示一个加载进度
        return const CircularProgressIndicator();
      } else {
        //加载完成后
        ui.Image? image = snapshot.data;
        //绘制图片并显示
        return CustomPaint(
          foregroundPainter: ImagePainter(image!),
```

```
            //画在 child 前面
            child: Container(
              width: double.infinity,
              height: 300,
              color: const Color.fromRGBO(155, 155, 155, 1),
            ),
          );
        }
      },
    );
  }
```

获取网络图片的代码如下：

```
//获取网络图片
Future<ui.Image>? _getNetImage() async {
  ui.Image uiImage = await getNetImage('图片地址', width: 200);
  return uiImage;
}

//获取网络图片,返回 ui.Image
Future<ui.Image> getNetImage(String url, {width, height}) async {
  ByteData data = await NetworkAssetBundle(Uri.parse(url)).load(url);
  ui.Codec codec = await ui.instantiateImageCodec(
    data.buffer.asUint8List(),
    targetWidth: width,
    targetHeight: height,
  );
  ui.FrameInfo fi = await codec.getNextFrame();
  return fi.image;
}
```

获取手机相册路径中的图片，代码如下：

```
//从本地路径中获取图片
Future<ui.Image> loadImageByFile(String path) async {
  Uint8List list = await File(path).readAsBytes();
  ui.Codec codec = await ui.instantiateImageCodec(list);
  ui.FrameInfo frame = await codec.getNextFrame();
  return frame.image;
}
```

8.3.3 绘制图片添加水印

在 8.3.2 节中，分别有本地资源目录下、网络图片、手机相册图片加载后转换为 ui.Image 的方法，这也是实现图片添加水印的第一步，第二步是向图片中添加文字水印，使用的原理是将图片与文字绘制到同一个画布上，然后将画布保存为一个新的图片，最后保存到设备本地存储路径中，代码如下：

```
//代码清单 8-12 图片加载水印文字
//lib/code8/example_821_Image_Wart.dart
Future<File> imageAddWaterMark(ui.Image image) async {

  //图片的宽度与高度
  int width = image.width;
  int height = image.height;

  //获得 Canvas
  ui.PictureRecorder recorder = ui.PictureRecorder();
  Canvas canvas = Canvas(recorder);

  //在 Canvas 上画 Image
  canvas.drawImage(image, const Offset(0, 0), Paint());

  //构建文本
  ui.ParagraphBuilder pb = ui.ParagraphBuilder(ui.ParagraphStyle(
    textAlign: TextAlign.start,
    fontWeight: FontWeight.normal,
    fontStyle: FontStyle.normal,
    fontSize: 30,
  ))
    ..pushStyle(ui.TextStyle(color: Colors.black))
    ..addText('早起的年轻人水印');

  ui.ParagraphConstraints pc =
      ui.ParagraphConstraints(width: width.toDouble());
  ui.Paragraph paragraph = pb.build()..layout(pc);
  //绘制文本
  canvas.drawParagraph(paragraph, const Offset(0, 0));

  ui.Picture picture = recorder.endRecording();
  //合成带水印的 Image
  ui.Image img = await picture.toImage(width.toInt(), height.toInt());
  //压缩为 ByteData
  ByteData? pngBytes = await img.toByteData(format: ui.ImageByteFormat.png);

  //保存到 SD 卡上
  final Directory _directory = await getTemporaryDirectory();
  final Directory _imageDirectory =
      await Directory('${_directory.path}/image/').create(recursive: true);
  String _targetPath = _imageDirectory.path;
```

```
    String localPath =
        '${_targetPath}_${DateTime.now().millisecondsSinceEpoch}.png';
    File file = File(localPath);
    file.writeAsBytesSync(pngBytes!.buffer.asInt8List());

    return file;
  }
```

在这里通过 getTemporaryDirectory 获取设备本地存储路径，需要添加的依赖如下：

```
path_provider: ^2.0.9
```

path_provider 是一个 Flutter 插件，支持 Android、iOS、Linux、macOS、Windows 系统，用于获取程序运行设备中的文件目录系统，使用方法如表 8-2 所示。

表 8-2　path_provider 获取文件目录概述

类　别	描　述
getTemporaryDirectory()	临时目录
getApplicationSupportDirectory()	应用程序支持目录
getLibraryDirectory()	应用程序持久文件目录
getApplicationDocumentsDirectory()	文档目录
getExternalStorageDirectory()	外部存储目录
getExternalCacheDirectories()	外部存储缓存目录
getExternalStorageDirectories()	外部存储目录(单独分区)
getDownloadsDirectory()	桌面程序下载目录

第9章 混合开发专题

CHAPTER 9

本书所描述的混合开发是指 Flutter 项目中加载 Android 原生与 iOS 原生功能及 Android 与 iOS 原生中加载 Flutter 模块功能。

在 Flutter 中,当应用遇到使用通知、应用生命周期、深链接、传感器、相机、电池、地理位置、声音、网络连接及与其他应用共享数据、打开其他应用、持久首选项、特殊文件夹、设备信息等相关数据时,就要使用到本章的知识点。

本章从开发 Flutter 应用程序、Flutter 依赖库、原生项目中集成 Flutter 模块三条线来分析讲解,在前面章节中已详细概述了 Flutter 应用程序的项目创建,在这里不做过多描述,下面概述一下 Flutter 依赖库项目的创建。

开发依赖库,首先应创建依赖库项目,创建方式有 3 种,前两种创建方式为可视化的操作,会打开如图 9-1 所示的创建项目的选择弹框。

(1) 在 Android Studio 的 Welcome 页面单击创建 Start New Flutter Project。

(2) 在 Android Studio 的工具栏中选中 File→New→New Flutter Project,如图 9-2 所示。

(3) 在命令行创建工程项目。

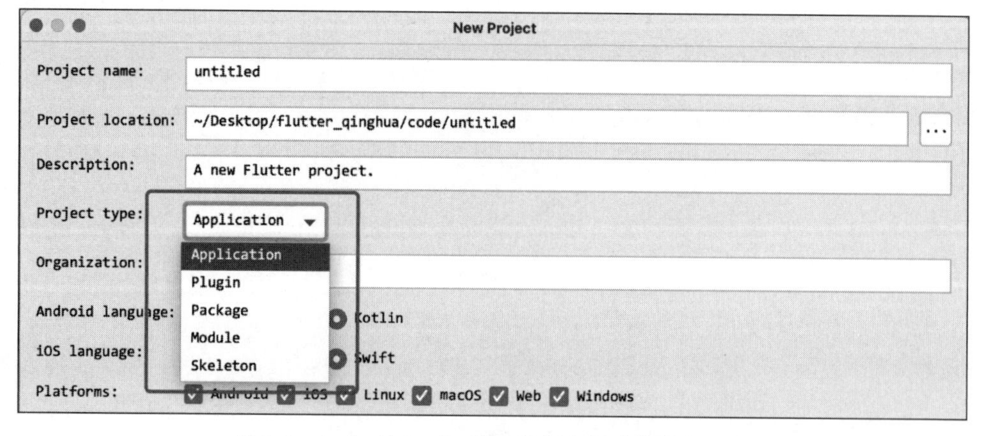

图 9-1 Android Studio 的工具栏中创建依赖库项目

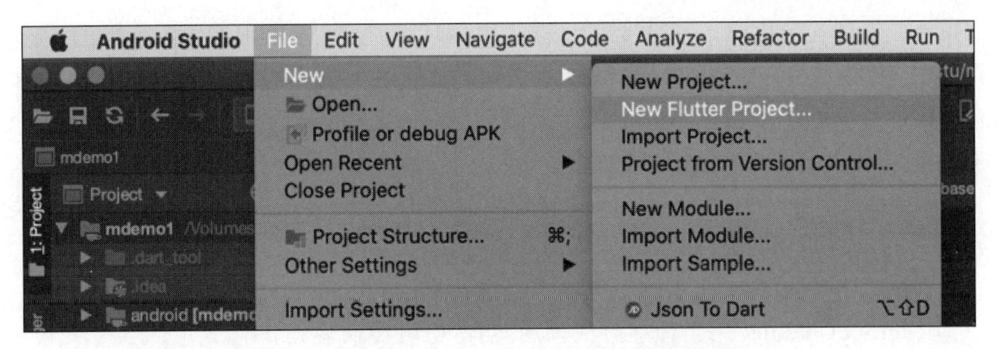

图 9-2　Android Studio 的工具栏中创建依赖库项目

Flutter 使用 Dart 语言开发移动应用，一套代码可以同时构建 Android 和 iOS 应用，但是 Dart 不会编译成 Android Dalvik 字节码，在 iOS 上也不会有 Dart/Objective-C 的绑定，也就意味着 Dart 代码并不会直接访问平台特定的 API，即 iOS Cocoa Touch 及 Android SDK 的 API。

在 Flutter 中，插件类型分为 3 种，Flutter Plugin、Flutter Package、Flutter Module。开发插件包，用来开发调用特定平台的 API 的包，这个插件包含针对 Android（Java 或 Kotlin 代码）或 iOS（Objective-C 或 Swift 代码）编写的特定于平台的实现（可以同时包含 Android 和 iOS 原生的代码），如加载 H5 的 WebView、调用自定义相机、录音、蓝牙等，需要创建 Flutter Plugin 项目，或者使用命令行工具，创建命令如下（Test 是创建的项目名称）：

```
flutter create -- org com.example -- template = plugin Test
```

Flutter Package 用来创建 Dart 语言类库，主要用于封装 Dart 语言一些共用的代码块，创建开发工程的方式可选择如图 9-2 所示的 Flutter Package，或者使用命令行，创建命令如下：

```
flutter create - t module Test
```

Flutter Module 用来创建 Flutter 的一些组件，以便在 Android、iOS 中嵌入使用，创建开发工程的方式可选择如图 9-2 所示的选项 Flutter Module，或者使用命令行，创建命令如下：

```
flutter create - t module Test
```

9.1　Flutter 与原生（Android、iOS）双向通信

所谓双向通信，在这里指的是在 Flutter 中调用 Android 与 iOS，反之在原生 Android 与 iOS 中调用 Flutter，指的是数据的双向传输。

在 Flutter 开发中，如果使用 Dart 语言结合 Widget 开发基本的业务，则可能不会用到这部分知识点，Flutter 框架及其底层图形引擎有足够的能力独立完成这些工作。

Flutter 编译代码由 Android 或 iOS 应用程序环境托管，应用程序的 Flutter 部分包含在标准的平台特定组件中，如 Android 上的 View 及 iOS 上的 UIViewController 中，所以在 Flutter 项目中也可以在宿主 App 中使用 Java/Kotlin 或者 Objective-C/Swift 执行尽可能多的操作，直接调用平台特定的 API。

在 Flutter 项目中，Platform Channels 提供了在 Dart 代码和宿主 App 的平台代码之间进行通信的方式，即双向通信方式，常用的通信方式有以下 3 种：

（1）BasicMessageChannel 用于传递字符串和半结构化的信息。

（2）MethodChannel 用于传递方法调用，当然也可传输参数。

（3）EventChannel 用于数据流 Event Streams 的通信。

双向通信应用于开发插件中的数据交互及 Flutter 项目中混合开发的双向互调，本节将会分别从这两种场景来论述。如在实际应用开发中在 Flutter 中调用系统相机，需要从 Flutter 将打开指令发送到 Android、iOS 中，然后执行调用操作，在获取照片后还需要将照片信息回传到 Flutter 中。

9.1.1 MethodChannel 实现消息监听与发送

在这里模拟将一组数据从 Flutter 中发送到 Android、iOS 中，然后做相应处理后再从 Android、iOS 将处理后的数据返回 Flutter 中。

对于 MethodChannel，官方 API 中解析为 A named channel for communicating with platform plugins using asynchronous method calls，也就是用来实现 Flutter 与原生 Android、iOS 通信，从 Flutter 中发送一条消息数据，经历的过程如图 9-3 所示。

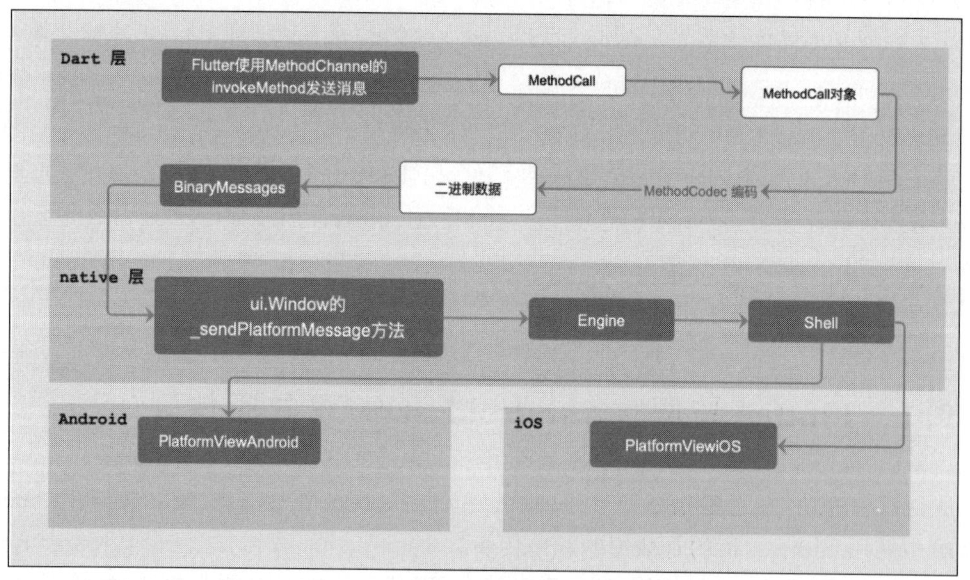

图 9-3 Flutter 消息发送流程示意图

如图 9-3 所示，当在 Flutter 中使用 MethodChannel 的 invokeMethod 方法发送一条消息时，invokeMethod 方法会将其入参 message 和 arguments 封装成一个 MethodCall 对象，并使用 MethodCodec 将其编码为二进制格式数据，再通过 BinaryMessages 将消息发出。

在实际项目开发中，使用 MethodChannel 时，可以考虑全局注册一个，在这里通过在一个自定义的 StatefulWidget 中进行开发概述，首先在 Flutter 中使用通信的 StatefulWidget，声明如下：

```
  //代码清单 9-1-1
//lib/MethodChannelPage.dart
  //创建 MethodChannel
  //参数一,flutter_and_native_101 为通信标识
  //参数二,codec 为参数传递的编码方式,非必选,默认为 StandardMessageCodec()
  //参数三,binaryMessenger 为使用的消息通道,默认为
  //[ServicesBinding.defaultBinaryMessenger]
  static const methodChannel = MethodChannel('flutter_and_native_101');

  //封装 Flutter 向原生中发送消息的方法
  static Future<dynamic> invokNative({
    required String method,                //method 为方法标识
    Map<String, dynamic>? arguments,       //arguments 为传递的参数
  }) async {
    if (arguments == null) {
      //无参数发送消息
      return await methodChannel.invokeMethod(method);
    } else {
      //有参数发送消息
      return await methodChannel.invokeMethod(method, arguments);
    }
  }
```

然后在 Flutter 中单击一个按钮调用 invokNative 方法向 Android、iOS 原生中发送一条数据，在收到回执消息后刷新页面进行显示，代码如下：

```
//代码清单 9-1-2
//lib/MethodChannelPage.dart
String _result = "";
buildTextButton() {
  return TextButton(
    onPressed: () async {
      //发送消息
      dynamic result = await invokNative(method: "test");
      //回执消息
      if (result != null) {
        _result = result['message'].toString();
```

```
            setState(() {});
          }
        },
        child: Text("发送消息 $_result"),
      );
    }
```

在 Android 平台，需要接收 Flutter 发送的消息，在如图 9-4 所示的 MainActivity 中编写 Android 方面的代码。

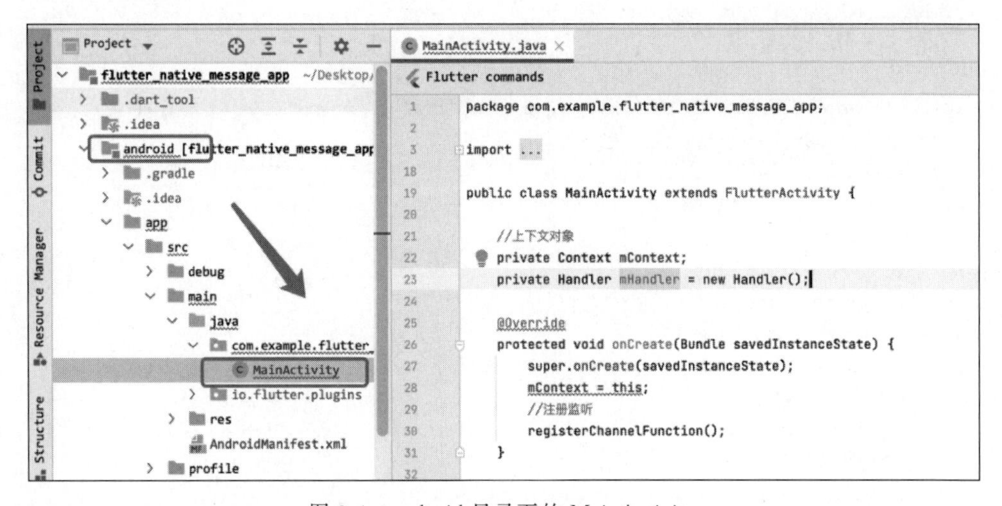

图 9-4　android 目录下的 MainActivity

在 MainActivity 文件上右击，可以将现有项目以 Android 工程目录的形式打开编辑，如图 9-5 所示。

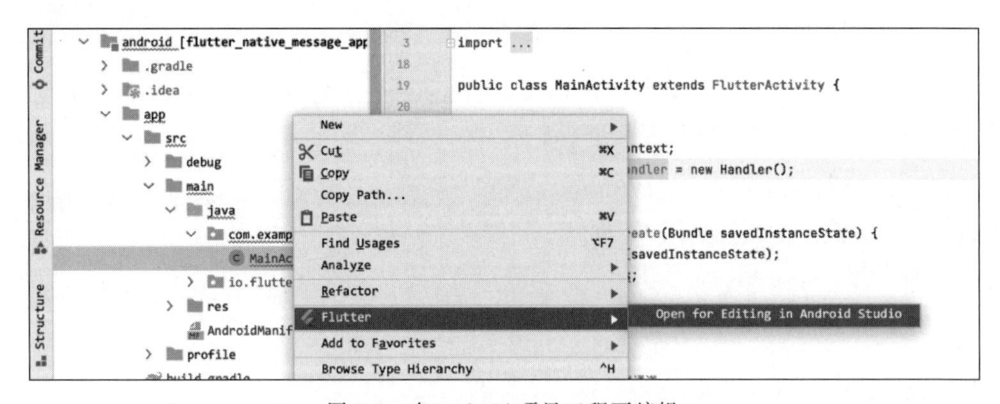

图 9-5　在 android 项目工程下编辑

```
//代码清单 9-2
//android/app/src/main/Java/com/example/flutter_native_message_app/MainActivity.Java
```

```
import io.flutter.embedding.android.FlutterActivity;
import io.flutter.embedding.engine.FlutterEngine;
import io.flutter.embedding.engine.dart.DartExecutor;
import io.flutter.plugin.common.BinaryMessenger;
import io.flutter.plugin.common.MethodChannel;

public class MainActivity extends FlutterActivity {

    //上下文对象
    private Context mContext;
    private Handler mHandler = new Handler();
    //消息通道
    private MethodChannel mMethodChannel;
    @Override
    protected void onCreate(Bundle savedInstanceState) {
        super.onCreate(savedInstanceState);
        mContext = this;
        //注册监听
        registerChannelFunction();
    }
    …
}
```

注册监听的代码如下：

```
//代码清单 9-2-1 注册监听
//android/app/src/main/Java/com/example/flutter_native_message_app/MainActivity.Java
//记着要在 onCreat 方法中调用
private void registerChannelFunction() {
    //Flutter 运行环境参数封装类
    FlutterEngine lFlutterEngine = getFlutterEngine();
    if (lFlutterEngine == null) {
        Log.e("ERROR", "注册消息通道失败 FlutterEngine = null");
        return;
    }
    //获取 Dart 缓存编译对象
    DartExecutor lDartExecutor = lFlutterEngine.getDartExecutor();
    //获取默认的 BinaryMessenger
    BinaryMessenger lBinaryMessenger = lDartExecutor.getBinaryMessenger();
    //消息通道名称
    String channelName = "flutter_and_native_101";
    //构建消息通道
    mMethodChannel = new MethodChannel(lBinaryMessenger, channelName);
    //设置监听,这里使用匿名内部类的方式
    mMethodChannel.setMethodCallHandler(methodCallHandler);
}
```

在 Flutter 中通过 methodChannel 来发送消息，对应地，在 Android 的 MethodCallHandler 监听中会接收到监听，可以获取 Flutter 中发送的方法名称与参数，代码如下：

```
//代码清单 9-2-2 监听回调
MethodChannel.MethodCallHandler methodCallHandler = (call, result) -> {
    //获取方法名称
    String lMethod = call.method;
    //获取参数
    Object lArguments = call.arguments;
    Map<String, Object> arguments = null;
    if (lArguments != null) {
        arguments = (Map<String, Object>) lArguments;
        //获取 Flutter 传递的参数
    }
    //处理消息
    if (lMethod.equals("test")) {
        Toast.makeText(mContext,
                "Flutter 调用到了 Android test",
                Toast.LENGTH_SHORT).show();
        Map<String, Object> resultMap = new HashMap<>();
        resultMap.put("message", "ABC");
        resultMap.put("code", 200);
        //将消息发送至 Flutter
        //此方法只能使用一次
        result.success(resultMap);

    } else {
        result.notImplemented();
    }
};
```

在这里从 Flutter 将数据发送到 Android 中，发送的数据格式要与 Android 中接收的数据类型一致，如 Flutter 中发送的是 Map<String,dynamic>类型，就需要在 Android 中使用 Map<String,Object>类型来接收，然后从 Android 向 Flutter 回传的数据为 Map<String,Object>类型，在 Flutter 中接收的数据类型应为 Map<dynamic, dynamic>。

上述已通过 MethodChannel 实现了数据从 Flutter 发送到 Android 原生，并通过 MethodChannel.Result(Android)实现了数据的回传，但是这两者只能回调一次。在实际项目开发中，在 Android 原生可通过 MethodChannel 实现主动向 Flutter 中多次发送数据，代码如下：

```
//代码块 9-2-3
Map<String, Object> resultMap2 = new HashMap<>();
resultMap2.put("message", "Android 主动调用 Flutter test 方法");
resultMap2.put("code", 200);
mMethodChannel.invokeMethod("test2", resultMap2);
```

对应地,在 Flutter 要接收消息,需要事先设置消息监听,代码如下:

```
  //代码块 9-1-3
//lib/MethodChannelPage.dart
  @override
  void initState() {
    super.initState();
    //设置监听
    methodChannel.setMethodCallHandler((MethodCall call) async{
      //处理原生 Android 和 iOS 发送过来的消息
      String method = call.method;
      Map arguments = call.arguments;
      int code = arguments["code"];
      String message = arguments["message"];
    });
  }
```

在 ios 目录下的 AppDelegate.m 文件上右击,以 iOS 项目目录的方式在 Xcode 开发工具中打开文件进行编辑,如图 9-6 所示。

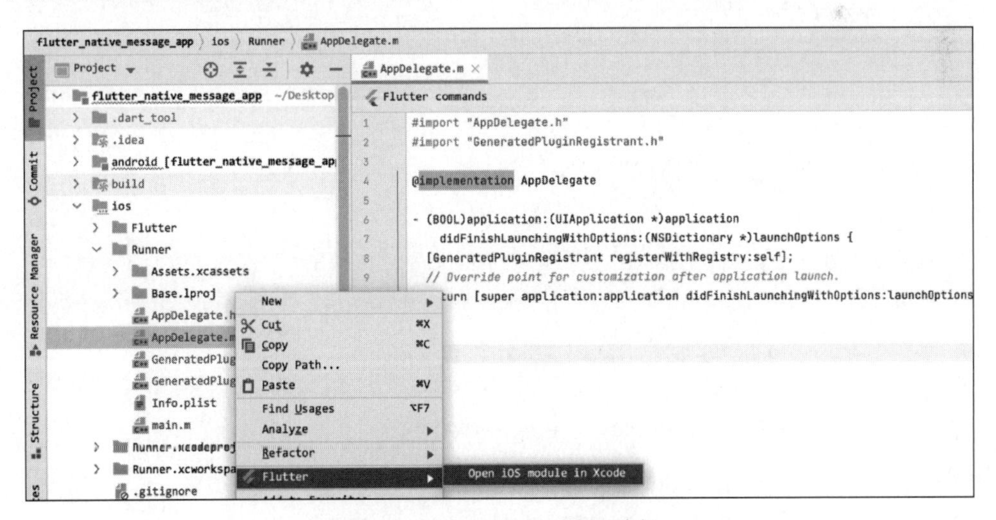

图 9-6　在 iOS 项目工程下编辑

同理在 iOS 中接收 Flutter 方面的消息,也需要先设置监听,在监听回调中处理消息,代码如下:

```
//代码清单 9-3 注册监听
#import "AppDelegate.h"
#import "GeneratedPluginRegistrant.h"


@implementation AppDelegate{
    FlutterMethodChannel * methodChannel;
}


- (BOOL)application:(UIApplication *)application
    didFinishLaunchingWithOptions:(NSDictionary *)launchOptions {
   [GeneratedPluginRegistrant registerWithRegistry:self];

    //FlutterMethodChannel 与 Flutter 之间的双向通信
    [self methodChannelFunction];


  return [super application:application didFinishLaunchingWithOptions:launchOptions];
}

- (void) methodChannelFunction{
    //获取当前的 ViewControlle
FlutterViewController * controller =
        (FlutterViewController *)self.window.rootViewController;
    //创建 FlutterMethodChannel
    //flutter_and_native_101 是通信标识
    methodChannel = [FlutterMethodChannel
                     methodChannelWithName:@"flutter_and_native_101"
                     binaryMessenger:controller.binaryMessenger];

    __weak typeof(self) weakSelf = self;

    //设置监听
[methodChannel setMethodCallHandler:^(FlutterMethodCall * call,
                            FlutterResult result) {
        //Flutter 传过来的方法名称
        NSString *method = call.method;
        //Flutter 传过来的参数
        NSDictionary * arguments = call.arguments;

        [weakSelf methodCallMethod:method
             andWithArguments:arguments andWithResult:result];
    }];
}
```

```
//消息方法处理
- (void)methodCallMethod: (NSString * ) method andWithArguments:(
    NSDictionary *  ) arguments andWithResult: (FlutterResult) result{

    if ([method isEqualToString:@"test"]) {

        NSLog(@"Flutter 调用到了 iOS test");
        NSMutableDictionary * dic = [NSMutableDictionary dictionary];
        [dic setObject:@"ABC" forKey:@"message"];
        [dic setObject: [NSNumber numberWithInt:200] forKey:@"code"];
        //FlutterResult 回调将消息发送至 Flutter 中
        //此方法只能调用一次
        result(dic);

    }
}

@end
```

上述的 FlutterResult 只可使用一次，在 iOS 原生中通过 FlutterMethodChannel 可多次向 Flutter 中回传数据，代码如下：

```
//代码清单 9-3-1 将消息发送至 Flutter 中
NSMutableDictionary * dics = [NSMutableDictionary dictionary];
[dics setObject:@" 原生中数据" forKey:@"message"];
[dics setObject: [NSNumber numberWithInt:200] forKey:@"code"];

//通过此方法可以主动向 Flutter 中发送消息
//可以多次调用
[methodChannel invokeMethod:@"test" arguments:dics];
```

9.1.2　BasicMessageChannel 实现消息监听与发送

对于 BasicMessageChannel，官方 API 中解析为 A named channel for communicating with platform plugins using asynchronous …，也就是用来实现 Flutter 与原生 Android、iOS 通信，与 MethodChannel 的区别是 MethodChannel 针对于方法调用，BasicMessageChannel 针对于消息数据。针对发送和接收字符串消息，首先在页面中注册消息通道，然后设置监听通道，用来实时获取 Android 与 iOS 原生回传的消息，代码如下：

```
//代码清单 9-4
//lib/BasicMessageChannelPage.dart
class BasicMessageChannelPageState extends State<BasicMessageChannelPage> {
  //创建 MethodChannel
```

```
  //参数一,flutter_and_native_101 为通信标识
  //参数二,codec 为参数传递的编码方式,必选
  //参数三,binaryMessenger 为使用的消息通道,默认为
  //[ServicesBinding.defaultBinaryMessenger]
  static const messageChannel =
      BasicMessageChannel('flutter_and_native_100', StandardMessageCodec());

  String _result = "";

  @override
  void initState() {
    super.initState();
    messageChannel.setMessageHandler((result) async {
      if (result != null&&result is Map) {
        //解析原生发给 Flutter 的参数
        int code = result["code"];
        String message = result["message"];
        setState(() {
          _result = "receiveMessage: code: $ code message: $ message";
        });
      }
      return 'Flutter 已收到消息';
    });
  }
  ...

}
```

然后在这个页面中,单击按钮,向 Android 与 iOS 原生发送一条消息,参数可以自定义,发送成功后再接收原生的回执消息进行显示,代码如下:

```
//代码清单 9-4-1
//lib/BasicMessageChannelPage.dart
buildTextButton() {
  return TextButton(
    onPressed: () async {
      //发送消息
      dynamic result = await sendMessage(arguments: {"method": "test"});
      //回执消息
      if (result != null) {
        _result = result['message'].toString();
        setState(() {});
      }
    },
    child: Text("发送消息 $ _result"),
```

```
    );
  }

  //使用 messageChannel 发送消息
  Future<Object?> sendMessage({required Map arguments}) async {
    Object? reply = await messageChannel.send(arguments);
    if (reply != null && reply is Map) {
      //解析原生发给 Flutter 的参数
      int code = reply["code"];
      String message = reply["message"];
      //更新 Flutter 中的页面显示
      setState(() {
        _result = "code: $ code message: $ message";
      });
    }
    return reply;
  }
```

在 Flutter 应用项目的 android 目录下的 MainActivity 中，一是要接收 Flutter 中发送过来的消息并回应消息，二是要主动多次地向 Flutter 中发送消息，首先是注册消息监听用于获取 Flutter 中发送的消息，代码如下：

```
  //代码清单 9-5-1 注册监听
  //BasicMessageChannel 实例对象
  private BasicMessageChannel<Object> mMessageChannel;

  //这种方法需要在 onCreate 方法中调用
  private void registerMessageChannelFunction() {

      //获取 BinaryMessenger
      BinaryMessenger lBinaryMessenger = getFlutterEngine()
              .getDartExecutor()
              .getBinaryMessenger();
      //消息接收监听
      //BasicMessageChannel (主要用于传递字符串和一些半结构体的数据)
      mMessageChannel = new BasicMessageChannel<Object>(
              lBinaryMessenger,
              "flutter_and_native_100",
              StandardMessageCodec.INSTANCE);

      //接收消息监听
      //监听参数 handlerArgument 是 Flutter 中传过来的消息
      //参数 reply 是 BasicMessageChannel.Reply<Object>类型
      mMessageChannel.setMessageHandler((handlerArgument, reply) -> {
          //获取传递的参数
```

```
        Map<String, Object> arguments = (Map<String, Object>) handlerArgument;
        //消息处理
        try {
            baseMessageChannelCallFunction(reply, arguments);
        } catch (Exception e) {
            e.printStackTrace();
            //异常回调
            Map<String, Object> resultMap = new HashMap<>();
            resultMap.put("message", "操作异常 " + e.getMessage());
            resultMap.put("code", 500);
            //回调此方法只能使用一次
            reply.reply(resultMap);
        }
    });

}
```

本实例中封装了一层处理参数，实际上就是根据不同的参数，来甄别不同的行为，从而实现不同的业务逻辑，代码如下：

```
//代码清单 9-5-2 处理参数
private void baseMessageChannelCallFunction(
        BasicMessageChannel.Reply<Object> reply,
        Map<String, Object> arguments) {
    if(arguments.containsKey("method")) {
        //处理参数
        //方法名标识
        String lMethod = (String) arguments.get("method");
        //测试 reply.reply()方法将消息发给 Flutter
        if (lMethod.equals("test")) {
            Toast.makeText(mContext, "调用到了 test",
                    Toast.LENGTH_SHORT).show();
            //回调 Flutter
            Map<String, Object> resultMap = new HashMap<>();
            resultMap.put("message", "reply.reply 返给 Flutter 的数据");
            resultMap.put("code", 200);
            //回调此方法只能使用一次
            reply.reply(resultMap);
            return;
        }
    }
    reply.reply("没有实现方法");
}
```

在这里的 Reply 只能回调一次，如果需要多次向 Flutter 发送消息，则需要通过

BasicMessageChannel 的 send 方法发送，发送的消息会在 Flutter 中的监听方法 MessageHandler 的回调中获取，代码如下：

```
//代码清单 9-5-3
//Android 中 BasicMessageChannel 向 Flutter 中主动发送消息
private void messageChannelSendFunction() {
    //构建参数
    Map<String, Object> resultMap = new HashMap<>();
    resultMap.put("message", "ABCSD");
    resultMap.put("code", 200);

    //向 Flutter 中发送消息
    //参数二可以再次接到 Flutter 中的回调
    //也可以直接使用 mMessageChannel.send(resultMap)
    mMessageChannel.send(resultMap, new BasicMessageChannel.Reply<Object>() {
        @Override
        public void reply(Object o) {
            //再次接收的回执消息
            Log.d("mMessageChannel", "mMessageChannel send 回调 " + o);
        }
    });
}
```

在 iOS 中通过 FlutterBasicMessageChannel 来处理 Flutter 中 BasicMessageChannel 的消息通道消息，首先应注册通道，然后设置监听，最后处理消息内容，代码如下：

```
//代码清单 9-6
#import "AppDelegate.h"
#import "GeneratedPluginRegistrant.h"

@implementation AppDelegate{
FlutterBasicMessageChannel * messageChannel;
}
- (BOOL)application:(UIApplication *)application
didFinishLaunchingWithOptions:(NSDictionary *)launchOptions {
    //默认需要保留
    [GeneratedPluginRegistrant registerWithRegistry:self];

    //FlutterBasicMessageChannel 与 Flutter 之间的双向通信
    [self basicMessageChannelFunction];

    return [super application:application
                    didFinishLaunchingWithOptions:launchOptions];
}
```

注册消息通道并设置回调监听，代码如下：

```
//代码清单 9-6-1
-(void) basicMessageChannelFunction{
    //获取当前的 controller
    FlutterViewController * controller
            = (FlutterViewController * )self.window.rootViewController;
    //初始化定义
    //flutter_and_native_100 j
    messageChannel = [FlutterBasicMessageChannel
                      messageChannelWithName:@"flutter_and_native_100"
                      binaryMessenger:controller.binaryMessenger];

     __weak typeof(self) weakSelf = self;
    //接收消息监听
    [messageChannel setMessageHandler:^(id message, FlutterReply callback) {
        //处理消息
        [weakSelf basicMessageChannelCall:message andWithReply:callback];
    }];
}
```

处理 Flutter 端发送的数据，并通过 FlutterReply 将一个回执消息发送到 Flutter 中，代码如下：

```
//代码清单 9-6-2 消息处理
-(void) basicMessageChannelCall:(id) message andWithReply:(FlutterReply )callback{
    NSString *method = message[@"method"];

    if ([method isEqualToString:@"test"]) {

        NSLog(@"Flutter 调用到了 iOS test");

        NSMutableDictionary *dic = [NSMutableDictionary dictionary];
        [dic setObject:@"ios 回调消息" forKey:@"message"];
        [dic setObject: [NSNumber numberWithInt:200] forKey:@"code"];

        //回调回执消息
        callback(dic);

    }
}
```

FlutterReply 只能回调一次，可通过 FlutterBasicMessageChannel 通道多次向 Flutter 端发送消息，代码如下：

```
//代码清单 9-6-3
//iOS 中通过 BasicMessageChannel 通道向 Flutter 中发送消息
-(void)baseMessageSendFunction{

    NSMutableDictionary *dic = [NSMutableDictionary dictionary];
    [dic setObject:@"这里是数据" forKey:@"message"];
    [dic setObject:[NSNumber numberWithInt:200] forKey:@"code"];

    //通过这种方法 iOS 可以主动多次向 Flutter 发送消息
    [messageChannel sendMessage:dic];

}
```

9.1.3 EventChannel 实现消息监听与发送

EventChannel 用于从原生向 Flutter 发送通知事件，例如 Flutter 通过其监听 Android、iOS 的重力感应变化，EventChannel 通道适用于频繁发送消息的情景，代码如下：

```
//代码清单 9-7
//lib/EventChannelPage.dart
class EventChannelPageState extends State<EventChannelPage> {
  //回传的消息标识
  String _reciverStr = "";
  //创建 EventChannel
  static const eventChannel = EventChannel('flutter_and_native_103');
  //消息流
  StreamSubscription? _streamSubscription;

  @override
  void initState() {
    super.initState();
    //设置监听
    _streamSubscription =
        eventChannel.receiveBroadcastStream().listen((Object? event) {
      //成功消息回调
      setState(() {
        _reciverStr = "回传数据：" + event.toString();
      });
    }, onError: (Object event) {
      //失败消息回调
      setState(() {
        _reciverStr = "回传数据：" + event.toString();
      });
    });
```

```
  }

  @override
  Widget build(BuildContext context) {
    return Scaffold(
      body: Center(
        child: Text(_reciverStr),
      ),
    );
  }
  @override
  void dispose() {
    super.dispose();
    //移除监听
    if (_streamSubscription != null) {
      _streamSubscription?.cancel();
      _streamSubscription = null;
    }
  }
}
```

然后在 Android 中注册通道，通过通道实时发送消息通知，代码如下：

```
//代码清单 9-8 EventChannel 通信
private EventChannel.EventSink mEventSink;

//需要在 MainActivity 的 OnCreate 方法中调用
private void registerEventChannerl() {

    //获取当前默认的 BinaryMessenger
    BinaryMessenger lBinaryMessenger = getFlutterEngine()
            .getDartExecutor()
            .getBinaryMessenger();
    //消息通道名称
    String channelName = "flutter_and_native_103";
    //创建 EventChannel 通道
    EventChannel lEventChannel = new EventChannel(lBinaryMessenger, channelName);
    //设置 StreamHandler
    lEventChannel.setStreamHandler(new EventChannel.StreamHandler() {
        //如果 onListen 可回调，则代表通道已经建好，可以发送数据了
        @Override
        public void onListen(Object o, EventChannel.EventSink eventSink) {
            //注意是通过 EventSink 发送消息
            mEventSink = eventSink;
```

```
        }

        //onCancel 表示 Flutter 端已取消接收消息
        @Override
        public void onCancel(Object o) {
            mEventSink = null;

        }
    });
}
```

当在 Android 中注册了 EventChannel 消息通道后，在 Flutter 中调用注册通道时，在 Android 中设置的 StreamHandler 的 onListen 函数就会回调，以通知 Android 原生消息通道建立，这样就可以使用 EventChannel.EventSink 进行数据回传了，回传方法如下：

```
//代码清单 9-8-1
private final Handler mHandler = new Handler();
//使用 EventChannel 通道向 Flutter 发送消息
//这里发送的消息为 String 类型
//可以在实际项目开发中发送其他数据类型
private void eventSendMessageFunction(String messsage) {
    if (mEventSink != null) {
        mHandler.post(new Runnable() {
            @Override
            public void run() {
                mEventSink.success(messsage);
            }
        });

    } else {
        Log.e("ERROR", "EventSink is null");
    }
}
```

在 iOS 中实现过程与此类似，首先应注册监听通道，然后发送消息，代码如下：

```
//代码清单 9-9
#import "AppDelegate.h"
#import "GeneratedPluginRegistrant.h"


@implementation AppDelegate{
    FlutterEventSink mEventSink;
    NSTimer * nsTime;
```

```
}


- (BOOL)application:(UIApplication * )application
didFinishLaunchingWithOptions:(NSDictionary * )launchOptions {
    //默认需要保留
    [GeneratedPluginRegistrant registerWithRegistry:self];

    [self registerEventChannerl];

    return [super application:application didFinishLaunchingWithOptions:launchOptions];
}
```

在 AppDelegate.h 文件中需要遵循协议 FlutterStreamHandler，代码如下：

```
#import <Flutter/Flutter.h>
#import <UIKit/UIKit.h>

@interface AppDelegate : FlutterAppDelegate <FlutterStreamHandler>

@end
```

然后注册 FlutterEventChannel 消息通道，代码如下：

```
//代码清单 9-9-1 注册
//EventChannel 通信
-(void)registerEventChannerl{
    //获取当前的 controller
    FlutterViewController * controller =
            (FlutterViewController * )self.window.rootViewController;
    //创建 EventChannel 通信
FlutterEventChannel * eventChannel = [FlutterEventChannel
            eventChannelWithName:@"flutter_and_native_103"
            binaryMessenger:controller.binaryMessenger];
    //添加监听
    [eventChannel setStreamHandler:self];
}

//EventChannel 注册成功回调方法
#pragma mark - FlutterStreamHandler
- (FlutterError * _Nullable)onListenWithArguments:
                (id _Nullable)arguments
                                        eventSink:(FlutterEventSink)eventSink{
    mEventSink = eventSink;
    NSLog(@"EventChannle 注册");
```

```
    //开启定时
    [self startTimer];
    return nil;
}

//EventChannel 取消回调方法
#pragma mark - FlutterStreamHandler
- (FlutterError * _Nullable)onCancelWithArguments:(id _Nullable)arguments {
    NSLog(@"EventChannle 取消");
    mEventSink = nil;
    [self stopTimer];
    return nil;
}
```

在本节中，通过计时器模拟数据的发送，代码如下：

```
//代码清单 9-9-2 模拟消息发送
//启动定时器
-(void)startTimer{
    //以 NSTime 的类方法创建一个定时器并且启动这个定时器
    //p1:每隔多长时间调用定时器函数,以秒为单位
    //p2:表示实现这个定时器函数的对象
    //p3:定时器函数对象
    //p4:可以传入定时器函数一个参数,无参数时传入 nil
    //p5:定时器是否重复操作,YES 表示重复,NO 表示只完成一次函数调用
    nsTime = [NSTimer scheduledTimerWithTimeInterval:1
                             target:self selector:@selector(updateTime:)
                                                 userInfo:nil repeats:YES];
}

-(void)updateTime:(NSTimer * ) timer{
    NSLog(@"参数为 %@",timer.userInfo);
    NSDate * datenow = [NSDate date];    /现在时间,输出后来看一下是什么格式

      NSString * timeSp = [NSString stringWithFormat:@" %ld",
                              (long)[datenow timeIntervalSince1970]];

    if(mEventSink!= nil){
        mEventSink([NSString stringWithFormat:@"定时执行 %@",timeSp]);
    }
}

//停止定时器
-(void)stopTimer{
    [nsTime invalidate];
}
```

9.2 Flutter 调用原生 View

Flutter 可将 Android 的 TextView 文本框、iOS 的 UIlabe 文本框内嵌在 Flutter 页面。Flutter 支持两种集成模式：虚拟显示模式（Virtual Displays）和混合集成模式（Hybrid Composition）。

（1）虚拟显示模式会将 android.view.View 实例渲染为纹理，因此它不会嵌入 Android Activity 的视图层次结构中（键盘处理和辅助功能），可能无法正常工作。

（2）混合集成模式会将原生的 android.view.View 附加到视图层次结构中，因此键盘处理和无障碍功能是开箱即用的，在 Android 10 之前，此模式可能会大大降低 Flutter UI 的帧吞吐量（FPS），虚拟显示模式将在 9.2.3 节开发插件中应用。

9.2.1 内嵌 Android 原生 View

第一步，创建 Android 原生的自定义 View，需要继承于 PlatformView，然后在 TestTextView 构造函数中创建 Anroid 原生的 View，在这里使用了 Android 原生文本组件 TextView，代码如下：

```
//代码清单 9-10
//com/example/flutter_native_message_app/TestTextView.Java
//创建 Android 原生自定义 View
public class TestTextView implements PlatformView {

    //这里使用的是一个 TextView
    private final TextView mTestTextView;

    TestTextView(Context context, int id, Map<String, Object> params) {
        //创建 TextView
        TextView lTextView = new TextView(context);
        //设置文字
        lTextView.setText("Android 的原生 TextView aas ");
        this.mTestTextView = lTextView;
        //Flutter 传递过来的参数
        if (params!= null&&params.containsKey("content")) {
            String myContent = (String) params.get("content");
            //设置文本显示
            lTextView.setText(myContent);
        }
    }
    @Override
    public View getView() {
        return mTestTextView;
```

```
    }

    @Override
    public void dispose() {}
}
```

第二步，创建一个 PlatformViewFactory 来关联上述自定义的 TestTextView，代码如下：

```
//代码清单 9-11
//com/example/flutter_native_message_app/TestViewFactory.Java
public class TestViewFactory extends PlatformViewFactory {


    public TestViewFactory() {
        super(StandardMessageCodec.INSTANCE);
    }

    /**
     * @param args args 是由 Flutter 传过来的自定义参数
     */
    @SuppressWarnings("unchecked")
    @Override
    public PlatformView create(Context context, int id, Object args) {
        //Flutter 传递过来的参数
        Map<String, Object> params = (Map<String, Object>) args;
        //创建自定义的 TestTextView
        return new TestTextView(context, id, params);

    }

    //需要在 MainActivity 的 onCreate 方法中调用
    public static void registerWith(FlutterEngine flutterEngine) {
        //通过 platformViewsController 获取 Registry
        PlatformViewRegistry registry =
                flutterEngine.getPlatformViewsController().getRegistry();
        //通过工厂类 PlatformViewRegistry 注册 Android 原生 View
        //参数一是设置标识
        //参数二是自定义的 Android 原生 View
        registry.registerViewFactory(
                "com.flutter_to_native_test_textview",
                new TestViewFactory());
    }
}
```

第三步，在 MainActivity 中绑定自定义的 TestViewFactory，代码如下：

```
//Flutter 项目中默认创建的 MainActivity
public class MainActivity extends FlutterActivity {

    @Override
    protected void onCreate(Bundle savedInstanceState) {
        super.onCreate(savedInstanceState);
        //新创建的插件
        TestViewFactory.registerWith(getFlutterEngine());
    }
}
```

最后，在 Flutter 中通过 AndroidView 来加载 Android 中自定义的原生文本组件 TextView，AndroidView 是虚拟显示模式，代码如下：

```
//代码清单 9 - 12
//lib/NaviteViewPage.dart
buildAndroidView() {
  return SizedBox(
    height: 200,
    child: AndroidView(
      //设置标识
      viewType: "com.flutter_to_native_test_textview",
      //参数
      creationParams: const {
        "content": " 34erw3 ",
      },
      onPlatformViewCreated: (int id) {
        //Android 原生的 View 创建后的回调
      },
      //参数的编码方式
      creationParamsCodec: const StandardMessageCodec(),
    ),
  );
}
```

9.2.2 内嵌 iOS 原生 View

第一步，创建 iOS 原生 View，需要继承于 FlutterPlatformView，代码如下：

```
//代码清单 9 - 13 FlutterIosTextLabel.h
#import <UIKit/UIKit.h>
```

```
#import <Flutter/Flutter.h>
NS_ASSUME_NONNULL_BEGIN

//实现协议 FlutterPlatformView
@interface FlutterIosTextLabel : NSObject <FlutterPlatformView>

- (instancetype)initWithWithFrame:(CGRect)frame
                   viewIdentifier:(int64_t)viewId
                        arguments:(id _Nullable)args
                  binaryMessenger:(NSObject<FlutterBinaryMessenger> *)messenger;
@end


NS_ASSUME_NONNULL_END
```

```
//代码清单 9-14 FlutterIosTextLabel.m
#import "FlutterIosTextLabel.h"

@interface FlutterIosTextLabel ()

@end

@implementation FlutterIosTextLabel{
    //FlutterIosTextLabel 创建后的标识
    int64_t _viewId;
    UILabel * _uiLabel;
    //消息回调
    FlutterMethodChannel * _channel;
}

//在这里只是创建了一个 UILabel
- (instancetype)initWithWithFrame:(CGRect)frame
                        viewIdentifier:(int64_t)viewId arguments:(id)args binaryMessenger:
(NSObject<FlutterBinaryMessenger> *)messenger{
    if ([super init]) {
        if (frame.size.width == 0) {
            frame = CGRectMake(frame.origin.x,
                               frame.origin.y,
                               [UIScreen mainScreen].bounds.size.width, 22);
        }
        _uiLabel = [[UILabel alloc] initWithFrame:frame];
        _uiLabel.textColor = [UIColor redColor];
        _uiLabel.text = @"ios 原生 uilabel ";
        _uiLabel.font = [UIFont systemFontOfSize:14];
        _uiLabel.textAlignment = NSTextAlignmentCenter;
```

```
        _uiLabel.backgroundColor = [UIColor grayColor];
        _viewId = viewId;

    }
    return self;
}

- (nonnull UIView * )view {
    return _uiLabel;
}

@end
```

第二步，定义 FlutterPlatformViewFactory，用来构建装载上述创建的 FlutterIosTextLabel，代码如下：

```
//代码清单 9-15 FlutterIosTextLabelFactory.h

#import <Foundation/Foundation.h>
#import <Flutter/Flutter.h>
NS_ASSUME_NONNULL_BEGIN

@interface FlutterIosTextLabelFactory : NSObject<FlutterPlatformViewFactory>

- (instancetype)initWithMessenger:(NSObject<FlutterBinaryMessenger> * )messager;

+ (void)registerWithRegistrar:(nonnull NSObject<FlutterPluginRegistrar> * )registrar ;
@end

NS_ASSUME_NONNULL_END
```

```
//代码清单 9-16 FlutterIosTextLabelFactory.m

#import "FlutterIosTextLabelFactory.h"
#import "FlutterIosTextLabelFactory.h"
#import "FlutterIosTextLabel.h"

@implementation FlutterIosTextLabelFactory{
     NSObject<FlutterBinaryMessenger> * _messenger;
}

//设置参数的编码方式
-(NSObject<FlutterMessageCodec> * )createArgsCodec{
    return [FlutterStandardMessageCodec sharedInstance];
}
```

```
//用来创建 iOS 原生 View
- (nonnull NSObject<FlutterPlatformView> *)createWithFrame:(CGRect)frame
                                            viewIdentifier:(int64_t)viewId
                                                 arguments:(id _Nullable)args {
    //args 为 Flutter 传过来的参数
    FlutterIosTextLabel * textLagel = [[FlutterIosTextLabel alloc]
                                          initWithWithFrame:frame
                                          viewIdentifier:viewId
                                          arguments:args binaryMessenger:_messenger];
    return textLagel;
}

+ (void)registerWithRegistrar:(nonnull NSObject<FlutterPluginRegistrar> *)registrar {
    //注册插件
    //注册 FlutterIosTextLabelFactory
    //com.flutter_to_native_test_textview 为 Flutter 调用此 textLabel 的标识
    [registrar registerViewFactory:[[FlutterIosTextLabelFactory alloc]
                                     initWithMessenger:registrar.messenger] withId:@"
com.flutter_to_native_test_textview"];
}

- (instancetype)initWithMessenger:(NSObject<FlutterBinaryMessenger> *)messager{
    self = [super init];
    if (self) {
        _messenger = messager;
    }
    return self;
}

@end
```

第三步，将自定义的 FlutterIosTextLabelFactory 注册关联，代码如下：

```
#import "AppDelegate.h"
#import "GeneratedPluginRegistrant.h"
#import "FlutterIosTextLabelFactory.h"

@implementation AppDelegate

- (BOOL)application:(UIApplication *)application
```

```
didFinishLaunchingWithOptions:(NSDictionary * )launchOptions {
    //默认需要保留
    [GeneratedPluginRegistrant registerWithRegistry:self];

[ FlutterIosTextLabelFactory   registerWithRegistrar:   [ self   registrarForPlugin:   @
"FlutterIosTextLabelPlugin"]];

    return [super application:application didFinishLaunchingWithOptions:launchOptions];
}
}
```

最后，在 Flutter 中加载自定义的这个 iOS 原生 UILabel，UiKitView 是混合集成模式，代码如下：

```
//代码清单 9-17
//lib/NaviteViewPage.dart
//通过 UiKitView 来加载 iOS 原生 View
buildUIKitView() {
  return const SizedBox(
    height: 200,
    child: UiKitView(
      //标识
      viewType: "com.flutter_to_native_test_textview",
      creationParams: {
        "content": "Flutter 传入的文本内容",
      },
      //参数的编码方式
      creationParamsCodec: StandardMessageCodec(),
    ),
  );
}
```

3min

9.3 原生项目内嵌 Flutter 模块

3min

在实际业务开发中，多数读者在已有的 Android、iOS 原生项目中内嵌 Flutter 模块来开发，如图 9-7 所示，如果是新的项目，则可使用 Flutter 项目来构建项目主体，如图 9-8 所示，可通过插件的方式来开发与平台相关的内容。读者可观看视频讲解【9.3-1 Android 中集成 Flutter 模块】和【9.3-2 iOS 中集成 Flutter 模块】。

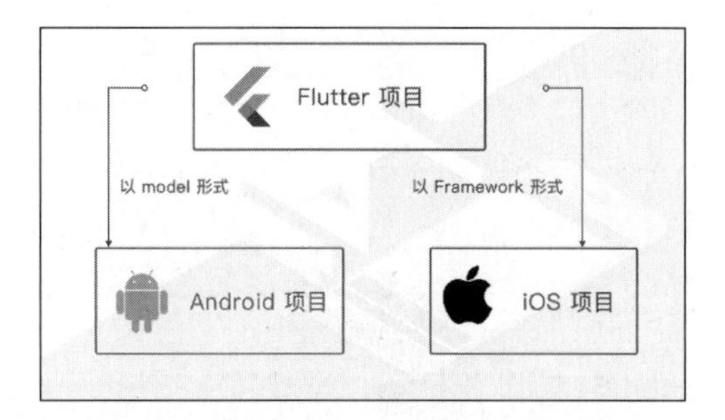

图 9-7 原生项目中使用 Flutter 插件

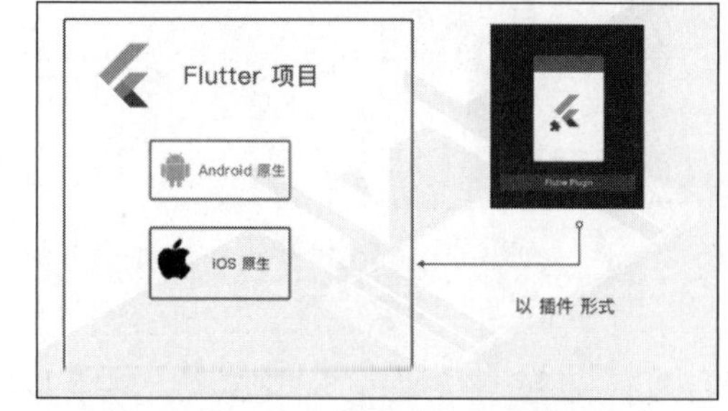

图 9-8 Flutter 项目中使用原生插件

首先创建 Flutter Module，如图 9-9 所示，然后分别创建 Android 原生项目与 iOS 原生项目，三者的项目目录空间结构如图 9-10 所示。

New Project
Project name: demo_flutter_module
Project location: ~/Desktop/f... demo_flutter_module
Description: A new Flutter project.
Project type: Module
Organization: com.example
Android language: Java Kotlin
iOS language: Objective-C Swift
Platforms: Android iOS Linux MacOS Web Windows

图 9-9 创建 Flutter Module

图 9-10 Flutter Module 与原生项目目录结构

9.3.1 Android 中集成 Flutter 模块

首先在 Android 项目的根目录下的 settings.gradle 配置文件中添加 Flutter Module 项目中的配置文件，代码如下：

```
//代码清单 9-18
//NativeAndFlutter/DemoAndroidApplication/settings.gradle
```

```
//加入下面的配置
setBinding(new Binding([gradle: this]))
//Flutter Module 项目中的配置文件
String path = "DemoFlutterModule/demo_flutter_module/.android/include_flutter.groovy"
evaluate(new File(
        settingsDir.parentFile,
        path
))
//引入 Flutter Module 项目的路径,这里使用的是相对路径
include ':demo_flutter_module'
project(':demo_flutter_module').projectDir =
        new File('../DemoFlutterModule/demo_flutter_module')
```

然后在 Android 项目的 app 目录下的 build.gradle 文件中加载依赖如下：

```
android { … }
dependencies {
    implementation project(':flutter')
}
```

在业务开发中，如单击一个按钮，打开 Flutter 页面，代码如下：

```
//创建打开 FlutterActivity 的意图
Intent intent = FlutterActivity.createDefaultIntent(this);
//打开 Flutter 的第 1 种方式
startActivity(intent);
```

FlutterActivity 会加载 Flutter Module 中 lib/main.dart 中的 main 方法，如果有多个 Flutter 页面，则可以动态指定路由，代码如下：

```
//打开 Flutter 的第 2 种方式
startActivity(FlutterActivity
            .withNewEngine()
            .initialRoute("/test")
            .build(this));
```

需要在 Flutter Module 中的 MaterialApp 中配置静态路由规则，代码如下：

```
MaterialApp(
  …
  routes: {
    'test': (context) {
      return SecPage();
    },
  },
)
```

当加载 FlutterActivity 页面时有一段时间明显会看到黑屏(有的手机卡着不动),这段时间主要是启动 Flutter 引擎(FlutterEngine),遇到这种情况时可以使用 Flutter 引擎(FlutterEngine)缓存,即在启动 FlutterActivity 前先启动 Flutter 引擎,然后使用缓存的引擎加载页面,代码如下:

```
//创建引擎
FlutterEngine lFlutterEngine = new FlutterEngine(this);
//设置一些参数,如初始加载显示的路由页面
lFlutterEngine.getNavigationChannel().setInitialRoute("/test");
//设置缓存
lFlutterEngine.getDartExecutor()
    .executeDartEntrypoint(DartExecutor.DartEntrypoint.createDefault());
//put 引擎
FlutterEngineCache.getInstance().put("test_engine", lFlutterEngine);
```

然后在启动 Flutter 页面时,使用缓存引擎来调用,代码如下:

```
//打开 Flutter 的第 3 种方式,前后的引擎 id 要对应起来
startActivity(FlutterActivity
        .withCachedEngine("test_engine")
        .build(this));
```

在 Android 开发中,一个 Fragment 代表了一块较大的模块化 UI,Flutter 提供了 FlutterFragment,用来在 Activity 中以模块的方式来加载 Flutter 内容,代码如下:

```
FlutterFragment lFlutterFragment = FlutterFragment
            .withCachedEngine("test_engine")
            .build();

getSupportFragmentManager()
            .beginTransaction()
            .add(R.id.fl_open_flutter, lFlutterFragment)
            .commit();
```

在 Android 中,Flutter Module 已集成并可加载显示基本的页面,其中避免不了需要实时地进行双向通信,可使用 9.1 节中的通信方式,本节也提供了 Demo,源码路径在本书配套源码中的 DemoAndroidApplication 的 FlutterMessageActivity.Java 文件中。

FlutterView 用于在 Android 设备上显示一个 Flutter UI,FlutterView 有两种不同的渲染模式,即 surface 和 texture,surface 模式的性能比较高,FlutterView 的默认构造器是 surface 模式,代码如下:

```
//代码清单 9-19
//FlutterViewActivity.Java
```

```
public class FlutterViewActivity extends AppCompatActivity {

    private FlutterEngine mFlutterEngine;
    @Override
    protected void onCreate(Bundle savedInstanceState) {
        super.onCreate(savedInstanceState);
        setContentView(R.layout.activity_flutter_view);
        //通过 FlutterView 引入 Flutter 编写的页面
        mFlutterEngine = new FlutterEngine(this);
        mFlutterEngine.getDartExecutor().executeDartEntrypoint(
    DartExecutor.DartEntrypoint.createDefault()
        );
        mFlutterEngine.getNavigationChannel().setInitialRoute("test");

        //通过 FlutterView 引入 Flutter 编写的页面
        FlutterView flutterView = new FlutterView(this);
        FrameLayout.LayoutParams lp = new FrameLayout.LayoutParams(
                ViewGroup.LayoutParams.MATCH_PARENT,
                ViewGroup.LayoutParams.MATCH_PARENT);

        //Android 页面布局
        FrameLayout flContainer = findViewById(R.id.fl_container);
        flContainer.addView(flutterView, lp);
        //关键代码,将 Flutter 页面显示到 FlutterView 中
        flutterView.attachToFlutterEngine(mFlutterEngine);

        //消息通道
    MethodChannel methodChannel = new
        MethodChannel(mFlutterEngine.getDartExecutor(),
                "com.example.flutter/native");
        //或
    MethodChannel methodChannel1 = new MethodChannel(
                    mFlutterEngine.getDartExecutor()
                        .getBinaryMessenger(),
    "com.example.flutter/native");
    }
    @Override
    protected void onResume() {
        super.onResume();
        mFlutterEngine.getLifecycleChannel().appIsResumed();
    }

    @Override
    protected void onPause() {
        super.onPause();
        mFlutterEngine.getLifecycleChannel().appIsInactive();
```

```
    }

    @Override
    protected void onStop() {
        super.onStop();
        mFlutterEngine.getLifecycleChannel().appIsPaused();
    }
}
```

9.3.2 iOS 中集成 Flutter 模块

本节全部源码在本书配套源码中的 DemoIosApplication 项目中。

iOS 与 Flutter 混合开发，首先应在 iOS 项目中的 Podfile 文件中添加与 Flutter 相关的配置，然后执行 pod install 命令加载依赖，代码如下：

```
#设置 Flutter 工程路径并添加脚本 podhelper.rb,此处使用的是相对路径
flutter_application_path = '../../DemoFlutterModule/demo_flutter_module'
load File.join(flutter_application_path, '.ios', 'Flutter', 'podhelper.rb')

target 'DemoiOSApplication' do
  use_frameworks!
  #arget 都需要绑定 Flutter
  install_all_flutter_pods(flutter_application_path)


  target 'DemoiOSApplicationTests' do
    inherit! :search_paths
  end

  target 'DemoiOSApplicationUITests' do
  end

end
```

然后创建测试页面 OpenFlutterControllerViewController，代码如下：

```
#import <UIKit/UIKit.h>

NS_ASSUME_NONNULL_BEGIN

@interface OpenFlutterControllerViewController : UIViewController

@end

NS_ASSUME_NONNULL_END
```

在这里模拟测试在 iOS 原生中打开 Flutter UI，以及将 Flutter UI 内嵌至 iOS 原生 View 中，首先初始化引擎，相当于提前缓存引擎，代码如下：

```
//代码清单 9-20
//OpenFlutterControllerViewController.m 创建的测试类

#import "OpenFlutterControllerViewController.h"
#import "Flutter/Flutter.h"
#import "FlutterPluginRegistrant/GeneratedPluginRegistrant.h"

@interface OpenFlutterControllerViewController ()

@property(nonatomic,strong) FlutterEngine *flutterEngine;

//用来将 Flutter 添加到本地的 View 窗口
@property (weak, nonatomic) IBOutlet UIView *otherView;

@end

@implementation OpenFlutterControllerViewController

- (void)viewDidLoad {
    [super viewDidLoad];
FlutterEngine *engine = [[FlutterEngine
            alloc]initWithName:@"native_flutter" project:nil];
    [engine runWithEntrypoint:nil];
    [GeneratedPluginRegistrant registerWithRegistry:engine];
    self.flutterEngine = engine;
}
...

@end
```

然后在使用的时候通过缓存的引擎去加载 Flutter UI，代码如下：

```
//单击按钮事件加载 Flutter
- (IBAction)openFlutterFunction:(id)sender {

    FlutterViewController *controller = [[FlutterViewController alloc]
                                          initWithEngine:self.flutterEngine
                                          nibName:nil bundle:nil];
    [self presentViewController:controller animated:true completion:nil];

}
```

也可以将 Flutter UI 加载到 iOS 中的 UIView 中，代码如下：

```
//单击按钮事件将 Flutter 加载到 UIView 中
- (IBAction)addFlutterToView:(id)sender {
    FlutterViewController * controller = [[FlutterViewController alloc]initWithEngine:
                                          self.flutterEngine
                                          nibName:nil bundle:nil];
    controller.view.frame = CGRectMake(0,0,100,200);
    [self.otherView addSubview:controller.view];
}
```

9.4 插件开发

插件按功能类型可分为两类，一类无 UI 页面，用于信息交互，如需要获取蓝牙信息、手机传感器信息等；另一类有 UI 页面，需要加载原生项目中的 View，本节中实现的是加载原生中的 View。

如图 9-11 所示，使用 AndroidStudio 创建一个用来解析 HTML 富文本的插件，使用原理是通过混合集成模式集成 Android 中的 TextView，TextView 通过 HTML 来解析文本，在 iOS 中通过 WKWebView 实现。

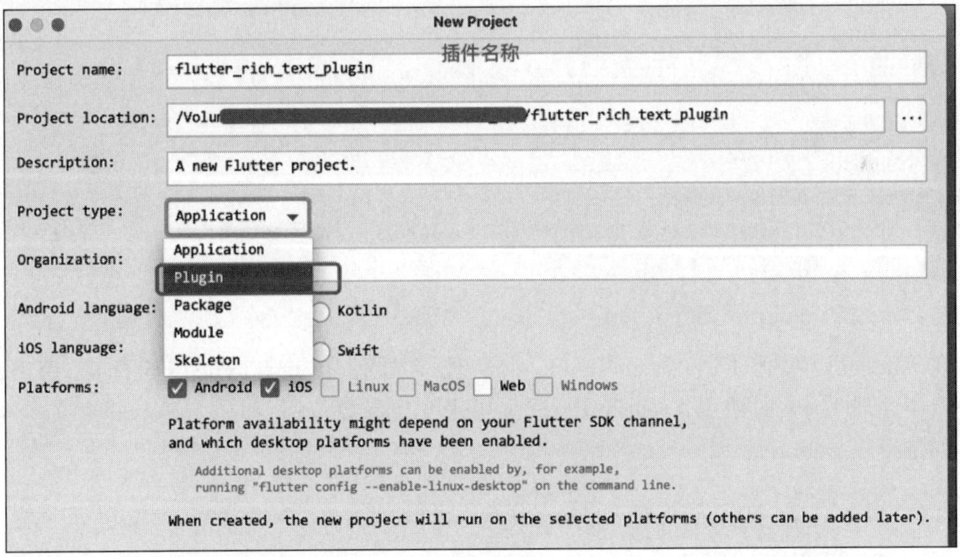

图 9-11 创建插件

9.4.1 Flutter 端开放 Widget

在插件项目中创建 src 文件夹目录，然后创建 nav_rich_text_view.dart 文件，如图 9-12 所示。

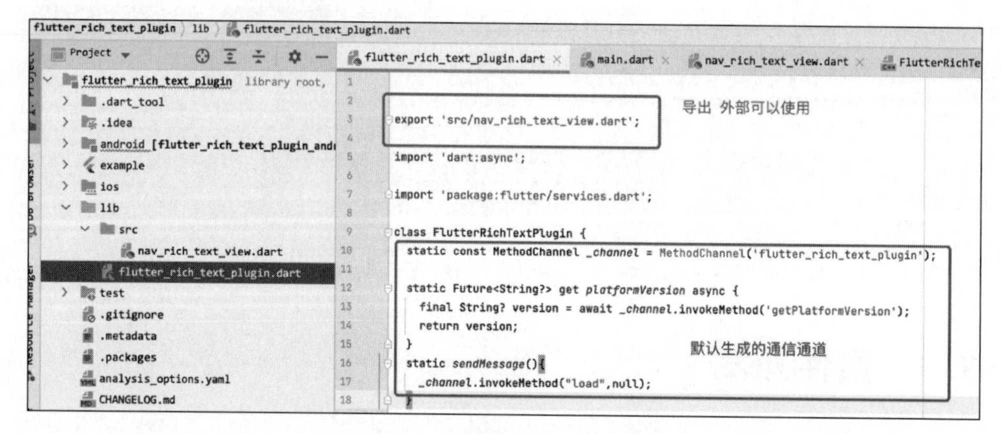

图 9-12 插件项目文件目录清单

在 nav_rich_text_view. dart 文件中，区分 Android 与 iOS 不同平台来处理，代码如下：

```
//代码清单 9 - 21 加载富文本组件
class NaviteRichWidget extends StatelessWidget {
  final String htmlText;

  NaviteRichWidget({Key? key, required this.htmlText}) : super(key: key);
  //通信加载 View 通道
  String viewType = 'com.flutter_to_native_test_textview';

  //内容构建
  @override
  Widget build(BuildContext context) {
    return Platform.isAndroid ? buildAndroidView() : buildUIKitView();
  }
}
```

在 Android 中，通过 PlatformViewLink 实现，需要使用 viewType 身份标识，用于在原生平台中注册 View 时使用，creationParams 用来实现参数传递，如初始显示的富文本内容，代码如下：

```
//代码清单 9 - 21 - 1 Android 中通过 PlatformViewLink 来加载
  buildAndroidView() {
    //参数
    Map<String, dynamic> creationParams = <String, dynamic>{};
    creationParams["content"] = htmlText;

    return PlatformViewLink(
      viewType: viewType,
      surfaceFactory:
```

```
        (BuildContext context, PlatformViewController controller) {
      return AndroidViewSurface(
        controller: controller as AndroidViewController,
        gestureRecognizers: const <Factory<OneSequenceGestureRecognizer>>{},
        hitTestBehavior: PlatformViewHitTestBehavior.opaque,
      );
    },
    onCreatePlatformView: (PlatformViewCreationParams params) {
      return PlatformViewsService.initSurfaceAndroidView(
        id: params.id,
        viewType: viewType,
        layoutDirection: TextDirection.ltr,
        creationParams: creationParams,
        creationParamsCodec: const StandardMessageCodec(),
        onFocus: () {
          params.onFocusChanged(true);
        },
      )
        ..addOnPlatformViewCreatedListener(params.onPlatformViewCreated)
        ..create();
    },
  );
}
```

在 iOS 中，通过 UiKitView 来加载使用，代码如下：

```
//代码清单 9-21-2 iOS 通过 UiKitView 来加载 iOS 原生 View
buildUIKitView() {
  return UiKitView(
    //标识
    viewType: viewType,
    creationParams: {
      "content": htmlText,
    },
    //参数的编码方式
    creationParamsCodec: StandardMessageCodec(),
  );
}
```

9.4.2　Android 端创建 TextView 并解析文本

如图 9-13 所示，在插件项目的 android 目录下，右击，以 Android 项目工程的方式来加载 Plugin。

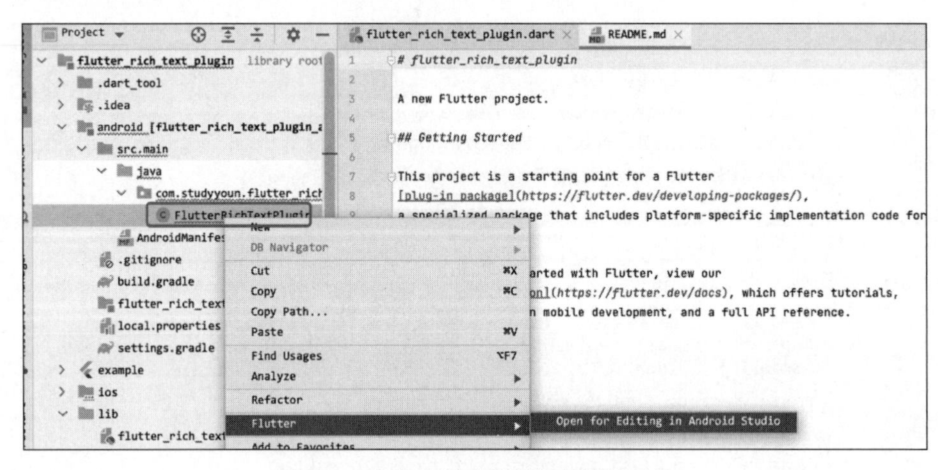

图 9-13　以安卓目录视图打开插件项目

如图 9-14 所示，在 Android 目录视图下，FlutterRichTextPlugin 为默认使用的插件，此插件可加载通信通道，在 onAttachedToEngine 回调方法中注册消息通道，在这里注册本节需要用到的 TextView，核心代码如下：

```
TestViewFactory.registerWith(flutterPluginBinding);
```

```
android > src > main > java > com > studyyoun > flutter_rich_text_plugin > FlutterRichTextPlugin > channel

import ...

/** FlutterRichTextPlugin */
public class FlutterRichTextPlugin implements FlutterPlugin, MethodCallHandler {
  /// The MethodChannel that will the communication between Flutter and native Andr
  ///
  /// This local reference serves to register the plugin with the Flutter Engine an
  /// when the Flutter Engine is detached from the Activity
  private MethodChannel channel;
  private FlutterPluginBinding flutterPluginBinding;

  @Override
  public void onAttachedToEngine(@NonNull FlutterPluginBinding flutterPluginBinding
    this.flutterPluginBinding = flutterPluginBinding;
    channel = new MethodChannel(flutterPluginBinding.getBinaryMessenger(), name: "f
    channel.setMethodCallHandler(this);
    TestViewFactory.registerWith(flutterPluginBinding);
  }
```

图 9-14　Android 项目工程视图

类 TestViewFactory 是创建的继承于 PlatformViewFactory 的子类，用来加载自定义 TestTextView 组件，代码如下：

```
public class TestViewFactory extends PlatformViewFactory {

    public TestViewFactory() {
        super(StandardMessageCodec.INSTANCE);
```

```
    }

    //args 是由 Flutter 传过来的自定义参数
    @Override
    public PlatformView create(Context context, int id, Object args) {
        //Flutter 传递过来的参数
        Map<String, Object> params = (Map<String, Object>) args;
        //创建自定义的 TestTextView
        return new TestTextView(context, id, params);

    }

    //需要在 MainActivity 的 onCreate 方法中调用
public static void registerWith(
                FlutterPlugin.FlutterPluginBinding flutterPluginBinding) {
        //通过 platformViewsController 获取 Registry
        PlatformViewRegistry registry =
            flutterPluginBinding.getPlatformViewRegistry();
        //通过工厂类 PlatformViewRegistry 注册 Android 原生 View
        //参数一是设置标识
        //参数二是自定义的 Android 原生 View
        registry.registerViewFactory(
                "com.flutter_to_native_test_textview",
                new TestViewFactory());
    }
}
```

在这里通过 PlatformViewRegistry 注册 View 用到的标识要与 Flutter 中 PlatformViewLink 用到的 viewType 标识一致，TestViewFactory 是自定义的 PlatformView，在其中创建了 TextView，代码如下：

```
public class TestTextView implements PlatformView {

    //这里使用的是一个 TextView
    private final TextView mTestTextView;
    private Context mContext;

    TestTextView(Context context, int id, Map<String, Object> params) {
        mContext = context;
        //创建 TextView
        TextView lTextView = new TextView(context);
        //设置文字
        lTextView.setText("");
        this.mTestTextView = lTextView;
        //Flutter 传递过来的参数
```

```
        if (params!= null&&params.containsKey("content")) {
            String myContent = (String) params.get("content");
            //设置文本显示
            lTextView.setText(Html.fromHtml(myContent));
        }
    }
    @Override
    public View getView() {
        return mTestTextView;
    }

    @Override
    public void dispose() {}
}
```

9.4.3 iOS端创建WKWebView并解析文本

如图9-15所示，在插件项目中，以Xcode工程视图打开项目，如图9-16所示，在Pods目录下，打开FlutterRichTextPlugin.m文件，在其中添加注册用到的iOS平台中的View，代码如下：

```
FlutterIosTextLabelFactory * factory =
      [[FlutterIosTextLabelFactory alloc]
                 initWithMessenger:registrar.messenger];
[registrar registerViewFactory:factory
                          withId:@"com.flutter_to_native_test_textview"];
```

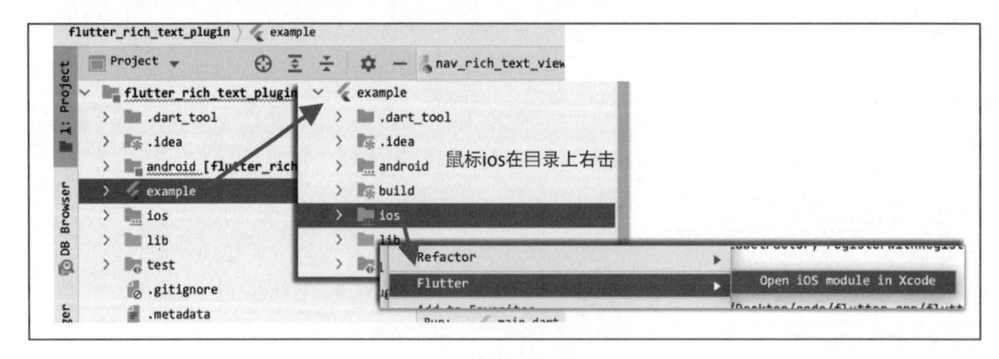

图9-15 以iOS Xcode工程打开项目目录视图

类FlutterIosTextLabelFactory是笔者创建的继承于FlutterPlatformViewFactory的子类，用来加载自定义TestTextView组件，代码如下：

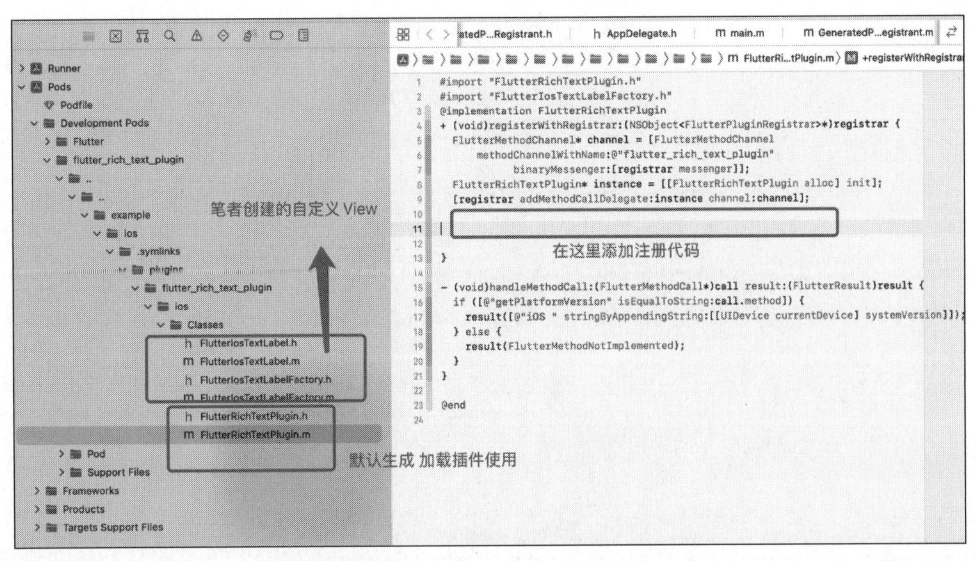

图 9-16 Xcode 工程目录视图

```
//FlutterIosTextLabelFactory.h

#import <Foundation/Foundation.h>
#import <Flutter/Flutter.h>
NS_ASSUME_NONNULL_BEGIN

@interface FlutterIosTextLabelFactory : NSObject<FlutterPlatformViewFactory>

- (instancetype)initWithMessenger:(NSObject<FlutterBinaryMessenger>*)messenger;

+ (void)registerWithRegistrar:(nonnull NSObject<FlutterPluginRegistrar> *)registrar ;
@end

NS_ASSUME_NONNULL_END
```

```
//FlutterIosTextLabelFactory.m

#import "FlutterIosTextLabelFactory.h"
#import "FlutterIosTextLabelFactory.h"
#import "FlutterIosTextLabel.h"

@implementation FlutterIosTextLabelFactory{
    NSObject<FlutterBinaryMessenger>*_messenger;
}

//设置参数的编码方式
```

```
-(NSObject<FlutterMessageCodec> *)createArgsCodec{
    return [FlutterStandardMessageCodec sharedInstance];
}

//用来创建 iOS 原生 View
- (nonnull NSObject<FlutterPlatformView> *)createWithFrame:(CGRect)frame
                                            viewIdentifier:(int64_t)viewId
                                                 arguments:(id _Nullable)args {
    //args 为 Flutter 传过来的参数
    FlutterIosTextLabel *textLagel = [[FlutterIosTextLabel alloc]
                                      initWithWithFrame:frame
                                      viewIdentifier:viewId
                                      arguments:args binaryMessenger:_messenger];
    return textLagel;
}

+ (void)registerWithRegistrar:(nonnull NSObject<FlutterPluginRegistrar> *)registrar {
    //注册插件
    //注册 FlutterIosTextLabelFactory
    //com.flutter_to_native_test_textview 为 Flutter 调用此 textLabel 的标识
    [registrar registerViewFactory:[[FlutterIosTextLabelFactory alloc]
                                    initWithMessenger:registrar.messenger] withId:@"
com.flutter_to_native_test_textview"];
}

- (instancetype)initWithMessenger:(NSObject<FlutterBinaryMessenger> *)messager{
    self = [super init];
    if (self) {
        _messenger = messager;
    }
    return self;
}

@end
```

FlutterIosTextLabel 是笔者创建的，用来构建 WKWebView 并以此加载解析 HTML 文本，代码如下：

```
#import <UIKit/UIKit.h>
#import <Flutter/Flutter.h>
NS_ASSUME_NONNULL_BEGIN

//实现协议 FlutterPlatformView
@interface FlutterIosTextLabel : NSObject<FlutterPlatformView>

-(instancetype)initWithWithFrame:(CGRect)frame
```

```
                    viewIdentifier:(int64_t)viewId
                         arguments:(id _Nullable)args
                   binaryMessenger:(NSObject<FlutterBinaryMessenger>*)messenger;
@end


NS_ASSUME_NONNULL_END
```

```
#import "FlutterIosTextLabel.h"
#import <WebKit/WebKit.h>
@interface FlutterIosTextLabel ()

@end

@implementation FlutterIosTextLabel{
    //FlutterIosTextLabel 创建后的标识
    int64_t _viewId;
WKWebView * _wKWebView;
    //消息回调
    FlutterMethodChannel * _channel;
}

- (instancetype)initWithWithFrame:(CGRect)frame
                   viewIdentifier:(int64_t)viewId arguments:(id)args binaryMessenger:
(NSObject<FlutterBinaryMessenger> *)messenger{
    if ([super init]) {
        if (frame.size.width == 0) {
            frame = CGRectMake(frame.origin.x,
                               frame.origin.y,
                               [UIScreen mainScreen].bounds.size.width, 22);
        }
        NSString *htmlString = args[@"content"];
        _wKWebView = [[WKWebView alloc] initWithFrame:self.view.bounds];
        [_wKWebView loadHTMLString:htmlString baseURL:nil];
        _viewId = viewId;

    }
    return self;
}

- (nonnull UIView *)view {
    return _wKWebView;
}

@end
```

9.4.4 插件发布

首先需要保证网络可以访问 https://www.google.com/，如图 9-17 所示，在终端目录下执行以下命令，查看插件开发是否符合规范。

```
flutter packages pub publish -- dry - run
```

检查结果如图 9-17 所示，检查不通过，提示需要在配置文件 pubspec.yaml 中配置 homepage 属性，这一般配置为 GitHub 仓库地址。

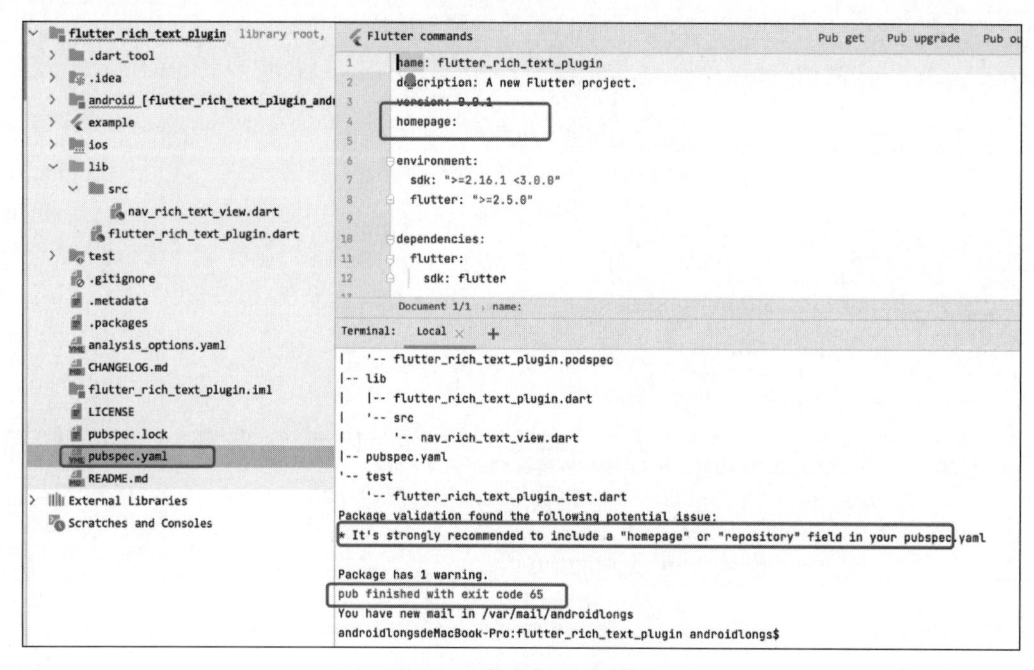

图 9-17 插件发布前准备

项目目录 LICENSE 用于配置开源协议，也是必填项，在文件 CHANGELOG.md 中配置插件版本及更新目录说明，与 pubspec.yaml 文件中的 version 要对应，配置好这些后，再次运行检查，如图 9-18 所示，检查无错误，然后执行以下命令进行发布：

```
dart pub publish
```

输入 y 确认发布，如果是第一次发布插件，Android Studio 会提示输入用户名登录认证，此处已登录认证过，所以可以直接发布，如图 9-19 所示。

发布成功后，在插件市场搜索 flutter_rich_text_plugin，然后在 Flutter 项目中添加依赖便可使用，代码如下：

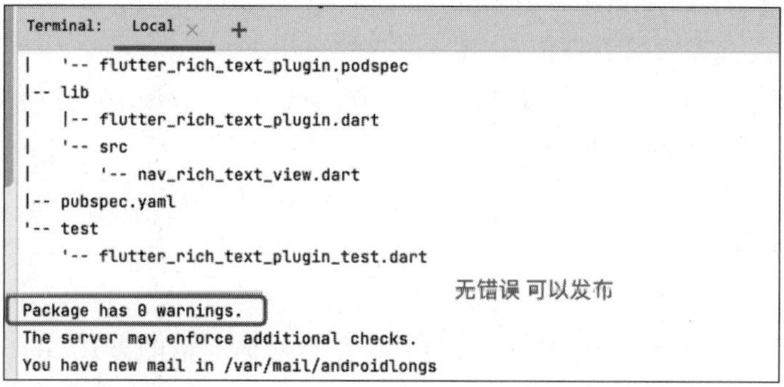

```
Terminal:   Local ×  +
|    '-- flutter_rich_text_plugin.podspec
|-- lib
|    |-- flutter_rich_text_plugin.dart
|    '-- src
|        '-- nav_rich_text_view.dart
|-- pubspec.yaml
'-- test
     '-- flutter_rich_text_plugin_test.dart
                                        无错误 可以发布
Package has 0 warnings.
The server may enforce additional checks.
You have new mail in /var/mail/androidlongs
```

图 9-18　插件发布无错误

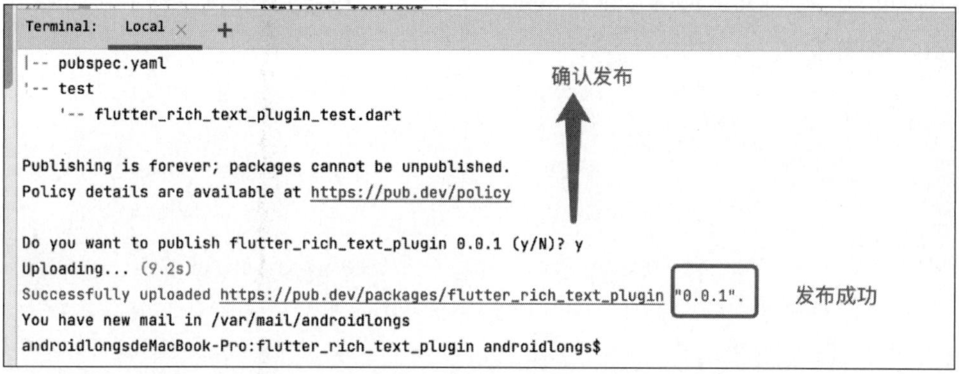

```
Terminal:   Local ×  +
|-- pubspec.yaml
'-- test                                       确认发布
    '-- flutter_rich_text_plugin_test.dart

Publishing is forever; packages cannot be unpublished.
Policy details are available at https://pub.dev/policy

Do you want to publish flutter_rich_text_plugin 0.0.1 (y/N)? y
Uploading... (9.2s)
Successfully uploaded https://pub.dev/packages/flutter_rich_text_plugin "0.0.1".    发布成功
You have new mail in /var/mail/androidlongs
androidlongsdeMacBook-Pro:flutter_rich_text_plugin androidlongs$
```

图 9-19　插件发布

```
//测试文本内容
String testText =
    "<strong>我的测试</strong><h1>标题标签</h1><p>p 标签在这里</></p><img src=\"ic
_launcher-web.png\"><img src=\"\">";

SizedBox(
  height: 200,
  child: NaviteRichWidget(
    htmlText: testText,
  ),
),
```

9.4.5　插件 API 升级

Flutter 1.12 版本发布后，Android 平台已可以使用新的 Android 插件 API，基于 PluginRegistry.Registrar 开发的插件目前已无法向 pub.dev 提交更新，需要升级 API，新版 API 的优点是为生命周期的相关组件提供了更简洁清晰的访问方式。

在插件的主类文件中（*Plugin.Java）实现FlutterPlugin接口，保留静态的registerWith()方法，来适配不兼容v2版本嵌入的应用。

当插件中使用Activity的引用时，需要实现ActivityAware接口，如果需要使用后台Service，则需要实现ServiceAware接口。

9.5 小结

本章概述了Flutter与原生Android与iOS平台双向通信的方式、在Flutter中嵌入原生View的方法，以及插件的开发。通过本章的内容，可以很容易地处理在Flutter项目开发中用到原生平台的相关功能，如自定义相机、视频、声频、蓝牙及各种第三方原生SDK的支持及使用，Android与iOS原生平台中集成了Flutter模块，很好地兼容了在旧项目中使用Flutter来开发。

第 10 章 文件操作与网络请求

CHAPTER 10

异步编程常用于网络请求、缓存数据加载、本地图片加载、定时与延时任务等，本章的内容如下：

（1）异步编程 Future、Timer 的使用分析，以及任务队列分析。

（2）文件文本的读写、SharedPreferences 轻量级数据缓存、sqflite 数据库操作。

（3）网络请求库 Dio 的详细使用分析，GET、POST、文件上传、文件下载、公共请求参数、请求代理、取消网络请求、JSON 数据组件等。

本章的源码存放在 flutter_base_widget 目录下：

```
flutter_base_widget/lib/code10
```

10.1 异步编程

大部分操作系统（如 Windows、Linux）的任务调度采用时间片轮转的抢占式调度方式，对于单核 CPU 来讲，并行执行两个任务，实际上是 CPU 在进行着快速切换，对用户来讲感觉不到有切换停顿，就好比 220V 交流电的灯光显示原理一样，也就是说一个任务执行一小段时间后强制暂停去执行下一个任务，每个任务轮流执行。任务执行的一小段时间叫作时间片，任务正在执行时的状态叫运行状态，任务执行一段时间后强制暂停去执行下一个任务，被暂停的任务就处于就绪状态等待下一个属于它的时间片的到来，任务的停与执行切换称为任务调度。

计算机的核心是 CPU，它承担了所有的计算任务，而操作系统是计算机的管理者，它负责任务的调度、资源的分配和管理，操作系统中运行着多个进程，每个进程是一个具有一定独立功能的程序在一个数据集上的一次动态执行的过程，是应用程序运行的载体。操作系统会以进程为单位，分配系统资源（CPU 时间片、内存等资源），进程是资源分配的最小单位，也就是操作系统的最小单位。

线程是进程中的概念，一个进程中可包含多个线程。任务调度采用的是时间片轮转的抢占式调度方式，进程是任务调度的最小单位。默认情况下，一般一个进程里只有一个线程，进程本身就是线程，所以线程可以被称为轻量级进程。

使用Flutter开发的App安装在手机上，当单击App图标启动时，手机操作系统会为当前App创建一个进程，然后在Flutter项目中通过main函数启动Flutter构建的项目。Dart是基于单线程模型的语言，所以在Flutter中一般的异步操作，实际上还是通过单线程通过调度任务优先级实现的，如图10-1所示。

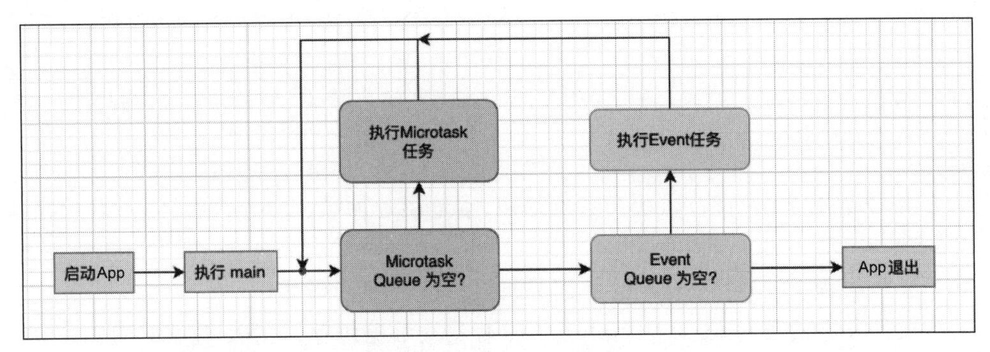

图10-1　消息队列机制

在Dart中的线程机制称为isolate，在Flutter项目中，运行中的Flutter程序由一个或多个isolate组成，默认情况下启动的Flutter项目，通过main函数启动就是创建了一个main isolate，在这里main isolate为Flutter的主线程，或称为UI线程。

单线程模型中主要在维护着一个事件循环（Event Loop）与两个队列（Event Queue和Microtask Queue），当Flutter项目程序触发如单击事件、I/O事件、网络事件时，它们就会被加入eventLoop中，eventLoop一直在循环之中，当主线程发现事件队列不为空时，就会取出事件，并且执行。

Microtask Queue只处理在当前isolate中的任务，优先级高于Event Queue，好比机场里的某个VIP候机室，总是VIP用户先登机了，才开放公共排队入口，如果在Event事件队列中插入Microtask，当前Event执行完毕即可插队执行Microtask事件，Microtask Queue队列为Dart提供了给任务队列插队的解决方案。

当事件循环正在处理Microtask事件时，Event Queue会被堵塞。这时候App就无法进行UI绘制，以及无法响应鼠标事件和I/O等事件。

10.1.1 async与await的基本使用

Future就是Event，每个被await标记的句柄也是一个Event，timer创建的任务也是一个Event，每创建一个Future就会把这个Future扔进Event Queue中排队。使用async和await组合，即可向Event Queue中插入Event实现异步操作。使用Future的Microtask方法可将任务添加到Microtask Queue任务队列中。

使用async关键字可开启一个异步处理，使用await关键字来等待处理结果，这个结果通常是一个 Future 对象。

如处理一个网络请求，或者加载一张图片、文件，需要异步加载，通过 async 与 await 的组合可以实现这个操作，如图 10-2 所示。

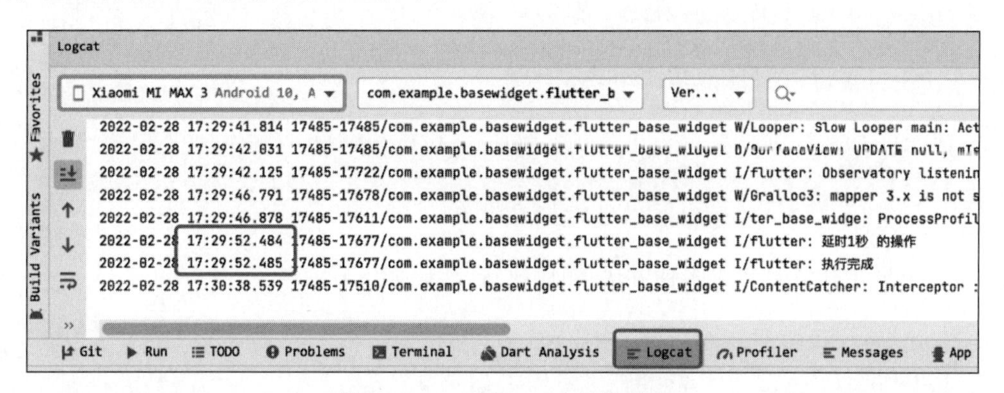

图 10-2　async 和 await 异步任务

代码如下：

```
//代码清单 10-1 async 和 await 基本使用
//lib/code10/example_1001_baseUse.dart
//async 关键字用于声明该函数内部有代码需要延迟执行
Future<bool> getData() async {
  //模拟一个耗时操作延时 1s
  //await 关键字声明运算为延迟执行,然后 return 运算结果
  //await 关键字声明后当前线程阻塞在这里
  await Future.delayed(const Duration(milliseconds: 1000), () {
    print("延时 1s 的操作");
  });
  print("执行完成");
  return true;
}
```

执行两个异步任务，这两个异步任务是串行的，异步任务 1 执行完毕后，获取结果 result，然后开启异步任务 2 执行，在实际项目中可使用第 1 个网络请求的结果动态加载第 2 个网络请求或者其他分类别的异步任务，代码如下：

```
//代码清单 10-1-1 async 多个 await
//lib/code10/example_1001_baseUse.dart
Future<bool> getData2() async {
  print("第 1 个耗时任务开始执行");
  //模拟一个耗时操作对异步任务 1 延时 1s
  await Future.delayed(Duration(milliseconds: 1000), () {
    print("第 1 个耗时任务执行完成");
```

```
  });
  print("第 2 个耗时任务开始执行");

  //模拟一个耗时操作,对异步任务 2 延时 2s
  await Future.delayed(Duration(seconds: 2), () {
    print("第 2 个耗时任务执行完成");
  });
  print("执行完成");
  return true;
}
```

将多个异步任务封装成多种方法,顺序执行,代码如下:

```
//代码清单 10 - 1 - 2 async 多个 await
//lib/code10/example_1001_baseUse.dart
Future<String> getData3() async {
  //await 关键字将运算声明为延迟执行,然后返回运算结果
  String result = await getDataA();
  String result2 = await getDataB();
  return result2;
}
Future<String> getDataA() async {
  //await 关键字将运算声明为延迟执行,然后返回运算结果
  await Future.delayed(Duration(milliseconds: 1000), () {
    print("第 1 个耗时任务执行完成");
  });
  return Future.value("执行完毕");
}
Future<String> getDataB() async {
  await Future.delayed(Duration(milliseconds: 1000), () {
    print("第 2 个耗时任务执行完成");
  });
  return Future.value("执行完毕");
}
```

也可以使用 then 函数获取对应的异步处理结果,代码如下:

```
//代码清单 10 - 1 - 3 async 多个 await
//lib/code10/example_1001_baseUse.dart
void getData4() {
  //await 关键字将运算声明为延迟执行,然后返回运算结果
  getDataA().then((valuea) {
    //valuea 为异步任务 getDataA 的返回值
    getDataB().then((valueb) {
      //valueb 为异步任务 getDataB 的返回值
```

```
    });
  });
}
```

10.1.2 Future 实现延时任务

在 Flutter 中实现延时操作有两种方式，一种是通过 Future；另一种是通过 Timer，通过 Future 实现延时 1s 的操作，代码如下：

```
//方式一
//参数一，延时的时间；参数二，延时执行的方法
Future.delayed(Duration(milliseconds: 1000), () {
      print("延时 1s 执行");
});
```

```
//方式二
Future.delayed(Duration(milliseconds: 1000)).whenComplete((){
      print("延时 1s 执行 whenComplete ");
});
```

```
//方式三
Future.delayed(Duration(milliseconds:1000)).then((value){
    print("延时 1s 执行 then ");
});
```

Future 的 then 函数的返回值类型为一个 Future 对象，所以支持链式调用，组合在一起就是串行方式调用，代码如下：

```
Future.delayed(Duration(milliseconds: 1000), () {
      print("延时 1s 执行");
      return Future.value("测试数据 1");
    }).then((value) {//函数 1
      print(" then $ value");
      //也可以再执行异步任务
      return Future.value("测试数据 2");
    }).then((value) {//函数 2
      print(" then $ value");
      return Future.value("测试数据 3");
    }).then((value) {//函数 3
      //value 是函数 1 回传的值
      print(" then $ value");
    });
```

假如在 then 函数中任何一个环节出现了异常，那么后续的函数将会被中断执行，对于手机界面来讲就是无响应或者黑屏显示。Future 的 whenComplete 方法，类似于 try-catch-finally 中的 finally 块，所以使用 whenComplete 来结束多个异步操作是一个合适的解决方案，代码如下：

```
Future.delayed(Duration(milliseconds: 1000), () {
      print("延时 1s 执行");
      return Future.value("测试数据 1");
    }).then((value) {//函数 1
      print(" then $ value");
      return Future.value("测试数据 2");
    }).then((value) {//函数 2
      print(" then $ value");
      throw 'Error!';
      return Future.value("测试数据 3");
    }).then((value) {//函数 3
      //value 是函数 1 回传的值
      print(" then $ value");
    }).catchError((err) {
      print('Caught $ err'); //Handle the error
    },test: (e){
      print('Caughte $ e'); //Han
      return e is String;
    }).whenComplete((){
      print("程序执行完成");
    });
```

catchError 函数，当 then 这几个函数中任何一个处理出现异常时，都会回调此方法，如这里函数 2 通过 throw 抛出一个异常，在 catchError 函数中捕捉到这个异常，然后回调 test 方法块，再回调 catchError 的参数一的函数处理，类似 try-catch-finally 中的 catch。

10.1.3 Timer 实现定时任务

通过 Timer 实现延时 2s 的操作，代码如下：

```
//延时 2s
Timer timer = new Timer(Duration(milliseconds: 2000), (){

});
```

从源码的角度来看，Future 中实现的延时操作也是通过 Timer 实现的，在实际开发中，如果只是一个单纯的延时操作，笔者建议使用 Timer，在当前 Widget 销毁时取消延时任务，可避免内存泄漏，代码如下：

```
//代码清单 10-2
class _TimerDelayedTestPageState extends State{
  //声明
  Timer ? _timer;
  @override
  void initState() {
    super.initState();
    //延时 2s
     _timer = new Timer(Duration(milliseconds: 2000), (){

    });
  }
  @override
  void dispose() {
//取消延时任务
if (_timer != null && _timer!.isActive) {
      _timer?.cancel();
      _timer = null;
  }

    super.dispose();
   }
    ...
}
```

通过 Timer 可以实现计时器功能，需要在页面的销毁函数中停止计时器，代码如下：

```
//代码清单 10-3-1 计时器
//lib/code10/example_1002_timer.dart
class _ExampleState extends State {
  //计时器执行的次数
  int _timNumber = 0;
  //计时器
  Timer? _timer;
  @override
  void initState() {
    super.initState();
    //创建计时器
    startTimer();
  }

  @override
  void dispose() {
    //关闭计时器
    stopTimer();
```

```
    super.dispose();
  }

  …

}
```

Timer 默认的构造函数实现的是一个倒计时功能(延时功能),Timer.periodic 实现的是一个间隔一定时间的计时器功能,代码如下:

```
  //代码清单 10-3-2 计时器创建
  void startTimer() {
    //安全校验
    stopTimer();
    //创建一个间隔 1000 毫秒的计时器
    //并开始执行
    _timer = Timer.periodic(
      const Duration(milliseconds: 1000),
      (timer) {
        _timNumber++;
        print("计时器 $_timNumber");
      },
    );
  }

  //关闭计时器
  void stopTimer() {
if (_timer != null && _timer!.isActive) {
      _timer?.cancel();
      _timer = null;
    }
  }
```

10.1.4 FutureBuilder 实现异步任务更新

FutureBuilder 用于将 Future 处理结果与 UI 数据刷新显示完美地结合在一起,代码如下:

```
//代码清单 10-4 FutureBuilder 的基本使用
//lib/code10/example_1004_FutureBuilder.dart
class _ExampleState extends State {
  Future<String>?_testFuture;
  @override
```

```
    void initState() {
      super.initState();
      //模拟一个异步任务,如读取文件、网络请求等
      _testFuture = Future.delayed(
        const Duration(milliseconds: 2000),
        () {
          return "模拟的数据";
        },
      );
    }
    @override
    Widget build(BuildContext context) {…}

    FutureBuilder<String> buildFutureBuilder() {
      return FutureBuilder<String>(
        //绑定 Future
        future: _testFuture,
        //默认显示的占位数据
        initialData: "",
        //需要更新数据对应的 Widget
        builder: (BuildContext context, AsyncSnapshot<String> snapshot) {
          return Text("${snapshot.data}");
        },
      );
    }
  }
```

需要注意的是必须在例如 initState、didUpdateWidget 或 didChangeDependencies 构建 Future 或构造 FutureBuilder 时,不能在 State.build 或 StatelessWidget.build 方法调用期间创建它,如果 Future 与 FutureBuilder 同时创建,则每次重新构建 FutureBuilder 的父类时都会重新启动异步任务。

10.2 文件的读写

App 安装包中会包含代码和 assets(资源)两部分,assets 会被打包到程序安装包中,可在运行时访问,常见类型包括静态数据(例如 JSON 文件、JS 文本)、配置文件、图标和图片(JPEG、WebP、GIF、动画 WebP/GIF、PNG、BMP 和 WBMP)及各种字体等。

在手机存储磁盘上也保存着一些共享的内容,如相册图片或者其他目录下的文件等,如选择手机中存储的 text 文本文件并上传到服务器,就需要获取这个文本文件,再如一些不常用大文件,需要保存在手机存储空间中,所以也要用到文件的保存操作。

10.2.1 资源目录 assets 文件读取

每个 Flutter 应用程序都有一个默认创建的 rootBundle 对象，通过它可以轻松访问主资源包，直接使用 package：flutter/services. dart 中全局静态的 rootBundle 对象来加载 asset 即可，如图 10-3 所示，在这里有一个 test. json 文件。

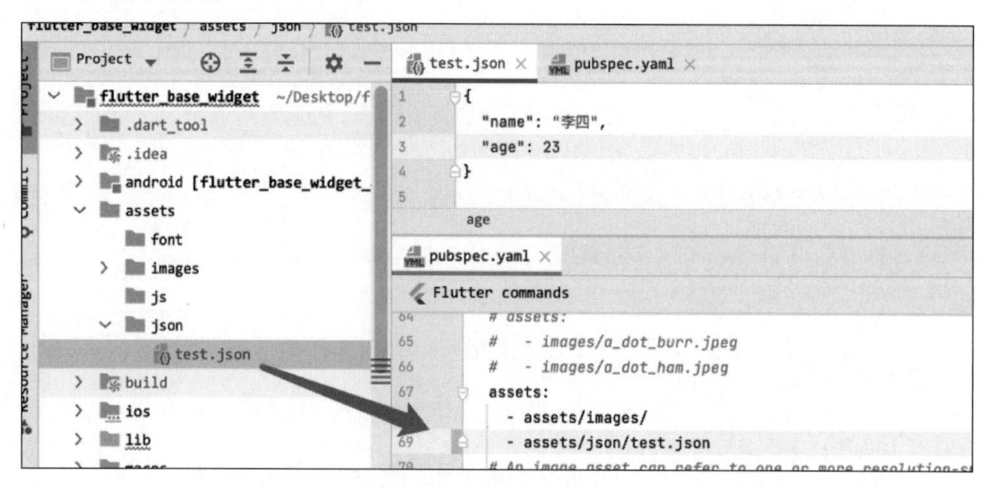

图 10-3 JSON 文件路径

如果需要在程序中获取这个文件中的内容，并转换为自定义对象，则可以这样来操作，代码如下：

```
//代码清单 10 - 5 加载 assets 路径下的文件
//lib/code/code10/example_1005_File.dart
import 'dart:convert' as convert;

void loadAssetTestJson() async {
  //文件在 Flutter 项目中的完整路径
  String json = await rootBundle.loadString('assets/json/test.json');
  print("加载完成 $json");
  //JSON 转 Map
  Map<String, dynamic> userMap = convert.jsonDecode(json);
  print("加载完成 userMap $userMap");
  //转换为自定义对象
  UserBean userBean = UserBean.fromMap(userMap);
}
```

运行结果如图 10-4 所示。

UserBean 为自定义数据体，代码如下：

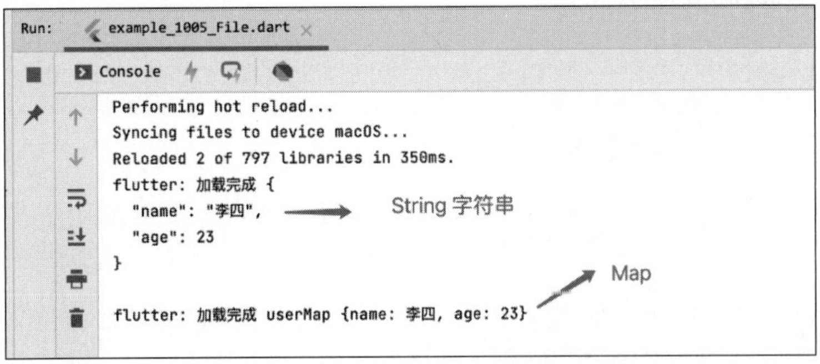

```
Run:  example_1005_File.dart
Console
Performing hot reload...
Syncing files to device macOS...
Reloaded 2 of 797 libraries in 350ms.
flutter: 加载完成 {
  "name": "李四",
  "age": 23
}

flutter: 加载完成 userMap {name: 李四, age: 23}
```

String 字符串

Map

图 10-4　JSON 文件日志输出

```
class UserBean {
  String? userName;
  int? userAge;
  UserBean.fromMap(dynamic map) {
    userName = map["name"] ?? "";
    userAge = map["age"] ?? 0;
  }
}
```

DefaultAssetBundle 获取当前 BuildContext 的 AssetBundle。这种方法没有使用应用程序构建的默认 asset bundle，而是使父级 Widget 在运行时动态地替换不同的 AssetBundle，一般可以使用 DefaultAssetBundle.of()在运行期间加载 asset 中的文件，而在 Widget 上下文之外，可以使用 rootBundle 直接加载这些 asset，上述获取 test.json 文件内容的代码可以改写，代码如下：

```
void loadAssetTestJson2() async {
  //文件在 Flutter 项目中的完整路径
  String json = await DefaultAssetBundle.of(context)
      .loadString('assets/json/test.json');
  print("加载完成 $json");
}
```

10.2.2 手机磁盘文件读写

Dart 的 I/O 库包含了文件读写的相关类，所以对于文件的读写操作需要导入 I/O 库，手机磁盘上的文件读写，首先要获取磁盘上开放的对应的目录空间，在这里使用插件 PathProvicer 实现这个操作，在配置文件 pubspec.yaml 中添加的依赖如下：

```
dependencies:
  path_provider: ^6.0.2
```

然后加载依赖，代码如下：

```
flutter pub get
```

在使用的文件中导入包，代码如下：

```
import 'package:path_provider/path_provider.dart';
```

Android 和 iOS 的应用存储目录不同，大体可分为两类，即临时目录和文档目录。临时目录：系统可随时清除的临时目录（缓存），在 Android 上，对应于 getCacheDir()，在 iOS 上，对应于 NSTemporaryDirectory()，使用 PathProvicer 获取临时目录，代码如下：

```
//代码清单 10-7 临时目录
//lib/code/code10/example_1005_File.dart
void getThemPath() async {
  //获取临时目录
  Directory dic = await getTemporaryDirectory();
  //临时目录的路径
  String path = dic.path;
  Uri uri = dic.uri;
//父级目录
  Directory parentDic = dic.parent;
  Directory absolute = dic.absolute;

}
```

文档目录，对应 Android 的 AppData 目录，在 iOS 上，对应于 NSDocumentDirectory 目录，该目录用于存储只有自己应用可以访问的文件，只有当应用程序被卸载时，系统才会清除该目录，获取方式如下：

```
//代码清单 10-8 文档目录
//lib/code/code10/example_1005_File.dart
void getDocumentPath() async {
  //获取文档目录
  Directory dic = await getApplicationDocumentsDirectory();

  String path = dic.path;
  Uri uri = dic.uri;
  //父级目录
  Directory parentDic = dic.parent;
  Directory absolute = dic.absolute;
}
```

File 文件适用于保存到这样的目录下，如应用程序必备的升级，需要将安装包数据先保存到文档目录中，再如音乐、视频一类的 App，需要将媒体数据通过 File 存到磁盘上，在这里提供一个将文本字符串信息保存到手机中的示例，代码如下：

```
//代码清单 10-9 保存文件
  //lib/code/code10/example_1005_File.dart
  void saveFile() async {
    //获取文档目录
    Directory documentsDir = await getApplicationDocumentsDirectory();
    //获取对应的路径
    String documentsPath = documentsDir.path;
    //构建保存文本的路径
    String filePath = '$documentsPath/test.json';
    //创建对应的文件
    File file = File(filePath);

    //如果文件不存在就创建
    if (!file.existsSync()) {
      file.createSync();
    } else {
      //否则就删除
      file.delete();
    }
    //向文件中写入字符串数据
    await file.writeAsString("测试数据");
    //直接调用 File 的 writeAs 函数时
    //默认文件打开方式为 WRITE:如果文件存在,则会将原来的内容覆盖
    //如果不存在,则创建文件
    //写入 String,默认将字符串以 UTF-8 进行编码
    //将数据内容写入指定文件中
    if (file.existsSync()) {
      print("保存成功");
    } else {
      print("保存失败");
    }
  }
```

然后获取这个文件中的文本，代码如下：

```
//代码清单 10-10 获取文件
//lib/code/code10/example_1005_File.dart
void getFile() async {
  //获取文档目录
  Directory documentsDir = await getApplicationDocumentsDirectory();
  //获取对应的路径
  String documentsPath = documentsDir.path;
```

```
    //构建保存文本的路径
    String filePath = '$documentsPath/test.json';
    //读取
    File file = File(filePath);
    if (file.existsSync()) {
      print("文件存在");
      //readAsString 读取文件,并返回字符串
      //默认返回的 String 编码为 UTF-8
      //相关的编解码器在 dart:convert 包中
      //包括以下编解码器:ASCII、LANTI1、BASE64、UTF-8、SYSTEM_ENCODING
      //SYSTEM_ENCODING 可以自动检测并返回当前系统编
      String data = await file.readAsString();
      print("文件内容 $data");

      //一行一行地读取
      List<String> lines = await file.readAsLines();
      lines.forEach((String line) => print(line));
    } else {
      print("文件不存在 ");
    }
  }
}
```

10.2.3 SharedPreferences 轻量级数据保存

上述提到的 File 文件磁盘读写操作，适用于大文件的存储，在应用开发中，还有一些轻量级的数据，如用户的基本信息、对应用的偏好设置信息，这些数据量小，权重比较高，可以使用数据存储 SharedPreferences，需要在 pubspec.yaml 文件中添加相关依赖，代码如下：

```
dependencies:?
  shared_preferences: ^2.0.13
```

这个插件在 Android 平台中，使用 SharedPreferences 存储机制，Sharedpreferences 是 Android 平台上一个轻量级的存储类，用来保存一些轻量级数据，如应用程序的各种配置信息，以键-值对的方式(key-value 的形式)保存数据的 XML 文件，其文件保存在/data/data/应用包名/shared_prefs 目录下，以 Map 形式存放简单的配置参数，在 iOS 平台中，使用 NSUserDefaults，NSUserDefaults 是 iOS 中的一个单例类，用来存储一些轻量级数据，如应用程序的基本配置和一些小数据类型等，以 key-value 的形式将数据存储到相应的 plist 文件中，存储路径为沙盒路径的 Library→Preferences 文件夹中。创建工具类 SPUtil，定义保存简单数据类型的方法，代码如下：

```
import 'dart:convert';
import 'package:shared_preferences/shared_preferences.dart';
class SPUtil {
  //静态实例
  static SharedPreferences? _sharedPreferences;

  //应用启动时需要调用
  static Future init() async {
    _sharedPreferences = await SharedPreferences.getInstance();
    return Future.value(true);
  }

  //异步保存基本数据类型
  static Future<bool> save(String key, dynamic value) async {
    if (_sharedPreferences == null) {
      return false;
    }
    if (value is String) {
      await _sharedPreferences!.setString(key, value);
    } else if (value is bool) {
      await _sharedPreferences!.setBool(key, value);
    } else if (value is double) {
      await _sharedPreferences!.setDouble(key, value);
    } else if (value is int) {
      await _sharedPreferences!.setInt(key, value);
    } else if (value is List<String>) {
      await _sharedPreferences!.setStringList(key, value);
    }
    return true;
  }

  static String getString(String key) {
    if (_sharedPreferences == null) {
      return "";
    }
    return _sharedPreferences!.getString(key) ?? "";
  }

  static int getInt(String key) {
    return _sharedPreferences!.getInt(key) ?? 0;
  }

  static bool getBool(String key) {
    return _sharedPreferences!.getBool(key) ?? false;
  }
```

```
  static double getDouble(String key) {
    return _sharedPreferences!.getDouble(key) ?? 0.0;
  }
```

在实际业务开发中，可以在main函数入口处调用SPUtil的init方法来初始化，在这里，通过单击按钮实现数据的保存与获取，代码如下：

```
TextButton(
      child: const Text("保存基本数据"),
      onPressed: () {
        SPUtil.save("testString", "张三");
        SPUtil.save("testInt", 33);
        SPUtil.save("是否生病了", false);
      },
    ),
    TextButton(
      child: const Text("获取基本数据"),
      onPressed: () {
        bool isBool = SPUtil.getBool("是否生病了");
        String userName = SPUtil.getString("testString");
      },
    ),
```

使用SPUtil保存简单数据在手机磁盘上的XML文件，在手机目录空间的data→data→应用包名→shared_prefs→FlutterSharedPreferences.xml中，如图10-5所示。

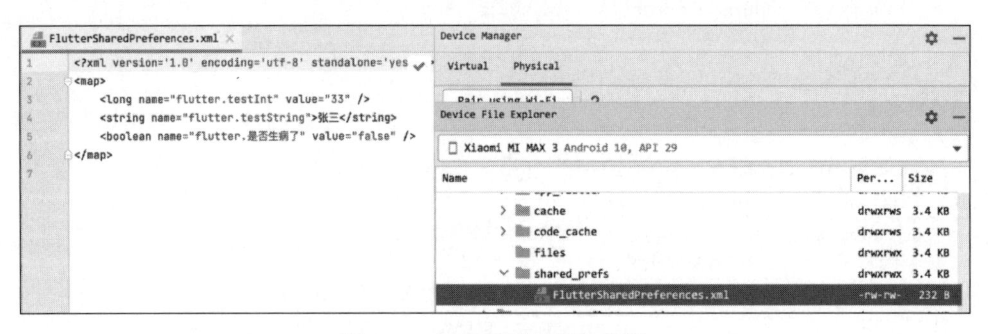

图10-5 SPUtil保存简单数据

10.2.4 sqflite数据库数据操作

在App的许多业务应用场景中，会生成很多列表数据，将这些列表数据适当地保存下来，当用户的手机无网络或者网络不好时，先使用这些数据填充应用的页面信息，可以很好地提高应用的用户体验，针对于列表数据，适合使用数据库来缓存。在这里使用插件sqflite实现这个操作，在配置文件pubspec.yaml中添加依赖，代码如下：

```
dependencies:
  sqflite: ^2.0.2
```

加载依赖,代码如下:

```
flutter pub get
```

在使用的文件中导包,代码如下:

```
import 'package:sqflite/sqflite.dart';
```

一个应用程序中可以对应多个库,每个库可以对应多个表,笔者创建了一个 test.db 数据库,使用一个用户数据表 t_user 来对用户数据进行增、删、改、查操作,在使用数据库时首先应创建数据库并且创建表,代码如下:

```
//代码清单 10 - 11
//lib/code10/example_1006_sqflite.dart
//数据库名称
final String _dbName = 'test.db';
//数据库版本
int version = 1;
//创建表的 SQL
String dbTables =
    "CREATE TABLE t_user (id INTEGER PRIMARY KEY, name TEXT,age INTEGER)";
//创建数据库 db [dbName]数据库名称 [version] 版本
void _createDb() async {
  //获取数据库路径
  String databasesPath = await getDatabasesPath();
  print(databasesPath);
  //获取数据库的完整路径
  //join 函数是 path 下的一个拼接路径的函数
  //导入 import 'package:path/path.dart';
  String path = join(databasesPath, _dbName);
  //打开数据库
  Database db = await openDatabase(path, version: version,
      onUpgrade: (Database db, int oldVersion, int newVersion) async {
    //数据库升级,只回调一次
    print("数据库需要升级!旧版:$ oldVersion,新版.$ newVersion");
  }, onCreate: (Database db, int version) async {
    //创建表,只回调一次
    await db.execute(dbTables,);
    await db.close();
  });
}
```

每当应用程序升级时，如表有改动或者有新的表需要创建，在这里就可以升级版本号，然后应用程序在升级后，如果本地已存在这个数据库，就会执行 onUpgrade 方法的回调，可以在这里创建新的表。

NSERT INTO 语句用于向表格中插入新的行，表中插入数据的 SQL 格式如下：

```
方式一
INSERT INTO 表名称 VALUES (值 1, 值 2, ….)
方式二
INSERT INTO table_name (列 1, 列 2, …) VALUES (值 1, 值 2, ….)
```

向表中添加一条数据，代码如下：

```
//代码清单 10 - 12 增
//lib/code10/example_1006_sqflite.dart
void _add() async {
  //获取数据库路径
  var databasesPath = await getDatabasesPath();
  String sql = "INSERT INTO t_user(name,age) VALUES('李四','22')";
  String path = join(databasesPath, _dbName);
  print("数据库路径: $ path");
  //打开数据 ylkd
  Database db = await openDatabase(path);
  //插入数据
  await db.transaction((txn) async {
    //可以根据插入的行数来判断是否插入成功
    int count = await txn.rawInsert(sql);
    print("插入数据 $ count");
  });
  await db.close();
  print("插入数据成功!");
}
```

DELETE 语句用于删除表中的行，SQL 格式如下：

```
DELETE FROM 表名称 WHERE 列名称 = 值
```

删除表中的一条数据，代码如下：

```
//代码清单 10 - 13 删
//lib/code10/example_1006_sqflite.dart
_delete() async {
  var databasesPath = await getDatabasesPath();
  String sql = "DELETE FROM t_user";
  String path = join(databasesPath, _dbName);
```

```
  Database db = await openDatabase(path);
  int count = await db.rawDelete(sql);
  await db.close();
  if (count > 0) {
    print("执行删除操作完成,该 sql 删除条件下的数目为 $count");
  } else {
    print("无法执行删除操作,该 sql 删除条件下的数目为 $count");
  }
}
```

当无删除条件时,代表删除这张表中的所有的数据。Update 语句用于修改表中的数据,SQL 格式如下:

```
UPDATE 表名称 SET 列名称 = 新值 WHERE 列名称 = 某值
```

修改表中的一条数据,代码如下:

```
//代码清单 10-14 改
//lib/code10/example_1006_sqflite.dart
void _update() async {
  var databasesPath = await getDatabasesPath();
  String sql = "UPDATE t_user SET name = ? WHERE id = ?";
  String path = join(databasesPath, _dbName);
  Database db = await openDatabase(path);
  //修改条件,对应参数值,参数二中的值要与 sql 中的"?"占位对应
  int count = await db.rawUpdate(sql, ["张三", 1]);
  await db.close();
  if (count > 0) {
    print("更新数据库操作完成,: $count");
  } else {
    print("无法更新数据库,: $count");
  }
}
```

SELECT 语句用于从表中选取数据,查数据是一门大学问,在这里只描述查询的基础用法,语法如下:

```
SELECT 列名称 FROM 表名称 WHERE 条件
```

查询表中所有的数据,代码如下:

```
//代码清单 10-15 查
//lib/code10/example_1006_sqflite.dart
```

```
void _queryAll() async {
  var databasesPath = await getDatabasesPath();
  String sql = "SELECT * FROM t_user";
  String path = join(databasesPath, _dbName);
  Database db = await openDatabase(path);
  //获取查询的条数
  int count = Sqflite.firstIntValue(await db.rawQuery(sql));
  //获取查询到的所有的结果集
  List<Map> list = await db.rawQuery(sql);
  print("查询完毕 count: $count 数据详情: $list");
  await db.close();

}
```

在实际项目开发中，数据库可以在适当的一个版块业务结束后再关闭，或者应用程序退出时再关闭，因为数据库的频繁打开与关闭也是一种性能消耗。

10.3 网络请求

Dart I/O 库中提供了用于发起 HTTP 请求的一些类，可以直接使用 HttpClient 来发起请求，Dio 是一个支持 RESTful API、FormData、拦截器、请求取消、Cookie 管理、文件上传/下载、超时等功能的封装网络框架。

10.3.1 HttpClient 网络请求操作

HttpClient 是 Dart I/O 库中的一个用于网络请求交互处理的类，使用 HttpClient 发起一个基本的 Get 请求，代码如下：

```
//代码清单 10-16 get 请求无参数
  //lib/code10/example_1007_HttpClient.dart

import 'dart:io'; //网络请求
import 'dart:convert'; //数据解析

  void getRequest() async {
    //定义请求 URL
    String url = 'http://192.168.40.167:8080/getUserList';
    //第一步，创建 Client
    HttpClient httpClient = HttpClient();
    String result;
    try {
      //第二步，构建 Uri
```

```
    Uri uri = Uri.parse(url);
    //第三步,发送 get 请求
    HttpClientRequest request = await httpClient.getUrl(uri);
    //第四步,获取响应同时关闭通道
    HttpClientResponse response = await request.close();

    if (response.statusCode == HttpStatus.ok) {
      //请求成功获取数据
      String json = await response.transform(utf8.decoder).join();
      //解析数据
      var data = jsonDecode(json);
      result = data.toString();
      print("请求到的数据为 ${data.toString()}");
    } else {
      result = '请求异常 ${response.statusCode}';
    }
  } catch (exception) {
    //异常数据处理
    result = 'Failed getting IP address';
  }
  setState(() {
    _netData = result;
  });
}
```

HttpClient 发起 Post 请求并提交 JSON 格式的数据,代码如下:

```
//代码清单 10-17 Post 请求参数为 JSON 格式
  //lib/code10/example_1007_HttpClient.dart
  void postRequest() async {
    HttpClient client = HttpClient();
    //定义请求 URL
    String url = 'http://192.168.40.167:8080/registerUser2';
    Uri uri = Uri.parse(url);
    //请求参数
    Map<String, dynamic> map = {"name": "张三", "age": 22};
    //发起网络请求
    final request = await client.postUrl(uri);
    //设置请求头
    request.headers
        .set(HttpHeaders.contentTypeHeader,
            "application/json; charset=UTF-8");
    //设置参数
    request.write(map.toString());
    //获取响应
```

```
    final response = await request.close();
    if (response.statusCode == HttpStatus.ok) {
      //请求成功获取数据,这里通过监听方式获取结果
      response.transform(utf8.decoder).listen((contents) {
        setState(() {
          _netData = contents;
        });
      });
    } else {
      //请求失败
      setState(() {
        _netData = "请求失败";
      });
    }
  }
```

10.3.2 网络请求库 Dio

Dio 是一个第三方库,所以在使用前需要在配置文件 pubspec.yaml 中添加依赖,代码如下:

```
dependencies:
  dio: ^4.0.4
```

加载依赖,代码如下:

```
flutter pub get
```

在使用的文件中导包,代码如下:

```
import 'package:dio/dio.dart';
```

Dio get 请求,获取列表 JSON 数据,并将 JSON 数据处理为自定义实例类,代码如下:

```
  //代码清单 10-18 Dio get 请求无参数
  //获取所有的用户信息
  //lib/code10/example_1008_Dio.dart
  void getRequest() async {
    //定义请求 URL 获取用户列表
    String url = 'http://192.168.40.167:8080/getUserList';
    //创建 Dio 对象
Dio dio = Dio();
//请求参数
```

```
Map<String, dynamic> map = Map();
    map["userId"] = userId;

    //发起 get 请求
    Response response = await dio.get(url,queryParameters: map);
    //响应数据,JSON 格式为 {"code": 200,"data": [],"message": "请求成功"}
    //这里第一层映射将 JSON 数据转换为 Map 类型
    Map<String, dynamic> responseMap = response.data;

    //然后在映射的这个 responseMap 中

    //解析第二层对应的 JSON 中的 data 是一个数组
    List<dynamic> dataList = responseMap["data"];
    //创建一个保存 UserBean 的集合 List
    List<UserBean> userBeanList = [];

    //解析第三层就是数组中对应的对象类型数据
    for (int i = 0; i < dataList.length; i++) {
      Map<String, dynamic> itemMap = dataList[i];
UserBean userBean = UserBean.fromJson(itemMap);
userBeanList.add(userBean);
    }

    setState(() {
      _netData = dataList.toString();
    });
  }
```

UserBean 是自定义的数据实体,代码如下:

```
//代码清单 10-19 用户信息数据模型
//lib/code10/example_1008_Dio.dart
class UserBean {
  String ? userName;
  String ?realName;
  int ?age;
  int ?id;

  static UserBean fromJson(Map<String, dynamic> data) {
    UserBean userBean = UserBean();
    userBean.id = data["id"];
    userBean.age = data["age"];
    userBean.userName = data["userName"];
    userBean.realName = data["realName"];
    return userBean;
  }
}
```

Dio post 请求提交 FormData 表单数据，FormData 将提交的参数 name 与 value 进行组合，实现表单数据的序列化，从而减少表单元素的拼接；从另一个角度描述，FormData 接口提供了一种表示表单数据的键-值对的构造方式，通过 FormData 发出的请求编码类型被设为 multipart/form-data，而在网络请求访问中，通过 Content-Type 来记录这个值，可以理解为 Content-Type 表示具体请求中的媒体类型信息，在实际开发中常用的 Content-Type 如下：

```
multipart/form-data
application/json                    //JSON 数据格式
application/x-www-form-urlencoded //表单数据格式
```

使用 Dio 来发起一个 post 请求，提交参数的格式为 FormData，代码如下：

```
//代码清单 10-20 Dio post 请求 [FormData]格式
//根据用户 ID 获取用户信息
//lib/code10/example_1008_Dio.dart
void postRequestFunction() async {
  //创建 Dio
  Dio dio = Dio();
  //发送 FormData
  FormData formData = FormData.fromMap(
    {"userName": "张三", "userAge": 22},
  );
  //请求 URL
  String url = "http://192.168.40.167:8080/registerUser";
  //发起 post 请求如这里的注册用户信息
  Response response = await dio.post(url, data: formData);
  Map<String, dynamic> data = response.data;
}
```

使用 Dio 发起一个 post 请求，提交 JSON 格式的参数，代码如下：

```
//代码清单 10-21 Dio post 请求发送 JSON 数据
//根据用户 ID 获取用户信息
//lib/code10/example_1008_Dio.dart
void postRequestFunction2() async {
  //请求 URL
  String url = "http://192.168.40.167:8080/registerUser2";

  //创建 Dio
  Dio dio = Dio();
  //创建 Map 封装参数
  Map<String, dynamic> map = {};
```

```
  map['userName'] = "小明";
  map['userAge'] = 44;

  //发起 post 请求
  Response response = await dio.post(url, data: map);
  //获取响应结果
  Map<String, dynamic> data = response.data;
}
```

Dio 文件上传并实现进度监听,代码如下:

```
//代码清单 10-22 Dio post 实现文件上传
//lib/code10/example_1008_Dio.dart
//手机中的图片
String localImagePath = "/storage/emulated/0/Download/17306285.jpg";
//上传的服务器地址
String netUploadUrl = "http://192.168.0.102:8080/fileupload";
//Dio 实现文件上传
void fileUplod() async {
  //创建 Dio
  Dio dio = Dio();

  Map<String, dynamic> map = {};
  map["auth"] = "12345";
  //对于图片数据一定要使用 await
  map["file"] = await MultipartFile.fromFile(
    localImagePath,
    filename: "xxx23.png",
  );
  //通过 FormData 传参数
  FormData formData = FormData.fromMap(map);
  //发送 post
  Response response = await dio.post(
    netUploadUrl, data: formData,
    //这里是发送请求回调函数
    //[progress] 当前的进度
    //[total] 总进度
    onSendProgress: (int progress, int total) {
      print("当前进度是 $progress 总进度是 $total");
    },
  );
  //服务器响应结果
  var data = response.data;
}
```

Dio 文件下载并实现进度监听,代码如下:

```
//代码清单 10 - 23 Dio post 实现文件下载
//lib/code10/example_1008_Dio.dart
//当前进度百分比,当前进度/总进度,即 0.0 ~ 1.0
double currentProgress = 0.0;

//下载文件的网络路径
String apkUrl = "";
//使用 Dio 下载文件
void downApkFunction() async {
  //申请写文件权限
  //
  //手机储存目录
  String? savePath = await getPhoneLocalPath();
  String appName = "rk.apk";

  //创建 Dio
  Dio dio = new Dio();

  //参数一,文件的网络储存 URL
  //参数二,下载的本地目录文件
  //参数三,下载监听
  Response response = await dio.download(apkUrl, " $ savePath $ appName",
      onReceiveProgress: (received, total) {
    if (total != -1) {
      //当前下载的百分比
      print((received / total * 100).toStringAsFixed(0) + " % ");
      //CircularProgressIndicator(value: currentProgress,) 进度 0 - 1
      currentProgress = received / total;
      setState(() {});
    }
  });
}

//获取手机的存储目录
//getExternalStorageDirectory() 获取的是 Android 的外部存储(External Storage)
//getApplicationDocumentsDirectory 获取的是 iOS 的 Documents 或 Downloads 目录
Future < String?> getPhoneLocalPath() async {
  final directory = Theme.of(context).platform == TargetPlatform.android
      ? await getExternalStorageDirectory()
      : await getApplicationDocumentsDirectory();
  return directory?.path;
}
```

10.3.3 选择图片插件

可通过 image_picker 插件实现相机拍照或者选择手机相册的图片来验证文件上传功能,添加的依赖如下:

```
#https://pub.flutter-io.cn/packages/image_picker 查看最新版本
dependencies:
image_picker: ^0.8.4+10
```

在 android 目录下的清单文件中添加配置,如图 10-6 所示。

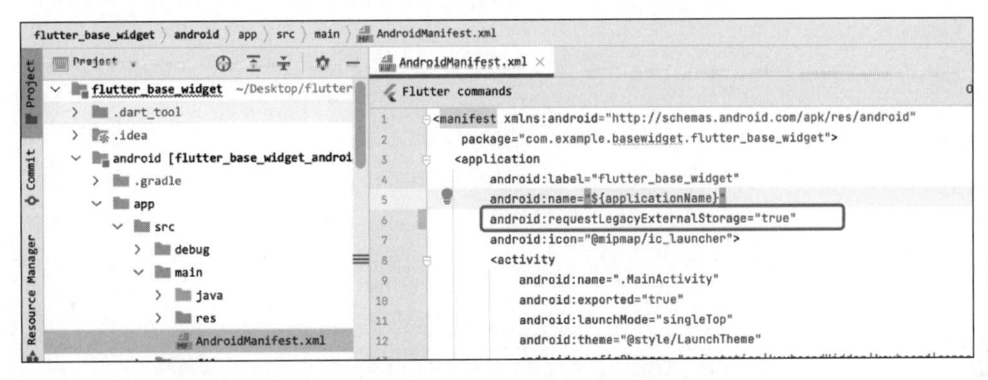

图 10-6　在 Android 清单文件中添加配置

代码如下:

```
import 'package:image_picker/image_picker.dart';

  void selectPhone() async{
    final ImagePicker _picker = ImagePicker();
    //打开相册,返回所选择的图片
final XFile? image = await _picker.pickImage(
                source: ImageSource.gallery);
    //打开相机,返回所拍照片
    final XFile? photo = await _picker.pickImage(source: ImageSource.camera);
    //打开相册,实现多选
    final List<XFile>? images = await _picker.pickMultiImage();
  }
```

10.4　小结

本章详细讲解了 Flutter、Timer 及 Flutter 中的消息队列任务模型,可以很好地帮助开发者理解 Flutter 中的异步机制,同时也可以一定程度上提高开发者异步任务开发上的调试协调,也详细介绍了文本数据持久化存储方案,可以帮助开发者很好地处理数据缓存方面的问题。

实战应用篇

第 11 章 Flutter 应用基础框架

CHAPTER 11

使用 Flutter 可以快速构建 Android、iOS 双平台的应用程序，从 0 到 1 的搭建过程会有很多细节要处理，本章不使用任何架构、框架及设计模式搭建，旨在帮助读者综合前 10 章所涉及内容，从最基础的搭建开始，读者可观看视频讲解【11 Flutter 应用基础架构】。本章的内容如下：

(1) Flutter 项目目录说明、Android build. gradle 及清单文件配置、iOS info. plist 配置概述。

(2) App 的图标配置、启动页面配置、打包发布流程概述。

(3) 基础 App 的启动、欢迎页面、引导页面、首页面，最终形成一个基础开发的脚手架。

本章最终开发完成的一个脚手架如图 11-1 所示，读者可以直接使用这个脚手脚来快速构建应用程序，当然按照本章的顺序也可以帮助读者形成 App 应用开发思维。

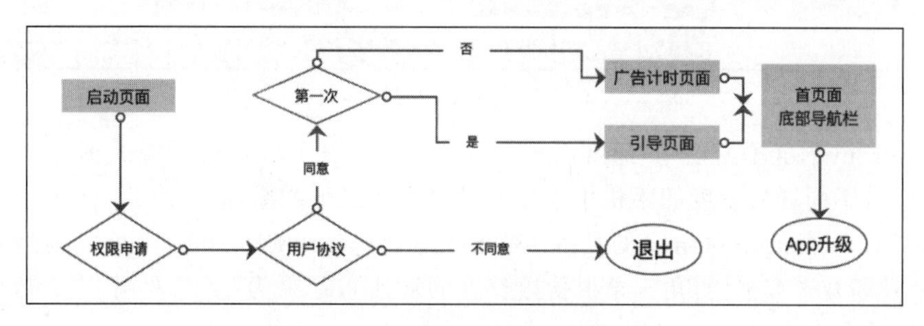

图 11-1　脚手架结构示意图

11.1　App 应用程序基本配置概述

一个完整的 App，初次呈现在用户面前的是手机桌面上的图标，在手机的设置中应用管理及快速搜索列表、手机通知栏都会有对应 App 的图标显示，在不同分辨率下的手机使用的图标大小不一样，所以需要分别配置不同大小的图标来适配。

对于打包生成发布的 App，在打包时需要配置应用版本信息、应用唯一标识信息、签名信息及其他针对不同平台的优化配置信息。

在 Flutter 项目工程开发中,也需要不同的配置,如添加依赖字体及一些原生 SDK 的接入等,本章构建的快速开发脚手架项目结构图如图 11-1 所示。

11.1.1 App 基本信息配置

在 Flutter 项目中的 android 目录(目录结构如图 11-2 所示)下的 res 目录中的各 mipmap-xxx 文件夹中放入不同尺寸的 App 图标,这些图标会分别在通知栏、快速搜索栏、手机桌面显示,在 android 目录下找到清单文件 AndroidManifest. xml,然后在清单文件中的 application 标签下配置 icon 属性即可。读者可观看视频讲解【11. 1. 1 App 基本信息配置(Android 平台)】。

2min

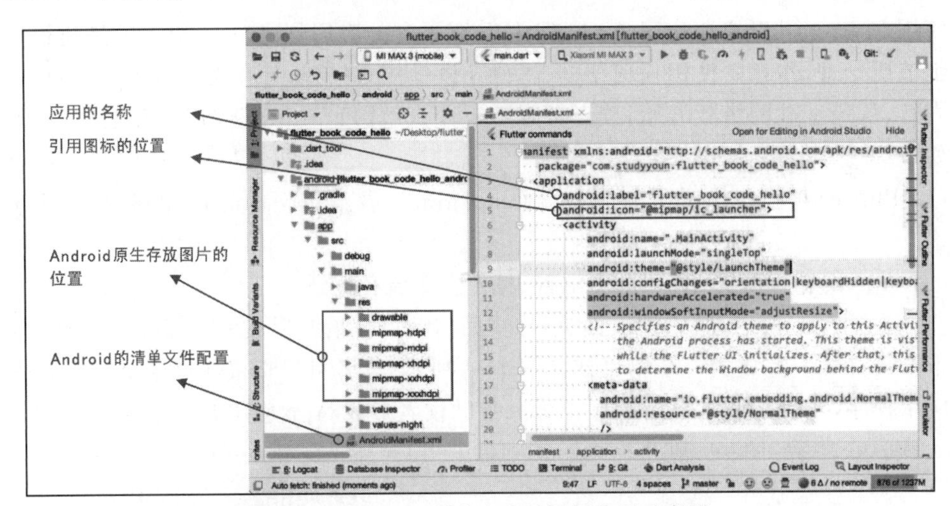

图 11-2　Android 原生中配置图标位置示意图

在 Android 的屏幕密度中,通常以 160dpi 为基准,每个目录所对应的图片尺寸描述如表 11-1 所示,不同屏幕密度的手机中采用的是对应目录下的图标。

屏幕的尺寸指的是对角线的长度,密度值是指(以 1920×1080,5 英寸为例)1920 和 1080 的平方和开平方根(直角三角形斜边长度的算法),结果为 2202. 9,除以 5 英寸约等于 441,441 接近表 11-1 中的 3 倍屏。

表 11-1　Android 目录下图标大小说明

文　件　夹	图标大小	密　度　值	分　辨　率
mipmap-mdpi	48px×48px	160	320px×480px 1 倍屏
mipmap-hdpi	72px×72px	240	480px×800px 1. 5 倍屏
mipmap-xhdpi	96px×96px	320	720px×1280px 2 倍屏
mipmap-xxhdpi	144px×144px	480	1080px×1920px 3 倍屏
mipmap-xxxhdpi	192px×192px	640	1440px×2560px 4 倍屏

对于 iOS 的图标配置，有两种配置方式，第一种是如图 11-3 所示，在 Flutter 工程项目中的 ios 目录下的 AppIcon. appiconset 中直接替换图标即可；第二种方式是将 Flutter 项目在 Xcode 中打开，然后在 Xcode 中进行图标配置，打开方式有以下 3 种：

（1）通过 Android Studio 的工具栏中的 Tools→Flutter→Open iOS module in Xcode 打开。

（2）在 Flutter 项目中的 ios 目录上单击，在弹出的对话框中选择 Flutter→Open iOS module in Xcode。

（3）在 Android Studio 的 Terminal 命令行工具中输入命令 open ios/Runner. xcworkspace 即可打开。

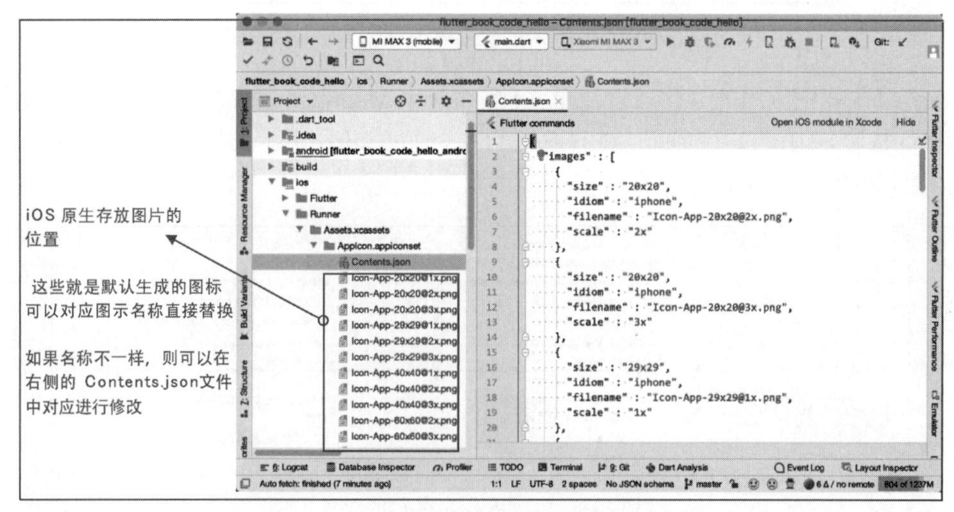

图 11-3　iOS 原生中配置图标位置示意图

在 Xcode 中打开的 Flutter 工程目录如图 11-4 所示，General 设置项中的 App Icons Source 用来配置 App 的图标，对应的 AppIcon 放在 Assets 资源文件夹中。AppIcon 用来配置各种机型上使用的不同的图标，如图 11-5 所示。读者可观看视频讲解【11. 1. 1 App 基本信息配置(iOS 平台)】。

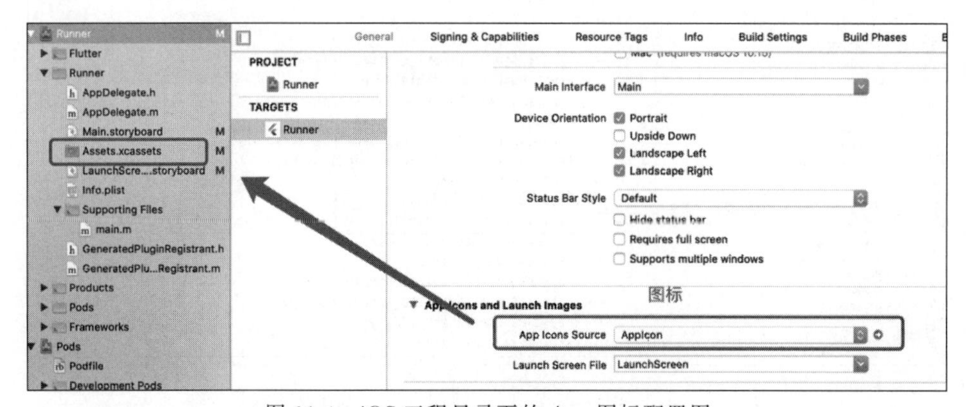

图 11-4　iOS 工程目录下的 App 图标配置图

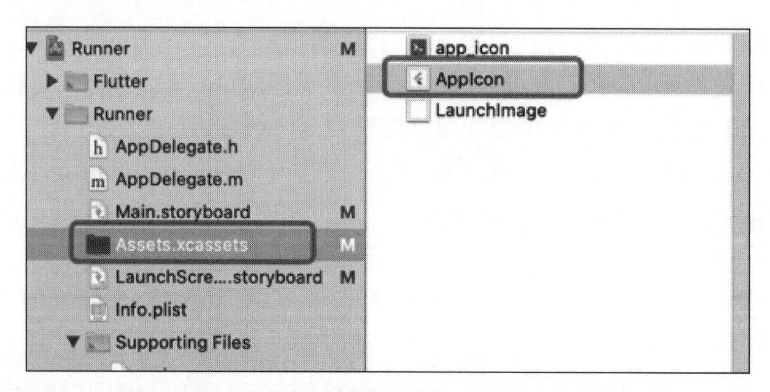

图 11-5　iOS 工程目录下的资源 Assets. xcassets 目录图

在 Android 中应用的名称通过清单文件 AndroidManifest. xml 中的 application 标签下 label 属性指定，如图 11-2 所示。在 iOS 平台中，通过 info. plist 文件中的 Bundle name 属性来配置，如图 11-6 所示。

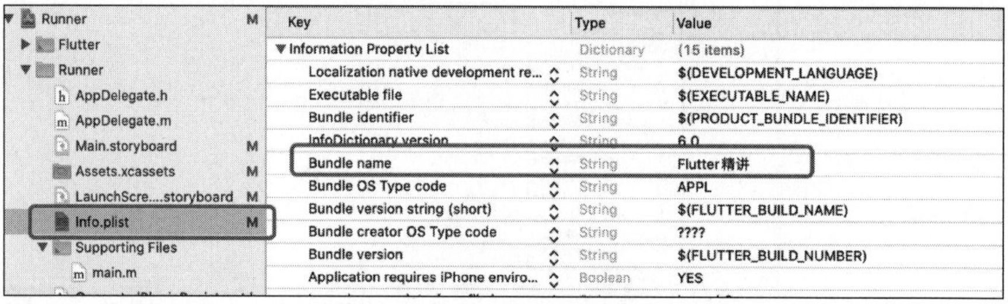

图 11-6　iOS 应用名称的配置

或者在 Flutter 工程项目中找到 ios 目录对应的文件直接修改，如图 11-7 所示。

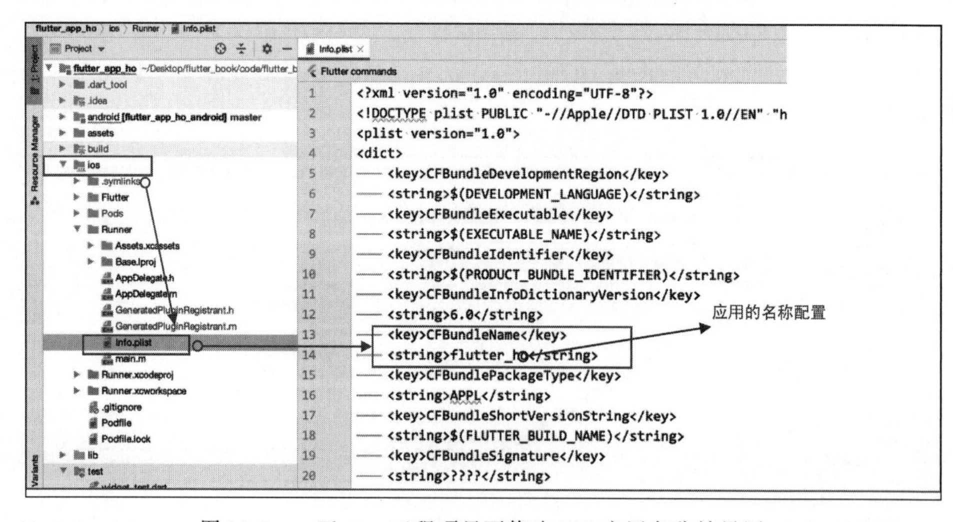

图 11-7　Flutter 工程项目下修改 iOS 应用名称效果图

11.1.2　Android 平台开发配置

4min

如图 11-8 所示，是 Android 平台的开发配置信息，图中所示的内容是比较全的一个配置，默认创建的项目缺少部分配置，这个可根据实际业务项目需要来配置，在本节中概述一下相关功能。读者可观看视频讲解【11.1.2 Android 平台开发配置】。

图 11-8　Android 工程项目中的 build.gradle 配置

首先在 defaultConfig 标签下配置一些基本的内容，描述如下：

```
defaultConfig {
    //应用的唯一标识
    applicationId " com.example.flutter_base_kit"
    //Android 依赖的 SDK 版本
    minSdkVersion 21
    targetSdkVersion 29
    //应用的版本信息
    versionCode flutterVersionCode.toInteger()
    versionName flutterVersionName
    //65536 方法限制分包打包
    multiDexEnabled true
    //为了减小 APK 体积,只保留 armeabi 和 armeabi - v7a 两个文件夹
    //并保证这两个文件夹中.so 数量一致
    //对只提供 armeabi 版本的第三方.so,原样复制一份到 armeabi - v7a 文件夹
    //不要修改,必须包含这三个
    ndk {
        abiFilters "armeabi - v7a", "arm64 - v8a", "armeabi"
    }
}
```

ndk 标签配置的是打包 Android 平台中 framework、so、JIN 库等相关的平台编译库，如在使用地图插件时，默认为编译支持的所有的平台 so 库，应用 APK 的体积是比较大的，通过这个标签减少一些不常用的平台编译，也是减少应用 APK 体积的一种有效方式。

同时也需要指定 JNI 的编译源位置，通过 sourceSets 标签来指定这个位置(需要在 app 目录下创建目录 libs 并将需要集成的第三方原生的库放在这个路径下)，代码如下：

```
 //目录指向配置
sourceSets {
        main {
            //指定 lib 库目录
            //可以在 Android Studio 的 Android 视图下生成 jniLibs 文件夹
            //可以方便存放 jar 包和库文件
            jniLibs.srcDirs = ['libs']
        }
    }
```

lintOptions 标签用来配置程序编译过程中对程序代码中可能存在的错误的一种检查，常用配置如下：

```
//程序在编译的时候会执行 lint
//当有任何错误提示时会停止 build,lintOptions {
    //即使报错也不会停止打包
    checkReleaseBuilds false
    //打包 release 版本的时候进行检测
    abortOnError false
    disable 'InvalidPackage'
}
```

signingConfigs 是自定义的方法块，用来配置加载的打包签名信息，这里配置的签名文件 test.jks 的位置是通过相对位置来引用的，如图 11-8 所示。

```
signingConfigs {
    //正式发布使用签名文件,名字自定义
    app_key {
        storeFile file('test.jks')          //文件位置
        storePassword '123456'              //文件密码
        keyAlias 'test'                     //别名
        keyPassword '123456'                //别名密码
    }

    //deBug 调试编译发布使用签名文件,名字自定义
    deBug_app_key {
```

```
        storeFile file('test.jks')
        storePassword '123456'
        keyAlias 'test'
        keyPassword '123456'
    }
}
```

配置好的 signingConfigs 会在 buildTypes 标签中使用，buildTypes 标签用于配置 Android 程序编译过程的一些信息，代码如下：

```
buildTypes {
    deBug {
        //设置签名信息
        signingConfig signingConfigs.deBug_app_key
        //是否对代码进行混淆
        minifyEnabled false
        //指定混淆的规则文件
        proguardFiles getDefaultProguardFile('proguard-android.txt'), 'proguard-rules.pro
'
        //是否在 APK 中生成伪语言环境,帮助国际化,一般不使用
        pseudoLocalesEnabled false
        //是否对 APK 包执行 ZIP 对齐优化,减小 ZIP 体积,增加运行效率
        zipAlignEnabled false
        //在 applicationId 中添加了一个后缀,一般不使用
        //applicationIdSuffix 'test'
        //在 applicationId 中添加了一个后缀,一般不使用
        //versionNameSuffix 'test'
        //是否支持断点调试
        deBuggable true
        //是否可以调试 NDK 代码
        jniDeBuggable true
        //是否开启渲染脚本,即 C 语言写的渲染方法
        renderscriptDeBuggable false
    }
    release {
        signingConfig signingConfigs.app_key
        //是否对代码进行混淆
        minifyEnabled false
        shrinkResources false
        //指定混淆的规则文件
        proguardFiles getDefaultProguardFile('proguard-android.txt'), 'proguard-rules.pro
'
        //是否在 APK 中生成伪语言环境,帮助国际化,一般不使用
        //pseudoLocalesEnabled false
```

```
            //是否对 APK 包执行 ZIP 对齐优化,减小 ZIP 体积,增加运行效率
            //zipAlignEnabled true
            //在 applicationId 中添加了一个后缀,一般不使用
            //applicationIdSuffix 'test'
            //在 applicationId 中添加了一个后缀,一般不使用
            //versionNameSuffix 'test'
            //是否开启渲染脚本,即 C 语言写的渲染方法
            //renderscriptDeBuggable false
        }
    }
```

清单文件 AndroidManifest.xml 如图 11-9 所示，其中包含了 App 的基本配置信息，系统可以根据里面的内容运行 App 的代码，显示界面，Android 中的四大组件、基本权限配置及一些集成第三方合作平台的 SDK 中配置的信息。

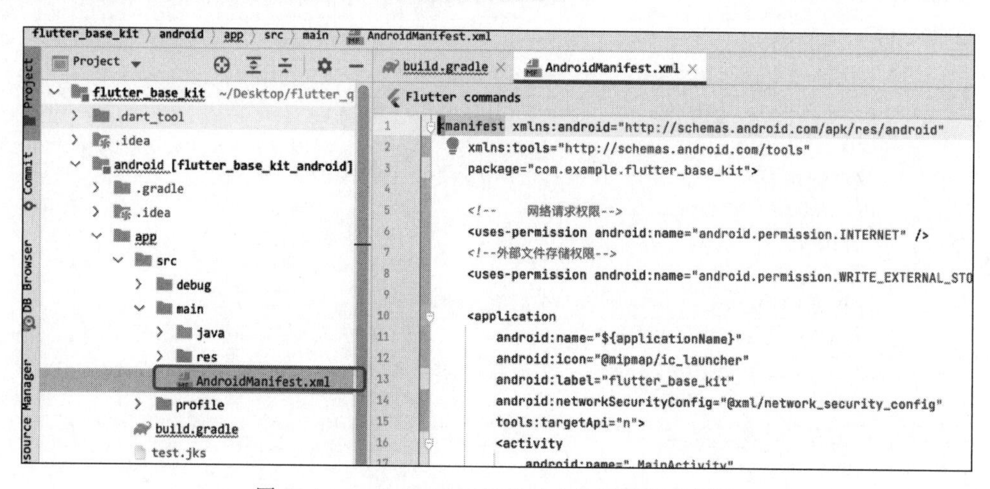

图 11-9　Android 工程项目中的清单文件调试

在业务项目开发中，Flutter 项目中可能会集成较多的插件，许多插件中会有相关项目示例，可能会导致 Flutter 项目编译打包 Android 平台的 APK 时清单文件合并冲突，如多个插件中配置了 label，此时就需要删除不需要的，可通过如图 11-10 所示的 Android 工程目录视图打开项目，然后在可视化区域查看冲突的文件并进行处理。

将 Flutter 项目工程以 Android 项目工程视图进行调试也是一种开发技巧，通常打开方式有以下 3 种：

(1) 通过 Android Studio 的工具栏中的 Tools→Flutter→Open for Editing Android Studio 打开调试页面。

(2) 在 Flutter 项目中的 android 目录上右击，在弹出的对话框中选择 Flutter →Open Androidmodule in Android Studio。

(3) 在 Android Studio 的顶部工具栏中选择 File→Open，在弹出的选项框中选中当前

Flutter 项目的 android 目录，然后打开。

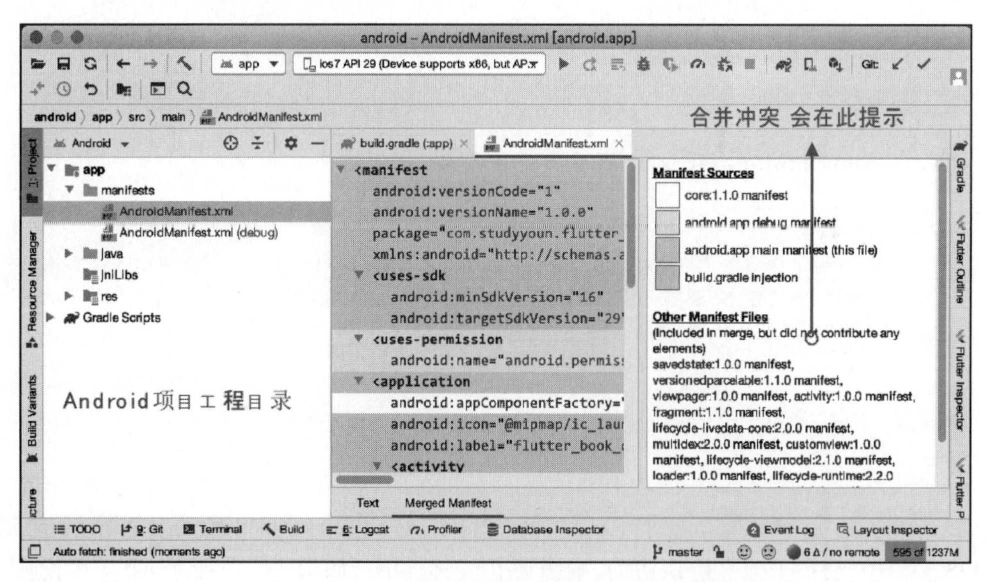

图 11-10　Android 工程项目中的清单文件调试

当在应用中使用相册、相机、传感器、蓝牙、位置等信息时，需要声明权限，首先需要在清单文件中声明，如这里在 AndroidManifest. xml 中配置了两个基本使用权限：

```
<!-- 网络请求权限 -->
<uses-permission android:name="android.permission.INTERNET" />
<!-- 外部文件存储权限 -->
<uses-permission android:name="android.permission.WRITE_EXTERNAL_STORAGE" />
```

在 Android P(9.0)以上的系统，默认禁止 App 使用 HTTP 协议访问网络，HTTP 明文传输协议不安全，同样在 iOS 9 和 OS X 10.11 以上的系统中，苹果引入了隐私保护功能 ATS(App Transport Security)，ATS 屏蔽了 HTTP 明文传输协议资源加载，在业务开发中，常会有一些资源使用 HTTP 访问，这种情况就需要应用程序支持 HTTP 传输。

在 Android 平台中，通常配置应用程序兼容 HTTP 协议有两种方式，第一种方式是在 AndroidManifest. xml 清单文件中，直接在 application 标签下配置，代码如下：

```
<uses-library
    android:name="org.apache.http.legacy"
    android:required="false" />
```

第二种方式是在 android 目录下的 res 目录下新建一个 xml 目录，然后创建一个名为 network_config. xml 的文件(文件名可以自定义)，内容如下：

```
<?xml version = "1.0" encoding = "utf - 8"?>
< network - security - config >
< base - config cleartextTrafficPermitted = "true" />
</network - security - config >
```

然后在 AndroidManifest.xml 清单文件中配置，代码如下：

```
< application
        android:label = "flutter_app_ho"
        android:networkSecurityConfig = "@xml/network_security_config"
        android:icon = "@mipmap/app_icon">

    ...

</application >
```

在 iOS 9 之后，苹果默认要求 App 访问的 URL 必须为 HTTPS 的安全链接，如果需要在 iOS 平台中支持 HTTP 协议访问，则需要在 info.plist（如图 11-11 所示）清单中配置，代码如下：

```
< key > NSAppTransportSecurity </key >
    < dict >
        < key > NSAllowsArbitraryLoads </key >
        < true/>
    </dict >
```

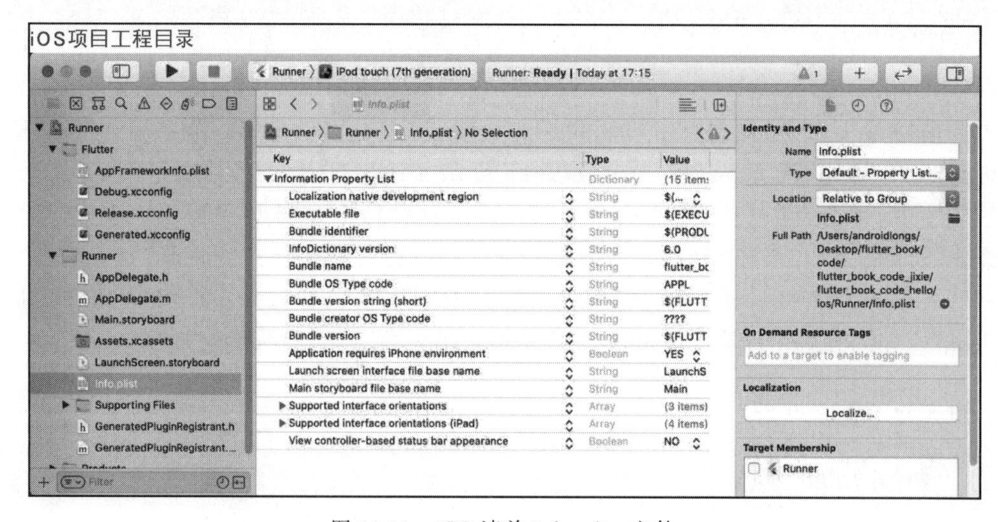

图 11-11　iOS 清单 info.plist 文件

11.1.3 Android 打包发布 APK

发布 App，就是将 Flutter 项目打包成 Android 平台使用的 APK 安装包，这个安装包是 release 模式的，通常打包方式有 3 种，第 1 种是在 Flutter 项目工程中通过命令打包，命令如下：

```
flutter run -- release
```

第 1 种打包方式通常需要在 Flutter 项目 android 目录下的 build. gradle 中配置签名信息，在 11. 1. 2 节中已讲解。

第 2 种是在 Android 工程目录视图下通过命令 aapt、Javac 等来综合生成 APK 安装包，在 Flutter 项目工程中不推荐这种方式。

第 3 种是在 Android 工程目录视图下通过工具栏的 Build→Generate Signed Bundle→APK，然后选择签名文件验证，最后选择打包 APK 路径、打包模式，然后完成打包，等待打包完成后就可以在如图 11-12 设置的保存路径中找到 APK 安装包，也就是可以用来发布到应用市场的安装包。读者可观看视频讲解【11. 1. 3 Android 打包发布 APK】。

3min

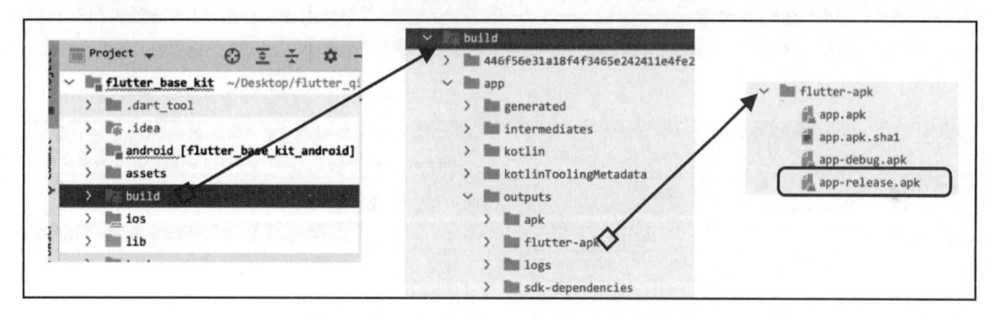

图 11-12 Android 打包 APK 文件路径

Android Studio 支持可视化生成签名，如图 11-13 所示，在 Android 工程目录下选择通过签名来打包 APK。

如果有签名文件，则可以在这里直接使用，如图 11-14 所示；如果没有签名文件，则可单击 Create new 按钮来打开如图 11-15 所示的生成新的签名文件的窗口。

11.1.4 iOS 打包发布

构建和发布一个 macOS 应用需要使用 Xcode，在发布之前，可以查看苹果公司的 App Store 审核指南，网址如下：

```
https://developer.apple.com/cn/app-store/review/
```

发布苹果应用，需要注册认证开发者账号，网址如下：

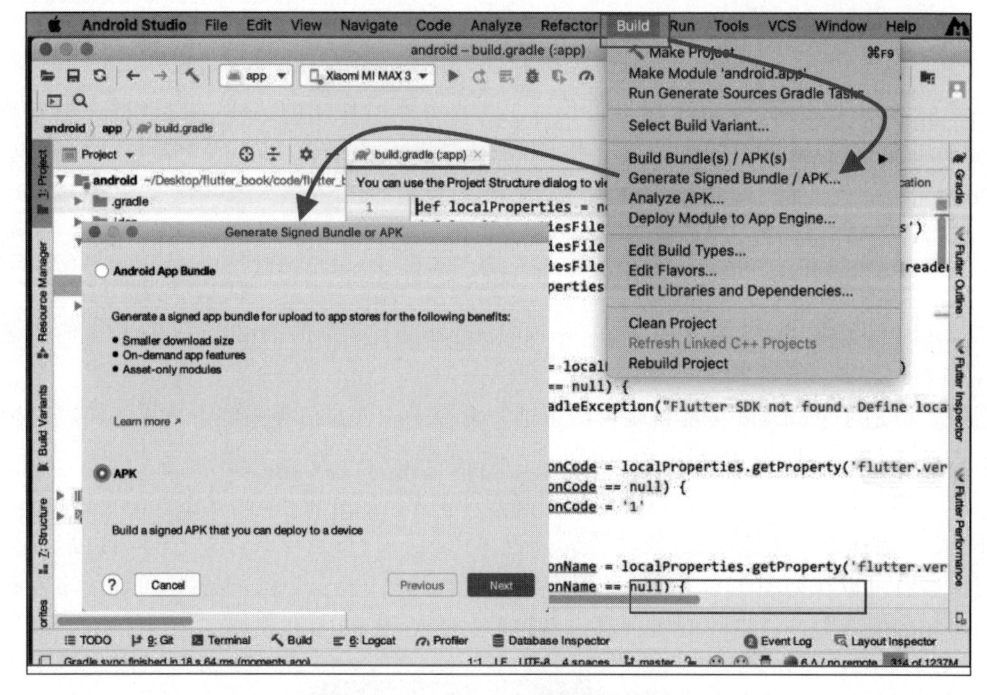

图 11-13　Android 工程目录下打包

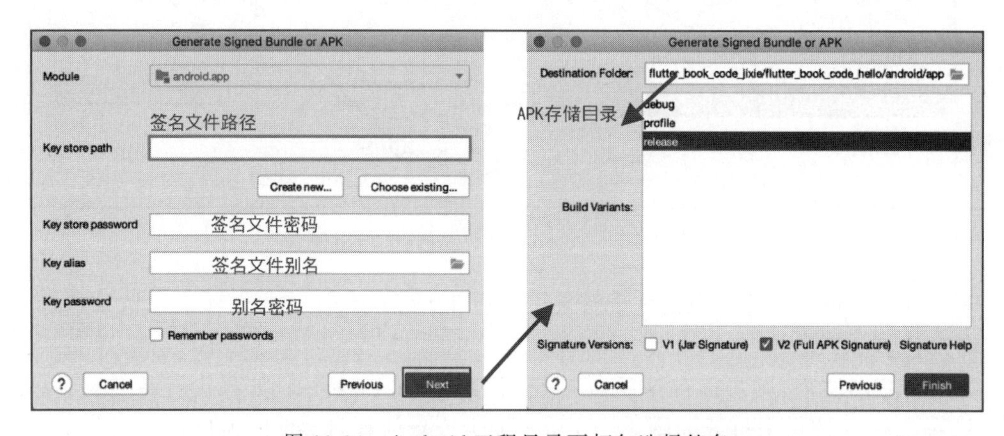

图 11-14　Android 工程目录下打包选择签名

```
https://developer.apple.com/cn/programs/
```

开发者帐号注册好后，需要在 App Store Connect 上注册相应的应用，App Store Connect 的使用介绍如下：

```
https://developer.apple.com/cn/support/app-store-connect/
```

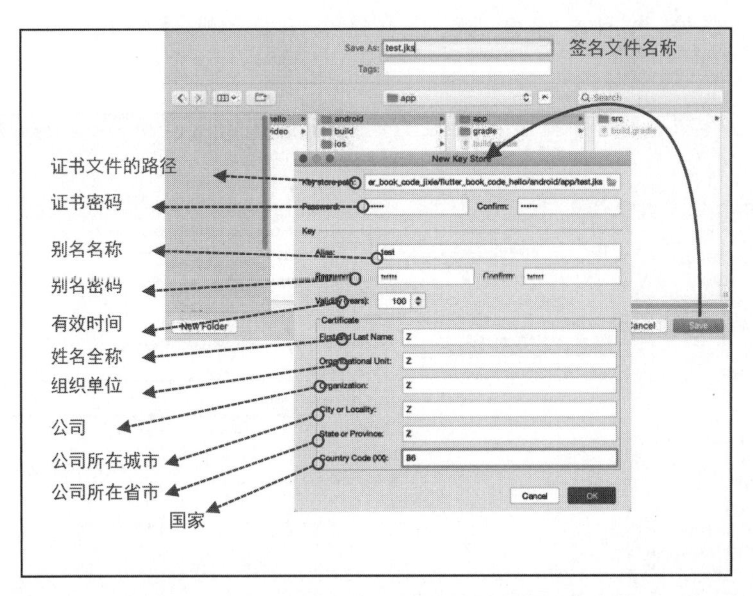

图 11-15 Android Studio 工程目录下生成证书文件

注册应用 Bundle ID，步骤如下：

（1）每个 iOS 应用都与一个唯一的套装 ID 关联，在开发者账号页面打开 App IDs 页面，网址如下：

```
https://developer.apple.com/account/ios/identifier/bundle
```

（2）在 App IDs 页面单击"+"来创建一个新的套装 ID。

（3）输入一个 App 名称，选择 Explicit App ID，然后输入一个 ID。

（4）选择 App 将要使用的服务，然后单击继续。

（5）在下一页，确认细节并单击注册来注册 Bundle ID。

在 App Store Connect 中注册应用的步骤如下。

（1）在浏览器里打开 App Store Connect，网址如下：

```
https://appstoreconnect.apple.com/
```

（2）在 App Store Connect 页面中，单击 My Apps。

（3）在我的 App 页面的顶部左侧，单击"+"，然后选择 New App。

（4）在出现的表单中填写 App 细节，在支持运行平台部分，确保 iOS 被选中。

（5）跳转到应用详情，从侧边栏选择 App Information。

（6）在基础信息部分，选择在前一步注册的套装 ID。

应用 Bundle ID 与 App Store Connect 完成应用注册后，在 Xcode 工程目录视图下打包发布，读者可观看视频讲解【11.1.4 iOS 打包发布】。首先选择发布账号，如果没有，

则应添加一个账号，并且本计算机已获取这个开发者账号的开发证书，如图 11-16 所示。

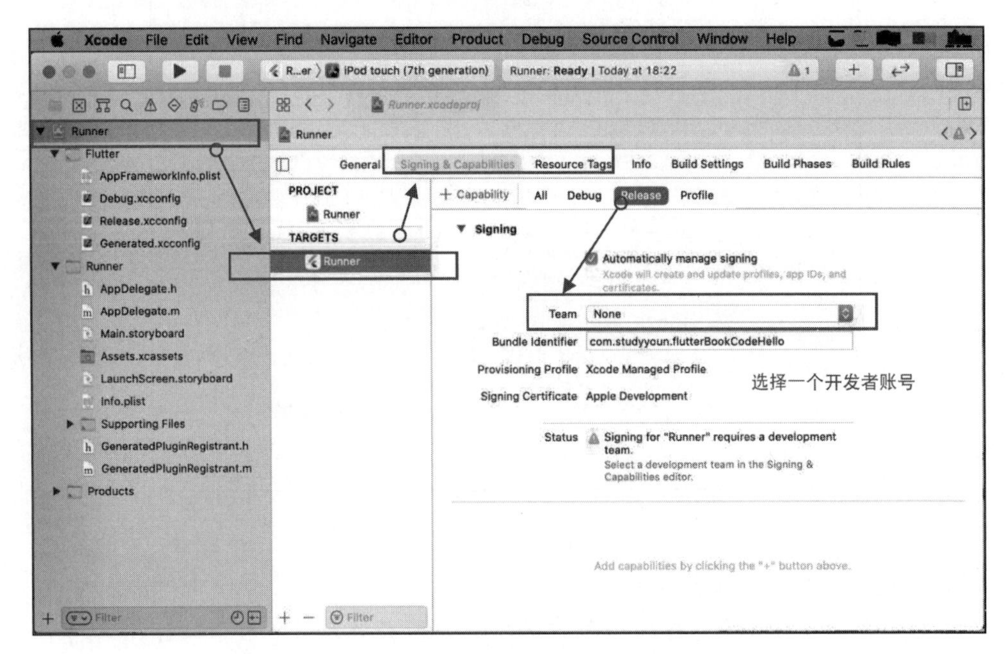

图 11-16　Xcode 工程目录下选择发布账号

然后在工具栏菜单中选择 Product→Archive 选项，如图 11-17 所示。

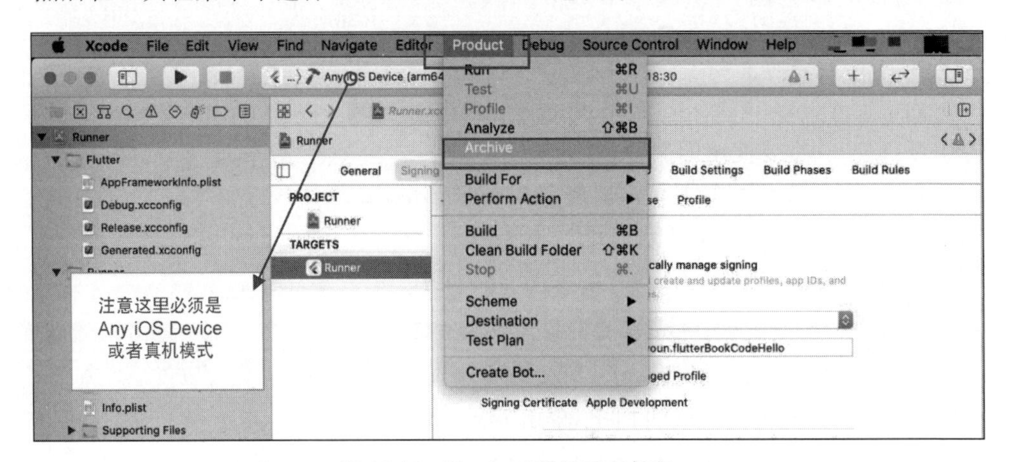

图 11-17　Xcode 工程目录下打包

等待编译完成后打开如图 11-18 所示的视图。

选择 Distribute App 进入如图 11-19 所示的选择操作页面，如果选择了上传 App Store，就需要在苹果开发者账号中创建对应的 Bundle Indentifier，以及本计算机的开发证书，同时需要在 App Store Connect 中创建对应的 App。

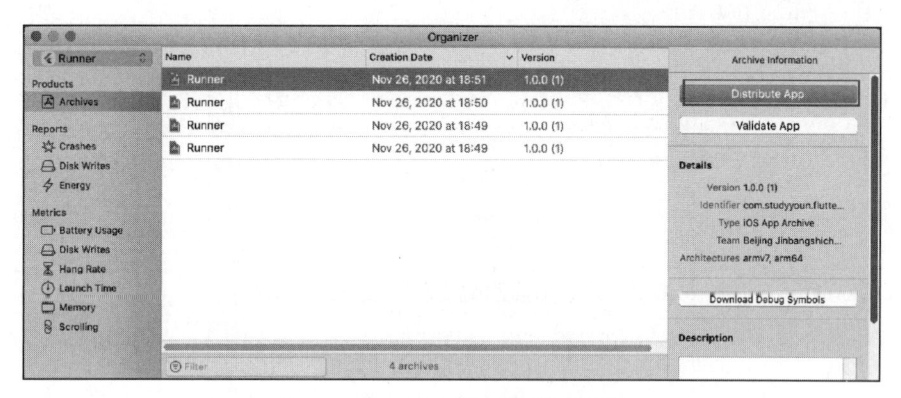

图 11-18　Xcode 工程目录下打包一

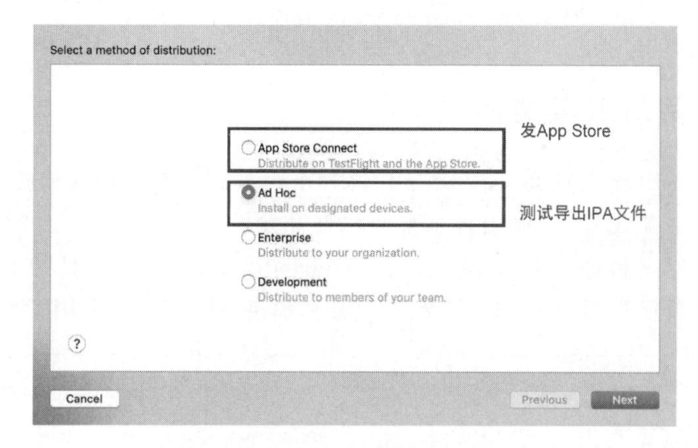

图 11-19　Xcode 工程目录下打包二

如果选择了 Ad Hoc 导出测试的 IPA 文件，则需要注意将测试 iOS 手机的 UDID 添加到开发者账号对应的证书设备中后，测试 iOS 手机才可以安装 IPA 文件。

当然从打包到 App Store 正式发布还有很多细节要处理，这不是本章的重点，在这里只提到了主要流程，读者可以根据相关提示信息灵活处理。

无论在哪个打包平台打包，在 Flutter 项目工程中，都应先清除缓存，再执行打包操作，这样更稳妥，清除命令如下：

```
flutter clean
```

11.2　App 开发工具类概述

日期的处理显示、字符串的判断分割、文件的读写、数据的缓存、数据的加密、消息提示框、数据网络请求等是开发中时常会用到的内容，通常会封装在固定的类中，以便在不同的业务代码块中使用，可以提升开发效率与代码的封装性，本章创建的项目目录结构如

图 11-20 所示,源码存放在本书配套源码 flutter_base_kit 中。

图 11-20　Android Studio 程序运行窗口日志

11.2.1　常用工具类概述

2min

本节概述的 Flutter 工具类是在 utils 包路径下。日志工具类用来向控制台输出日志信息,可以辅助开发者调试程序,输出日志有两种查看方式,第一种如图 11-21 所示,在程序运行窗口查看日志,第二种如图 11-22 所示,在 Andriod Studio 开发工具的 LogCat 日志窗口查看,可实现多条件筛选过滤日志。读者可观看视频讲解【11.2.1 常用工具类概述】。

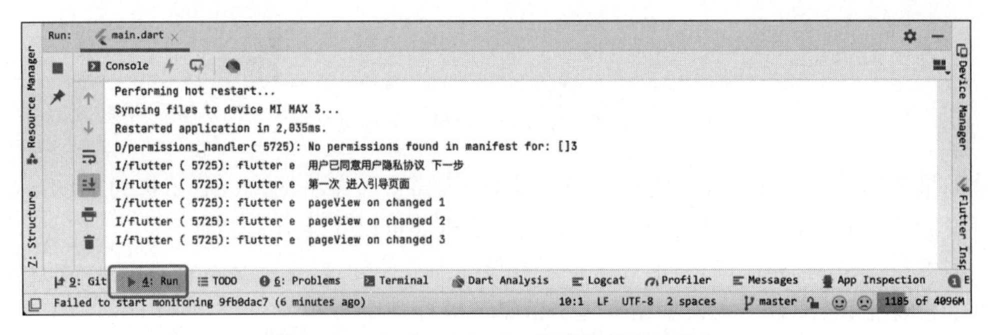

图 11-21　Android Studio 程序运行窗口日志

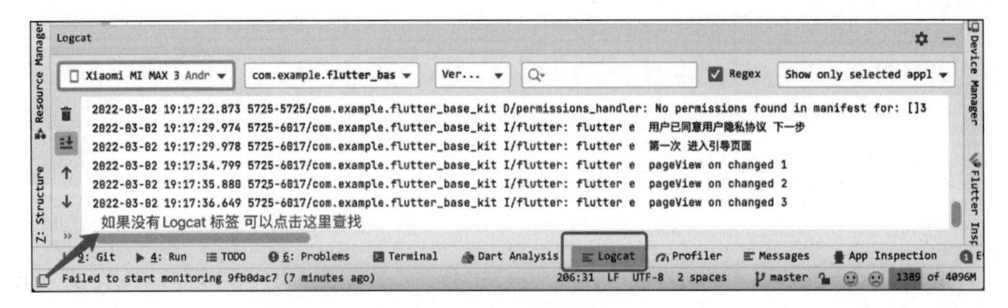

图 11-22　Android Studio Logcat 日志

日志封装工具类的实现代码如下：

```
//lib/utils/log_util.dart
//日志输出工具类
class LogUtil {
  //打印 log 的标签
  static const String _defaultLogTag = "flutter_log";
  //是否是 deBug 模式,true: log 不输出
  static bool _deBugMode = false;
  //log 日志的长度
  static int _maxLogLength = 130;

  //可以在程序启动的时候调用此方法来修改日志标识
  static void init({
    String tag = _defaultLogTag,
    bool isDeBug = false,
    int maxLen = 130,
  }) {
    _deBugMode = isDeBug;
    _maxLogLength = maxLen;
  }
  //外部调用输出日志的方法
  static void e(Object object, {String tag = "flutter"}) {
    if(_deBugMode){
      _printLog(tag, 'e', object);
    }
  }
  //内部实现日志输出方法
  static void _printLog(String tag, String stag, Object object) {
    String da = object.toString();
    if (da.length <= _maxLogLength) {
      deBugPrint("$tag$stag $da");
      return;
    }
    //日志过长,需要分行输出
    deBugPrint(
        '$tag$stag —————— st ——————————————————');
    while (da.isNotEmpty) {
      if (da.length > _maxLogLength) {
        deBugPrint("$tag$stag| ${da.substring(0, _maxLogLength)}");
        da = da.substring(_maxLogLength, da.length);
      } else {
        deBugPrint("$tag$stag| $da");
        da = "";
      }
    }
    deBugPrint('$tag$stag —— ed ——————————');
  }
}
```

在 App 业务开发时，如对订单商品的价格进行简单计算，通常需要保留两位小数，可通过 num 的 toStringAsFixed 方法来转换，在这里将这个方法放在 StringUtils 中，代码如下：

```
//字符串操作工具类
//lib/utils/string_utils.dart
class StringUtils {
  //判断 String 是否为空,为空时返回 true
  static bool isEmpty(String? tagText) {
    return tagText == null || tagText.isEmpty;
  }

  //对小数点的取舍,默认保留小数点后两位
  static String getDecimalPoint({
    required double tagNumber,
    int fractionDigits = 2,
  }) {
    //返回一个字符串
    return tagNumber.toStringAsFixed(fractionDigits);
  }
}
```

人与人通过手机应用沟通信息，密码学的加密手段保证了核心敏感数据在目前不安全的网络传输过程中的安全，添加的依赖如下：

```
dependencies:
  encrypt: ^5.0.1
  crypto: ^3.0.1
  convert: ^3.0.1
```

常用的有 Base64 编码、MD5 加密、AES 加密等，代码如下：

```
import 'package:encrypt/encrypt.dart' as encrypt;
import 'dart:convert';
import 'package:crypto/crypto.dart';
import 'package:convert/convert.dart';

class EncryptUtils {
  //MD5 加密
  static String encodeMd5(String data) {
    var content = const Utf8Encoder().convert(data);
    var digest = md5.convert(content);
    return hex.encode(digest.Bytes);
  }
```

```
  //Base64 编码加密
  static String encodeBase64(String data) {
    var content = utf8.encode(data);
    var digest = base64Encode(content);
    return digest;
  }

  //Base64 编码解密
  static String decodeBase64(String data) {
    List<int> Bytes = base64Decode(data);
    String result = utf8.decode(Bytes);
    return result;
  }
}
```

AES 加密是开发中常用的对称加密手段,核心代码如下:

```
//AES 加密
static String encodeAes(String aesKey, String data) {
  if(aesKey.length!= 32){
    throw "key 必须是 32 位";
  }
  final key = encrypt.Key.fromUtf8(aesKey);
  final iv = encrypt.IV.fromLength(16);           //偏移量
  final encrypter = encrypt.Encrypter(encrypt.AES(
    key,
    mode: encrypt.AESMode.cbc,
  ));
  //加密
  encrypt.Encrypted encrypted = encrypter.encrypt(data, iv: iv);
  return encrypted.base64;
}

//AES 解密
static String decodeAes(String aesKey, String data) {
  final key = encrypt.Key.fromUtf8(aesKey);
  final iv = encrypt.IV.fromLength(16);           //偏移量
  final encrypter = encrypt.Encrypter(encrypt.AES(
    key,
    mode: encrypt.AESMode.cbc,                    //加密模式
  ));
  //解密
  final decrypted = encrypter.decrypt64(data, iv: iv);
  return decrypted;
}
```

单元测试 AES 加密结果如图 11-23 所示。

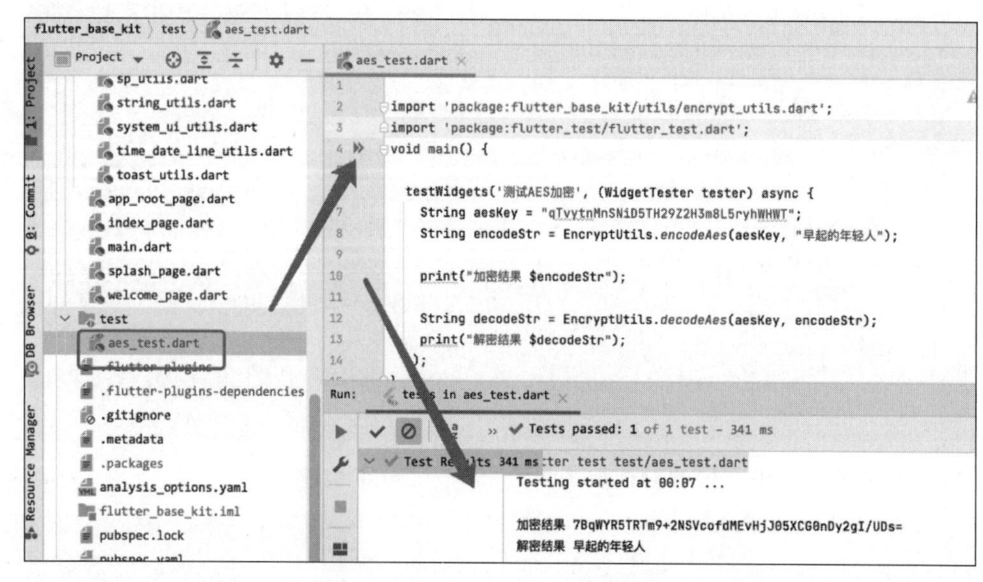

图 11-23　AES 单元测试运行效果图

路由功能封装是程序开发所必备的工具类封装，本脚手架提供了 NavigatorUtils 路由工具类，封装方法如下（全部代码读者可查看本书配套源码）：

```
class NavigatorUtils {
  //关闭当前页面
  static pop(){ … }

  //静态路由
  static pushName(){ … }

  //动态路由方法封装
  static pushPage(){ … }

  //以透明过渡的方式打开新的页面
  static openPageByFade(){ … }

  //以圆形过渡的方式打开新的页面
  static pushPageAboutCircle(){ … }

  //从下向上打开页面
  static openPageFromBottom(){ … }
}
```

核心跳转功能封装如下：

```
static pushPage(
    {required BuildContext context,
    required Widget page,
    String? routeName,
    paramtes,
    bool isReplace = false}) async {
  PageRoute pageRoute;
  //导入 io 包
  if (Platform.isIOS) {
    //iOS 平台使用支持滑动关闭页面的路由控制
    pageRoute = CupertinoPageRoute(
      builder: (_) => page,
      settings: RouteSettings(name: routeName, arguments: paramtes),
    );
  } else {
    //Android 等其他平台使用 Material 风格的路由控制
    pageRoute = MaterialPageRoute(
      builder: (_) => page,
      settings: RouteSettings(name: routeName, arguments: paramtes),
    );
  }
  if (isReplace) {
    //替换当前页面
    return Navigator.of(context).pushReplacement(pageRoute);
  }
  //压栈
  return Navigator.of(context).push(pageRoute);
}
```

颜色处理也是 UI 效果交互中常用的功能，本脚手架提供了 ColorUtils，可以用来生成随机颜色，代码如下：

```
class ColorUtils {
  //获取随机颜色
  //十六进制记法，#FF000000，取值范围为 00～FF
  //RGB 色彩通过对红(R)、绿(G)、蓝(B)
  //三种颜色通道的变化和它们相互之间的叠加来得到各式各样的颜色
  //ARGB 色彩模式与 RGB 相同，只是在 RGB 模式上新增了 Alpha 透明度
  static Color getRandomColor() {
    return Color.fromARGB(
      255,
      Random.secure().nextInt(200),
      Random.secure().nextInt(200),
      Random.secure().nextInt(200),
    );
  }
}
```

不同格式样式的日期时间处理，在 App 开发中是时常用到的功能，有时接口提供给 App 的日期数据是年月日形式的，在 App 中会进行简单比较，或者在不同格式的显示及切换时，需要处理日期时间，本脚手架提供了 DateFormatUtils，用来将日期时间对象 DateTime 格式化为常用日期格式，以及将 String 类型的常用日期格式解析为 DateTime，代码如下：

```
//代码路径 lib/utils/date_format_utils.dart
import 'package:intl/intl.dart';

class DateFormatUtils {
  //获取当前时间
  static String getCurrentTime({format = "HH:mm:ss"}) {
    DateTime dateTime = DateTime.now();
    //格式化时间 import 'package:intl/intl.dart';
    String _formatTime = DateFormat(format).format(dateTime);
    return _formatTime;
  }

  //将 DateTime 格式化为 String
  static String dateTimeToString({
    required DateTime dateTime,
    format = "yyyy - MM - dd HH:mm:ss",
  }) {
    String _formatTime = DateFormat(format).format(dateTime);
    return _formatTime;
  }
  //将 String 类型的日期转换为 DateTime
  static DateTime stringToDate({
    required String strTime,
    format = "yyyy - MM - dd HH:mm:ss",
  }) {
    DateTime _dateTime = DateFormat(format).parse(strTime);
    return _dateTime;
  }
}
```

单元测试用例如下：

```
//test/date_test.dart
void main() {
  testWidgets('测试日期工具类', (WidgetTester tester) async {

    String currentTime = DateFormatUtils.getCurrentTime();
    print("当前时间 $currentTime");
```

```
    //格式化时间
    String currentTimeFr = DateFormatUtils.dateTimeToString(
      dateTime: DateTime.now(),
    );
    print("当前时间 $currentTimeFr");

    //解析时间
    String strTime = "2022-03-03 00:36:25";
    DateTime datetime = DateFormatUtils.stringToDate(strTime: strTime);
    //再将 datetime 转换为其他格式
    String otherTime = DateFormatUtils.dateTimeToString(
      dateTime: datetime,
      format: "yyyy年MM月dd日",
    );
    print("其他时间格式 $otherTime");
  });
}
```

测试结果如图 11-24 所示。

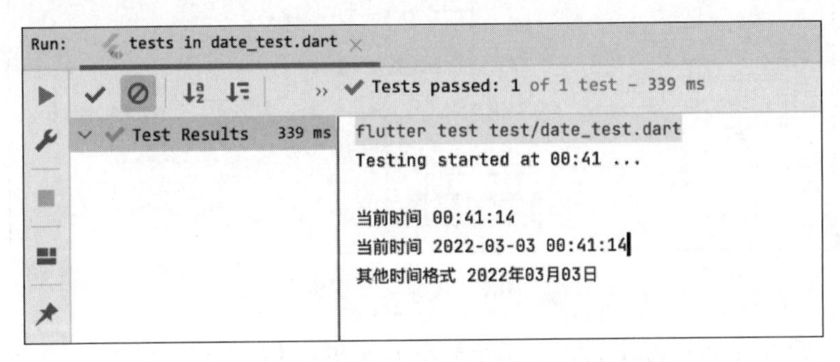

图 11-24　日期格式化工具类单元测试效果图

11.2.2　网络请求与页面交互状态显示

为便捷统一管理请求 URL 网址，所以定义了 http_helper.dart 帮助类，loading_statues.dart 用于对已定义的网络请求状态进行便捷管理，主要用于页面交互，是一个枚举，代码如下：

```
enum LoadingStatues {
  success,        //加载成功有数据
  noData,         //加载成功无数据
  faile,          //加载失败
  none,           //默认无状态
  loading,        //加载中
}
```

网络请求封装代码结构图如图 11-25 所示。

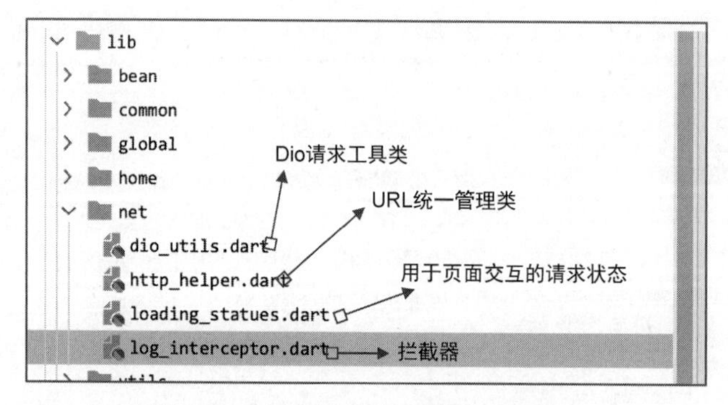

图 11-25 网络请求工具类代码结构图

结合网络请求数据状态实现的页面显示效果如图 11-26 所示。

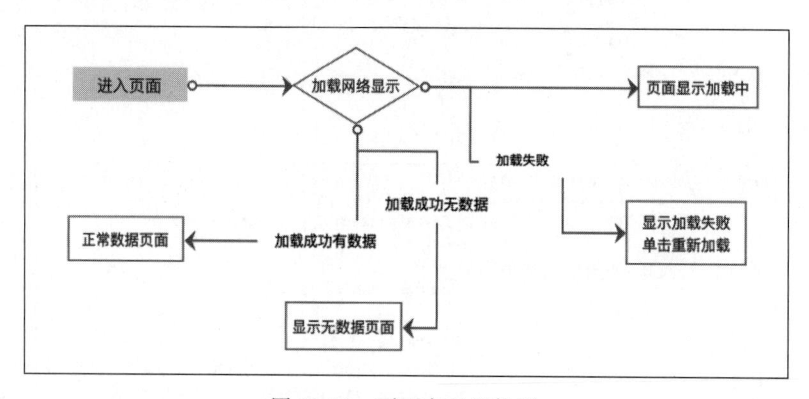

图 11-26 页面交互逻辑图

实现宗旨为在任何情况下，App 页面都会有足够的信息反馈用户，在这里需要定义无数据通用页面组件与加载中页面组件，代码如下：

```
//代码路径 lib/common/loading_widget.dart
class LoadingWidget extends StatelessWidget {
  const LoadingWidget({Key? key}) : super(key: key);
  @override
  Widget build(BuildContext context) {
    return const Center(
      child: CupertinoActivityIndicator(
        radius: 22,
      ),
    );
  }
}
```

更新优美的页面 UI，读者可根据实际 UI 设计师的设计来排版，在本章节以实现业务交互为主，定义无数据显示页面组件的代码如下：

```
//代码路径 lib/common/no_data_widget.dart
class NoDataWidget extends StatelessWidget {
  final GestureTapCallback? onTap;
  final String noDataText;
  const NoDataWidget({
    Key? key,
    this.onTap,
    this.noDataText = "暂无数据",
  }) : super(key: key);

  @override
  Widget build(BuildContext context) {
    if (onTap == null) {
      return Center(
        child: Text(noDataText),
      );
    }
    return Center(
      child: InkWell(
        onTap: onTap,
        onDoubleTap: onTap,
        child: Text(noDataText),
      ),
    );
  }
}
```

11.2.3 网络请求工具类封装概述

本节封装的是 Dio 网络请求工具类，所以需要添加的依赖如下：

```
dependencies:
  #用来加载网络数据
  dio: ^4.0.4
```

log_interceptor.dart 是封装的一个拦截器，可以监控网络的请求及响应等状态，在本节中用到的功能只是拦截显示日志，代码如下：

```
class LogsInterceptors extends InterceptorsWrapper {
  @override
  Future onRequest(
```

```
    RequestOptions options,
    RequestInterceptorHandler handler,
  ) {
    super.onRequest(options, handler);
    print("\n================== 请求数据 ==========================");
    print("|请求 url: ${options.path}");
    print('|请求头: ' + options.headers.toString());
    print('|请求参数: ' + options.queryParameters.toString());
    print('|请求方法: ' + options.method);
    print("|contentType =  ${options.contentType}");
    print('|请求时间: ' + DateTime.now().toString());
    if (options.data != null) {
      print('|请求数据: ' + options.data.toString());
    }

    return Future.value(options);
  }

  @override
  onResponse(
    Response response,
    ResponseInterceptorHandler handler,
  ) {
    super.onResponse(response, handler);
    print("\n| ================== 响应数据 ==========================");
    if (response != null) {
      print("|url =  ${response.realUri.path}");
      print("|code =  ${response.statusCode}");
      print("|data =  ${response.data}");
      print('|返回时间: ' + DateTime.now().toString());
      print("\n");
    } else {
      print("|data = 请求错误 E409");
      print('|返回时间: ' + DateTime.now().toString());
      print("\n");
    }
  }

  @override
  onError(
    DioError e,
    ErrorInterceptorHandler handler,
  ) {
    super.onError(e, handler);
    print("\n================== 错误响应数据 ======================");
    print("|url =  ${e.requestOptions.path}");
```

```
    print("|type = ${e.type}");
    print("|message = ${e.message}");

    print('|response = ${e.response}');
    print("\n");

  }
}
```

dio_utils.dart 文件用来封装 Dio 实际操作的功能，本实例提供的网络请求封装工具类采用单例模式来管理工具类，代码如下：

```
class DioUtils {
  Dio ?_dio;

  //单例模式
  factory DioUtils() => _getInstance();
  static DioUtils get instance => _getInstance();
  static DioUtils? _instance;
  static DioUtils _getInstance() {
    return _instance ??= DioUtils._internal();
  }
  //配置代理标识, false 表示不配置
  bool isProxy = false;
  //网络代理地址
  String proxyIp = "";
  //网络代理端口
  String proxyPort = "";

  DioUtils._internal() {
    BaseOptions options = new BaseOptions();
    //请求时间
    options.connectTimeout = 2000;
    options.receiveTimeout = 2 * 60 * 1000;
    options.sendTimeout = 10 * 1000;
    //初始化
    _dio = Dio(options);
    //当 App 运行在 Release 环境时, inProduction 为 true
    //当 App 运行在 DeBug 和 Profile 环境时, inProduction 为 false
    bool inProduction = bool.fromEnvironment("dart.vm.product");
    if (!inProduction) {
      deBugFunction();
    }
  }
}
```

可配置网络请求代理，用于在开发测试阶段进行网络请求抓包，协助开发测试，代码如下：

```
void deBugFunction() {
  //添加 log
  _dio!.interceptors.add(LogsInterceptors());
  //配置代理
  if (isProxy) {
    _setupPROXY();
  }
}

//配置代理
void _setupPROXY() {
  (_dio!.httpClientAdapter as DefaultHttpClientAdapter).onHttpClientCreate =
      (HttpClient client) {
    client.findProxy = (uri) {
      //proxyIp 为地址, proxyPort 为端口
      return 'PROXY $proxyIp : $proxyPort';
    };
    client.badCertificateCallback =
        (X509Certificate cert, String host, int port) {
      //忽略证书
      return true;
    };
  };
}
```

Get 请求方法封装，代码如下：

```
Future<ResponseInfo> getRequest(
    {required String url,
    Map<String, dynamic>? queryParameters,
    CancelToken? cancelTag}) async {
  //发起 get 请求
  try {
    Response response = await _dio!.get(url,
        queryParameters: queryParameters, cancelToken: cancelTag);
    //响应数据
    dynamic responseData = response.data;
    //数据解析
    if (responseData is Map<String, dynamic>) {
      //转换
      Map<String, dynamic> responseMap = responseData;
      //
```

```
        int code = responseMap["code"];
        if (code == 200) {
          //业务代码处理正常
          //获取数据
          dynamic data = responseMap["data"];
          return ResponseInfo(data: data);
        } else {
          //业务代码异常
          return ResponseInfo.error(code: responseMap["code"]);
        }
      } else {
        return ResponseInfo.error();
      }
    } catch (e, s) {
      //异常
      return errorController(e, s);
    }
  }
```

数据解析将在11.3.3节讲解，App启动获取基本配置信息时会使用。Post请求方法封装，代码如下：

```
  Future<ResponseInfo> postRequest(
      {required String url,
      Map<String, dynamic>? formDataMap,
      Map<String, dynamic>? jsonMap,
      CancelToken? cancelTag}) async {
    FormData? form;
    if (formDataMap != null) {
      form = FormData.fromMap(formDataMap);
    }

    //发起 post 请求
    try {
      Response response =
          await _dio!.post(url, data: form ?? jsonMap, cancelToken: cancelTag);
      //响应数据
      dynamic responseData = response.data;
      if (responseData is Map<String, dynamic>) {
        Map<String, dynamic> responseMap = responseData;
        int code = responseMap["code"];
        if (code == 200) {
          //业务代码处理正常
          //获取数据
          dynamic data = responseMap["data"];
```

```
        return ResponseInfo(data: data);
      } else {
        //业务代码异常
        return ResponseInfo.error(
            code: responseMap["code"], message: responseMap["message"]);
      }
    }
    return ResponseInfo.error();
  } catch (e, s) {
    return errorController(e, s);
  }
}
```

错误异常处理 errorController 封装，代码如下：

```
Future<ResponseInfo> errorController(e, StackTrace s) {
  ResponseInfo responseInfo = ResponseInfo();
  responseInfo.success = false;
  //网络处理错误
  if (e is DioError) {
    DioError dioError = e;
    switch (dioError.type) {
      case DioErrorType.connectTimeout:
        responseInfo.message = "连接超时";
        break;
      case DioErrorType.sendTimeout:
        responseInfo.message = "请求超时";
        break;
      case DioErrorType.receiveTimeout:
        responseInfo.message = "响应超时";
        break;
      case DioErrorType.response:
        //响应错误
        responseInfo.message = "响应错误";
        break;
      case DioErrorType.cancel:
        //取消操作
        responseInfo.message = "已取消";
        break;
      case DioErrorType.other:
        //默认自定义其他异常
        responseInfo.message = dioError.message;
        break;
    }
  } else {
```

```
      //其他错误
      responseInfo.message = "未知错误";
    }
    return Future.value(responseInfo);
  }
```

11.3 App 应用搭建

App 最初展示的效果就是将一张图标展示在手机桌面上,当单击 App 的这个图标时,会启动 App,应用启动的一刹那,手机会先白屏或者黑屏一段时间,然后进入应用程序的主页,但当退出应用后再次打开 App,却会发现白屏时间极短或者压根感觉不出来,前者称为冷启动,后者称为热启动。

冷启动：当启动某个应用程序时,如果手机系统中后台没有该应用程序的进程,则会先创建一个该应用程序的进程,这种方式叫冷启动。

热启动：当启动某个应用程序时,如果系统后台已经有一个该应用程序的进程,则不会再创建一个新的进程,这种方式叫热启动,通常按返回键退出应用,按 Home 键回到桌面,该应用进程还一直存活,再次启动应用都叫热启动。

通常情况下,白屏现象会在冷启动情况下出现,如果白屏时间过长,就会给人一种 App 很卡顿的感觉,这对于应用的用户体验是极为不好的。

11.3.1 Android 与 iOS 双平台的闪屏页面

2min

在 Flutter 项目中的 android 目录(目录结构如图 11-27 所示)下配置 Android 应用启动时显示图片功能,读者可观看视频讲解【11.3.1 Android 与 iOS 双平台的闪屏页面】。

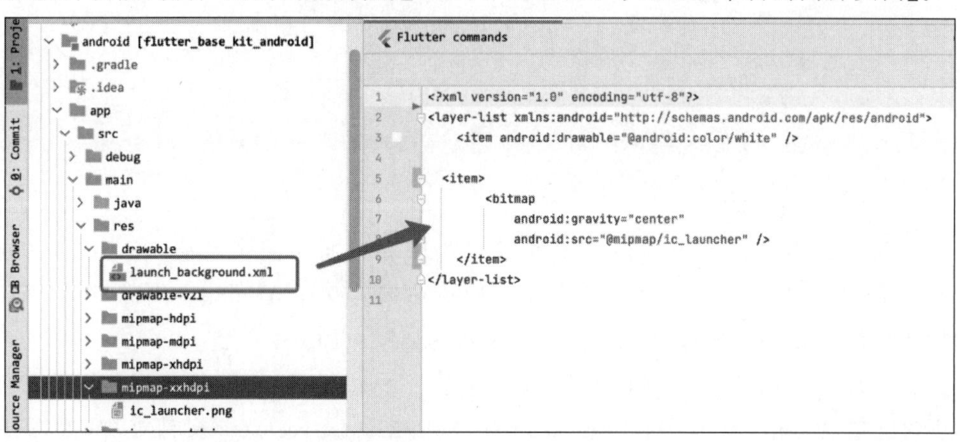

图 11-27 Anddroid 闪屏页面配置

第一步，将启动图片放到图片文件夹 mipmap 中。

第二步，将这张图片放到一个 xml 文件中，如图 11-26 所示的 launch_background. xml 文件，这个文件在 Flutter 项目中会默认创建好，可以只进行必要的修改。

第三步，将创建的 launch_background. xml 文件在样式中引用，在 res 的 styley 目录下的 styles. xml 文件中引用，配置如下：

```
<?xml version = "1.0" encoding = "utf - 8"?>
< resources >
<!-- 启动页面背景样式 -->
< style name = "LaunchTheme" parent = "@android:style/Theme.Black.NoTitleBar">
< item name = "android:windowBackground">@drawable/launch_background </item >
</style >
<!-- 其他情况 Activity 的背景样式 -->
< style name = "NormalTheme" parent = "@android:style/Theme.Black.NoTitleBar">
< item name = "android:windowBackground">@android:color/white </item >
</style >

</resources >
```

(4) 在清单文件 AndroidManifest. xml 中给启动 Activity 配置使用这个样式，Flutter 工程项目默认已配置好，代码如下：

```
< application
      …
< activity
            android:name = ".MainActivity"
            #这里配置的样式
            android:theme = "@style/LaunchTheme"
            android:WindowsoftInputMode = "adjustResize">
            …
</application >
```

对于 iOS 的启动白屏瞬间也需要通过配置启动页面功能来解决，在 Xcode 中打开 Flutter 工程目录，如图 11-28 所示，在 Launch Screen File 中配置了启动页面，在 Flutter 项目中默认配置的 LaunchScreen. storyboard 文件为启动页面。

在 LaunchScreen. storyboard 文件中创建一个 Image View，并设置居中的约束，然后将显示图片配置为 app_icon，图片 app_icon 存放在 Assets. xcassets 文件夹中，如图 11-29 所示。

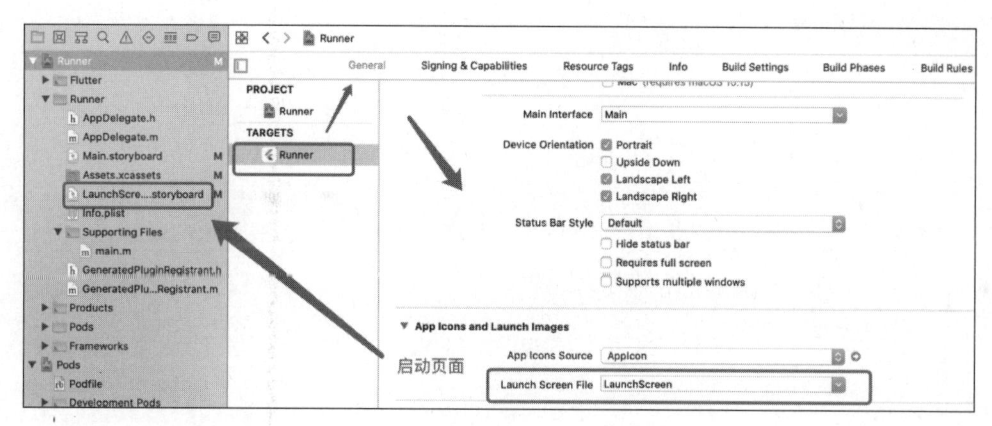

图 11-28　Xcode 中的 Flutter 工程项目图

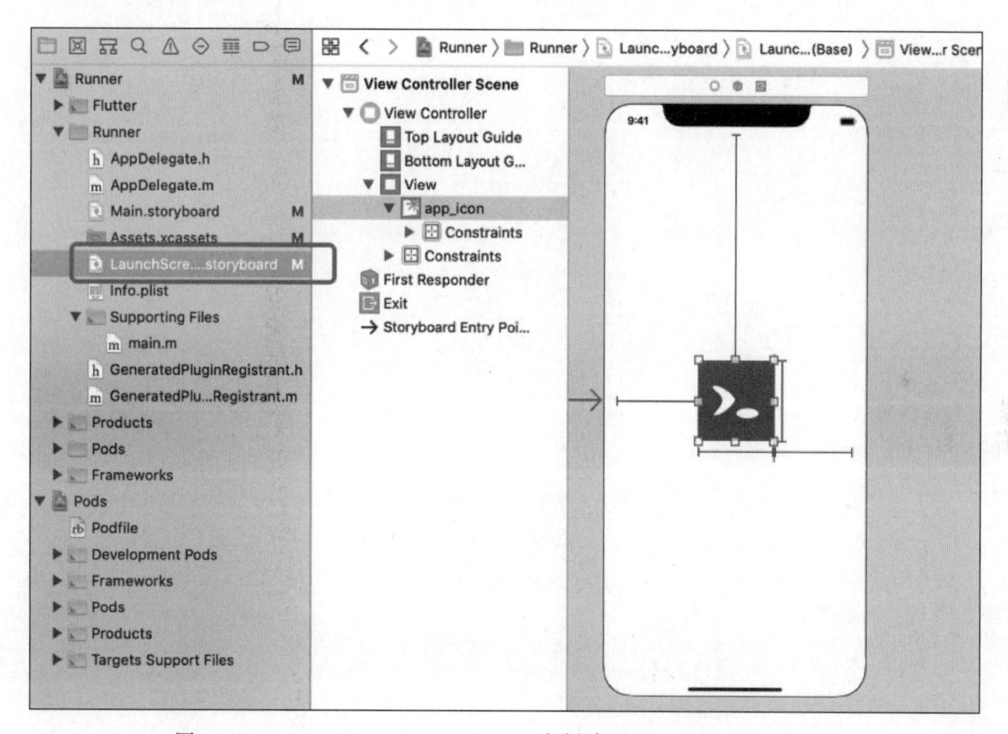

图 11-29　LaunchScreen. storyboard 中创建的图片设置约束说明图

11.3.2　应用根视图基本配置

首先在 Flutter 项目工程中默认的 main. dart 文件中配置启动的根视图及 Flutter 项目运行 App 报错信息捕捉 UI 显示功能，读者可观看视频讲解【11. 3. 2 应用根视图基本配置】，代码如下：

```
//代码清单 11-1
//程序的入口 lib/main.dart
void main() {
  //启动根目录
  runApp(const AppRootPage());
  //自定义报错页面
  ErrorWidget.builder = (FlutterErrorDetails flutterErrorDetails) {
    //deBug 模式下输出日志
    deBugPrint(flutterErrorDetails.toString());
    return Scaffold(
      body: SizedBox(
        width: double.infinity,
        child: Column(
          mainAxisAlignment: MainAxisAlignment.center,
          children: [
            Text(
              "App 错误,快去反馈给作者! ${flutterErrorDetails.exception}",
              maxLines: 4,
            ),
          ],
        ),
      ),
    );
  };
}
```

启动函数 runApp,打开一个根视图 AppRootPage,配置最初始的数据,此处并没有具体的页面 UI 构建,代码如下:

```
//代码清单 11-2
//根视图配置 lib/app_root_page.dart
class _AppRootPageState extends State<AppRootPage> {
  //默认过滤的颜色
  Color _defaultFilterColor = Colors.transparent;
  @override
  Widget build(BuildContext context) {
    return StreamBuilder(
      stream: rootStreamController.stream,
      initialData: GlobalBean(
        data: Colors.transparent,
        messageType: GlobalMessageType.mainThemColor,
      ),
      builder: (BuildContext context, AsyncSnapshot<GlobalBean> snapshot) {
        //获取数据
        if (snapshot.data != null &&
```

```
            snapshot.data!.messageType == GlobalMessageType.mainThemColor) {
          _defaultFilterColor = snapshot.data!.data;
        }
        return ColorFiltered(
          colorFilter: ColorFilter.mode(
              //动态生成过滤颜色
              _defaultFilterColor,
              //过滤模式
              BlendMode.color),
          //构建 MaterialApp 根组件
          child: buildMaterialApp(),
        );
      },
    );
  }
  @override
  void dispose() {
    //注销全局流订阅
    rootStreamController.close();
    super.dispose();
  }
}
```

ColorFiltered 可实现 App 整体颜色过滤，如整体显示为黑灰色，在根视图中使用 StreamBuilder 与 StreamController 实现全局数据刷新控制，StreamController 创建的是一个多订阅流，可实现多次监听，存放在 global.dart 文件中，代码如下：

```
//全局通信配置页面 lib/global/global_config.dart
class GlobalBean {
  GlobalMessageType messageType;
  dynamic data;
  GlobalBean({required this.messageType, this.data});
}

//全局数据更新流控制器
//多订阅流
StreamController<GlobalBean> rootStreamController =
    StreamController.broadcast();
//全局路由导航 Key
GlobalKey<NavigatorState> globalNavigatorKey = GlobalKey();
//注册 RouteObserver,作为 navigation observer
final RouteObserver<PageRoute> routeObserver = RouteObserver<PageRoute>();
//全局通信类型目前只定义了一种颜色过滤
enum GlobalMessageType { mainThemColor }
```

MaterialApp 一般用来构建应用的根视图，在其中配置主题、路由信息、国际化语言支持及路由观察者等基本信息，代码如下：

```
//代码清单 11-3
//MaterialApp 组件 lib/app_root_page.dart
MaterialApp buildMaterialApp() {
  return MaterialApp(
    //应用的主题
    theme: ThemeData(
      //主背景色
      primaryColor: Colors.blue,
      //Scaffold 脚手架的背景色
      scaffoldBackgroundColor: Color(0xffeeeeee),
    ),
    //应用程序默认显示的页面
    home: IndexPage(),
    //duBug 模式下不显示 deBug 标签
    deBugShowCheckedModeBanner: false,
    //国际化语言环境
    localizationsDelegates: const [
      //初始化默认的 Material 组件本地化
      GlobalMaterialLocalizations.delegate,
      //初始化默认的通用 Widget 组件本地化
      GlobalWidgetsLocalizations.delegate,
      //支持使用 CupertinoAlertDialog 的代理
      GlobalCupertinoLocalizations.delegate,
    ],
    navigatorKey: globalNavigatorKey,
    //路由导航观察者配置
    navigatorObservers: [routeObserver],
    //配置程序语言环境
    locale: const Locale('zh', 'CN'),
  );
}
```

11.3.3 启动页面动态权限申请

启动页面一般用来执行初始化程序配置操作，如图 11-30 所示，如请求一个配置接口，从而获取最新的协议，展示给用户，目前隐私格外重要，所以在这个页面需要提示隐私协议与用户协议及必要的基础权限申请。读者可观看视频讲解【11.3.3 启动页面初始化配置】。

启动页面 UI 方面，显示一张背景图与一个进度小圆圈，本节实现的是当网络请求加载失败时，显示加载失败原因，然后单击按钮重新发起加载请求。

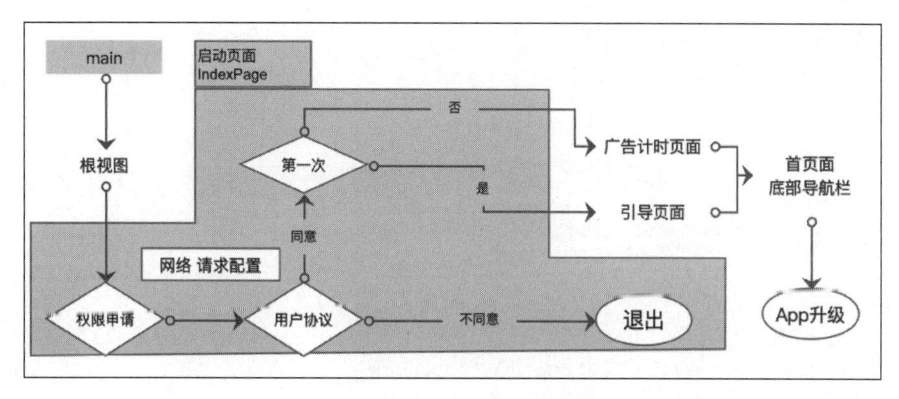

图 11-30 启动页面内容图

可以在这里进行缓存优化，如果加载失败，则查询是否有缓存的基础配置信息并且在可用时效内，如果有就使用缓存配置，如果没有缓存，则显示加载失败，代码如下：

```
//代码清单 11 - 4 启动页面 UI 构建 lib/index_page.dart
class _IndexPageState extends State<IndexPage> {

  @override
  Widget build(BuildContext context) {
    return Scaffold(
      //层叠布局
      body: SizedBox(
        width: double.infinity,
        height: double.infinity,
        child: Stack(
          alignment: Alignment.center,
          children: [
            //构建背景
            Positioned(
              left: 0,
              right: 0,
              top: 0,
              bottom: 0,
              child: Image.asset(
                "assets/images/welcome_bg.jpeg",
                fit: BoxFit.fill,
              ),
            ),
            Positioned.fill(
              child: Container(
                color: const Color(0x50ffffff),
              ),
            ),
```

```
            buildLoading(),
          ],
        ),
      ),
    );
  }
  //网络请求加载
  LoadingStatues _loadingStatues = LoadingStatues.loading;
  String _errMsg = "";
  buildLoading() {
    if (_loadingStatues == LoadingStatues.loading) {
      //显示一个加载小圆圈
      return const CupertinoActivityIndicator();
    }
    //加载失败时显示原因,单击重新发起加载
    if (_loadingStatues == LoadingStatues.faile) {
      return TextButton(
        onPressed: () {
          _loadingStatues = LoadingStatues.loading;
          setState(() {});
          initData();
        },
        child: Text(_errMsg),
      );
    }
  }
}
```

一般应用程序可能会在最开始需要获取手机的基本信息(如 IMEI 和 Mac 地址),在获取一些敏感隐私数据时,需要动态申请权限(如获取手机信息、传感器、相机、录音、蓝牙、位置等),本节使用 permission_handler 插件实现权限动态请求,需要添加的依赖如下:

```
dependencies:
  #权限请求框架
  permission_handler: ^9.2.0
```

在 Android 9 及更低版本中,用户在向应用授予位置信息访问权限时会做出永久性选择,可以拒绝或允许访问;选择"允许"后,应用将始终具有访问权限(无论是在前台还是后台运行)。Android 10 中的三态位置权限针对是否允许应用访问设备的位置信息为用户提供了三个选项。当应用请求权限时,系统会提示用户授予或拒绝权限级别,三态权限描述如下。

(1) 始终允许:即使用户并未使用应用(应用在后台运行),应用也知道设备的位置信息。此选项等同于 Android 9 及更低版本中的允许权限。

(2) 仅在使用该应用期间允许：(仅限在前台运行的应用)只有当应用正在运行时，设备的位置信息才会向应用显示。

(3) 拒绝：设备的位置信息绝不会向应用显示。此选项与 Android 9 及更低版本中的拒绝权限相同。

本节通过位置权限来讲解这部分内容的应用，对于位置信息，需要在 Flutter 项目的 android 目录下的清单文件中声明权限，代码如下：

```
<!-- 位置后台位置权限 -->
<uses-permission android:name="android.permission.ACCESS_BACKGROUND_LOCATION" />
<!-- 位置设备估算位置权限 -->
<uses-permission android:name="android.permission.ACCESS_COARSE_LOCATION" />
<!-- 位置精确位置权限 -->
<uses-permission android:name="android.permission.ACCESS_FINE_LOCATION" />
```

当用户拒绝后，需要弹框提示用户使用权限的目的，当用户拒绝权限并选择了不再提示时，只能去设置中心再开启，此时需要提供跳转设置中心的功能，无轮权限申请是否通过，应用程序都应可以正常使用部分功能，代码如下：

```
//代码清单 11-4-1 动态权限申请 lib/index_page.dart
@override
void initState() {
  super.initState();
  Future.delayed(Duration.zero, () {
    //检查权限
    checkPermissonFunction();
  });
}

void checkPermissonFunction() async {
  //权限请求封装功能
  dynamic result = await showPermissionRequestPage(
    context: context,
    permissionList: [
      Permission.location,
      Permission.storage,
    ],
    noPassMessageList: ["获取位置权限", "获取文件存储权限"],
  );
  //初始化数据
  initData();
}
```

动态权限申请弹框是通过自定义路由的方式打开一个普通的 StatefulWidget 页面，代码如下：

```
//代码清单 11-5 动态权限申请弹框
//lib/common/permission_request_page.dart
showPermissionRequestPage({
  required BuildContext context,
  required List<Permission> permissionList,
  required List<String> noPassMessageList,
}) async {
  //以透明的方式打开权限请求 Widget
  dynamic result = await NavigatorUtils.openPageByFade(
    context,
    PermissionRequestPage(
      permissionList: permissionList,
      noPassMessageList: noPassMessageList,
    ),
    opaque: false,
  );
  return result;
}
```

当用户跳转到设置中心给应用授权后返回当前 App 中时,可通过 WidgetsBindingObserver 观察者来监听应用的前后台切换,代码如下:

```
//代码清单 11-5-1 动态权限申请实现类
class _PermissionRequestState extends State<PermissionRequestPage>
    with WidgetsBindingObserver {
  @override
  void initState() {
    super.initState();
    LogUtil.e("权限请求页面");
    WidgetsBinding.instance!.addObserver(this); //添加观察者
    //检查权限
    checkPermissonFunction();
  }

  @override
  void dispose() {
    //销毁观察者
    WidgetsBinding.instance!.removeObserver(this);
    super.dispose();
  }
  //是否打开设置中心
  bool isOpenSetting = false;
  //生命周期变化时回调
  //resumed:应用可见并可响应用户操作
  //inactive:用户可见,但不可响应用户操作
```

```
  //paused:已经暂停了,用户不可见、不可操作
  //suspending:应用被挂起,此状态 iOS 永远不会回调
  @override
  void didChangeAppLifecycleState(AppLifecycleState state) {
    super.didChangeAppLifecycleState(state);
    if (state == AppLifecycleState.resumed && isOpenSetting) {
      isOpenSetting = false;
      checkPermissonFunction();
    }
  }
  @override
  Widget build(…)
}
```

本页面的动态权限申请,通过两个集合分别记录用户拒绝授权与用户拒绝并选择不再提示的权限,在页面中通过 ListView 显示(UI 构建源码可查看本书配套源码),动态权限申请需要执行调用,代码如下:

```
//代码清单 11-5-2 权限申请过程
List<String> noPassList = [];           //未通过
List<String> goSettingList = [];        //未通过,需要跳转到设置中心

void checkPermissonFunction() async {
  noPassList.clear();
  goSettingList.clear();
  //动态权限申请发起
  Map<Permission, PermissionStatus> statuses =
      await widget.permissionList.request();
  //返回结果集 statuses
  Iterable<Permission> keys = statuses.keys;
  //遍历循环检查,查看权限是否申请通过
  for (Permission keyName in keys) {
    //获取权限申请结果
    PermissionStatus? keyValue = statuses[keyName];
    //判断未通过的权限
    if (keyValue != null && !keyValue.isGranted) {
      //获取权限在原集合中的索引
      int index = widget.permissionList.indexOf(keyName);
      //根据索引获取权限描述
      String message = widget.noPassMessageList[index];
      //判断权限是否是用户的选择(拒绝且不再提示)
      if (keyValue.isPermanentlyDenied||keyValue.isDenied) {
        goSettingList.add(message);
      } else {
        noPassList.add(message);
```

```
            }
          }
        }
        if (noPassList.isEmpty && goSettingList.isEmpty) {
          Navigator.of(context).pop(true);
        } else {
          setState(() {});
        }
      }
```

权限申请，用户可能会拒绝或者同意，申请结果状态如表 11-2 所示。

表 11-2　权限申请结果状态表

类　别	描　述
isDenied	用户选择拒绝使用权限
isPermanentlyDenied	仅在 Android 平台下有此状态，用户选择拒绝使用权限并选择不再提示，此情景下需要用户去手机设置中心的应用权限中手动开启权限
isRestricted	仅在 iOS 平台下有此状态，如手机设备禁用位置权限，或者用户手动关闭应用位置授权
isLimited	仅在应用程序使用中使用权限
isGranted	权限申请通过

权限申请结束后，接下来进行一些程序初始化配置，如工具类初始化、缓存信息加载、配置信息请求等，代码如下：

```
//代码清单 11-6-1 初始化功能
//用户是否第一次使用
bool _userFirst = false;
void initData() async {
    //获取当前的运行环境
    //当 App 运行在 Release 环境时，inProduction 为 true
    //当 App 运行在 DeBug 和 Profile 环境时，inProduction 为 false
    const bool inProduction = bool.fromEnvironment("dart.vm.product");
    //为 ture 时输出日志
    const bool isLog = !inProduction;
    //初始化统计
    //初始化本地存储工具
    await SPUtil.init();
    //初始化日志工具
LogUtil.init(tag: "flutter_log", isDeBug: isLog);
//网络请求加载配置
bool isSuccess = await requestNet();
    if(!isSuccess){
```

```
        return;
      }
      //获取用户是否为第一次登录
      _userFirst = SPUtil.getBool(spUserIsFirstKey);
      //获取用户隐私协议的状态
      bool _userProtocol = SPUtil.getBool(spUserProtocolKey);
      //记录
      UserHelper.getInstance.userProtocol = _userProtocol;
      //初始化用户的登录信息
      UserHelper.getInstance.init();
      //下一步
      openUserProtocol();
    }
```

网络请求加载使用的是11.2.3节封装的DioUtils,如果加载失败,则更新页面并显示加载失败,代码如下:

```
//代码清单 11-6-2 网络请求加载配置
Future<bool> requestNet() async{
  ResponseInfo responseInfo = await requestSetting();
  if (!responseInfo.success) {
    _errMsg = responseInfo.message;
    _loadingStatues = LoadingStatues.faile;
    setState(() {});
    return false;
  }
  return true;
}
Future<ResponseInfo> requestSetting() async {
  //网络请求获取 App 的配置信息
  ResponseInfo responseInfo =
  await DioUtils.instance.getRequest(url: HttpHelper.SETTING_URL);
  if (responseInfo.success) {
    //解析数据
    AppSettingBean settingBean = AppSettingBean.fromMap(responseInfo.data);
    //配置 App 主题
    if (settingBean.appThemFlag == 1) {
      //将 App 设置成灰色主题
      rootStreamController.add(GlobalBean(
        messageType: GlobalMessageType.mainThemColor,
        data: Colors.grey,
      ));
    }
  }
  return responseInfo;
}
```

ResponseInfo 为数据的第一层解析，一般接口会有统一的规范，第一层解析一般是固定的，在 11.2.3 节中网络请求封装处有解析数据。本实例中 App 配置网络请求的 JSON 格式数据的映射关系如图 11-31 所示。

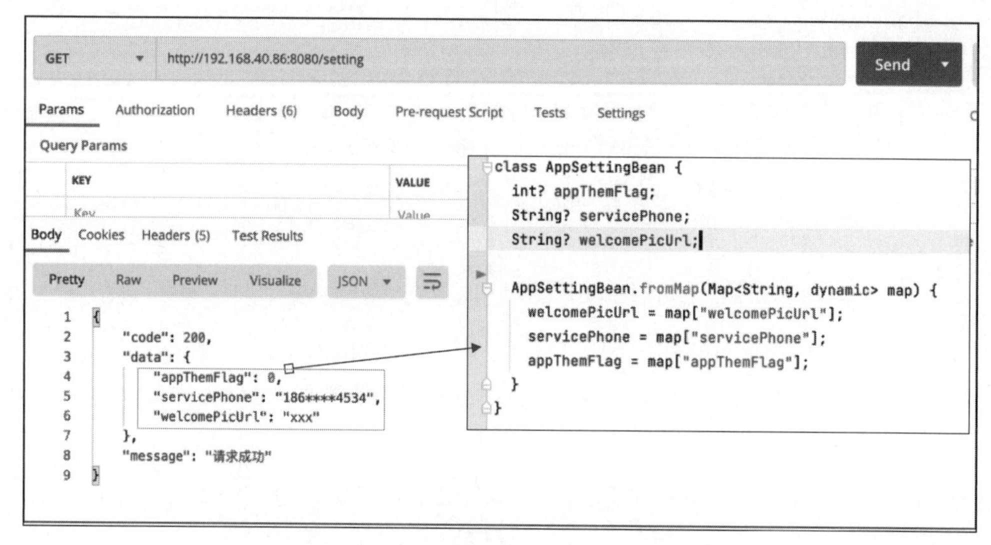

图 11-31　App 配置数据 JSON 映射实体

一般在基础配置信息中包含用户协议相关信息，接下来判断用户是否查看并同意协议，代码如下：

```
//代码清单 11-6-3 用户协议与页面跳转说明
  void openUserProtocol() async {
    //已同意用户隐私协议,下一步
    if (UserHelper.getInstance.isUserProtocol) {
      LogUtil.e("用户已同意用户隐私协议,下一步");
      openNext();
    } else {
      LogUtil.e("未同意用户协议,弹框显示");
      //未同意用户协议,弹框显示
      await showUserProtocolPage(context: context);
      openNext();
    }
  }
  void openNext() {
    //获取配置信息
    if (_userFirst == false) {
      LogUtil.e("第一次进入引导页面");
      //第一次隐藏 logo 显示左右滑动的引导
      NavigatorUtils.openPageByFade(context, SplashPage(), isReplace: true);
    } else {
```

```
        LogUtil.e("非首次进入首页面");
        //非第一次隐藏 logo 显示欢迎
        NavigatorUtils.openPageByFade(context, WelcomePage(), isReplace: true);
      }
    }
```

11.3.4 加载PDF文件显示

查看用户协议弹框定义方式与显示用户隐私协议弹框方式类似，也是通过自定义路由以透明渐变过渡动画方式打开一个普通页面，其中核心代码使用了本书2.3.3节中RichText显示的组合样式文本，详细UI构建读者可查看本书配套源码，协议一般是以链接或者PDF的方式加载，加载PDF文件，在Flutter中可使用插件flutter_pdf_plugin，加载网页链接可使用webView插件，读者可观看视频讲解【11.3.4 加载PDF文件显示】，添加的依赖如下：

2min

```
dependencies:
flutter_pdf_plugin: ^1.0.0
  #搜索查看最新版本
  #https://pub.flutter-io.cn/packages/flutter_fai_webview
  flutter_fai_webview: ^1.1.6
```

在这里使用的是PdfNetScaffold组件，此组件是插件中封装好的加载组件，调用下载功能，将PDF文件下载到手机磁盘中，再预览，代码如下：

```
//代码清单 11-7 打开用户协议
//lib/common/user_protocol_page.dart

import 'package:flutter_pdf_plugin/flutter_pdf_viewer.dart';

void openPrivateProtocol() {
  NavigatorUtils.pushPage(
    context: context,
    page: PdfNetScaffold(
        pdfUrl: "http://www.jinbangshichuang.com/20201209105650.pdf"),
  );
}
```

可以通过PdfScaffold组件打开手机SD卡中的PDF文件，代码如下：

```
//代码清单 11-8 打开 PDF 文件本地路径

import 'package:flutter_pdf_plugin/pdf_viewer_scaffold.dart';
```

```
class PdfLocalPage extends StatelessWidget {
  String localPath = "";
  PdfLocalPage(this.localPath);

  @override
  Widget build(BuildContext context) {
    return PdfScaffold(
      appBar: AppBar(title: Text("加载本地路径")),
      path: localPath,
    );
  }
}
```

11.3.5 滑动引导页面与倒计时页面

1min

App 的开屏页面，一般可以用来展示 3～5s 的广告或者重要消息，核心代码采用倒计时的方式实现，读者可观看视频讲解【11.3.5 滑动引导页面与倒计时页面】，代码如下：

```
//代码清单 11-9 倒计时页面
class _WelcomePageState extends State<WelcomePage> {
  //时间计时器
  Timer? _timer;
  //初始的时间
  double progress = 1000;
  //倒计时时间
  double totalProgress = 6000;

  @override
  void initState() {
    super.initState();
    //初始化时间计时器，每 100ms 执行一次
    _timer = Timer.periodic(const Duration(milliseconds: 100), (timer) {
      //进度每次累加 100，
      progress += 100;
      //计时完成后进入首页面
      if (progress >= totalProgress) {
        //完成计时后取消计时进入首页面
        _timer!.cancel();
        goHome();
      }

      setState(() {});
    });
```

```
  }
  //生命周期函数页面销毁时执行一次
  @override
  void dispose() {
    //取消定时
    if(_timer!= null&&_timer!.isActive) {
      _timer!.cancel();
    }
    super.dispose();
  }

  @override
  Widget build(BuildContext context) {…}
}
```

倒计时显示的是一个进度小圆圈,小圆圈中间显示的是剩余时间,单击小圆圈可以跳过当前计时页面,整体使用 Stack 层叠布局来构建,核心代码如下:

```
//代码清单 11-9-1 倒计时进度构建
@override
Widget build(BuildContext context) {
  return Scaffold(
    body: SizedBox(
      width: double.infinity,
      height: double.infinity,
      //层叠布局
      child: Stack(
        children: [
          //背景图片、其他适用的广告信息或重要提醒

          //倒计时使用的进度圆圈
          buildTimerProgress(),
        ],
      ),
    ),
  );
}

//可单击的倒计时进度圆圈在中间显示
Positioned buildTimerProgress() {
  return Positioned(
    //右上角对齐
    right: 20, top: 60,
    //单击事件回调
    child: InkWell(
```

```
        onTap: () {
          goHome();
        },
        child: Stack(
        alignment: Alignment.center,
        children: [
          CircularProgressIndicator(
            value: progress / totalProgress,
          ),
          Text(
            "${progress ~/ 1000}",
            style: TextStyle(color: Colors.white),
          )
        ],
      ),
    ),
  );
}
```

在用户第一次安装应用程序或者 App 有大版本修改时，为快速介绍新功能，常用的一个做法是在开屏页面设计一个左右滑动的引导页面来加载图片显示，核心代码如下：

```
//代码清单 11-10 底部可滑动的图片
//lib/splash_page.dart
Widget buildPageView() {
  //PageView 用于整屏切换效果
  //默认情况下是左右切换
  return PageView.builder(
    //滑动视频滑动结束时回调
    //参数 value, PageView 当前显示的页面索引
    onPageChanged: (value) {
      LogUtil.e("pageView on changed $value");
      //修改右上角的页面角标
      buildTopText(value);
    },
    //构建条目的总个数,如这里的 4
    itemCount: 4,
    //每一页的显示 Widget
    itemBuilder: (BuildContext context, int postion) {
      return Center(
        child: CachedNetworkImage(
          imageUrl: '',
          fit: BoxFit.fitHeight,
        ),
      );
```

```
    },
  );
}
```

CachedNetworkImage 是一个可以缓存图片的插件，通过缓存策略，可以有效地降低 App 流量的消耗，需要添加的依赖如下：

```
dependencies:
  #加载图片的缓存组件
  cached_network_image: ^3.2.0
```

11.3.6 应用首页

应用的首页面由底部导航栏与内容主区域构成，内容主区域可使用 PageView 来加载，读者可观看视频讲解【11.3.6 应用首页】，本节实现应用首页，代码如下：

```
//代码清单 11 - 11 应用的首页面
class _HomeMainState extends State < HomeMainPage > {
  //当前显示页面的标签
  int _tabIndex = 0;
  //[PageView]使用的控制器
  final PageController _pageController = PageController();
  @override
  Widget build(BuildContext context) {
    //Scaffold 用来搭建页面的主体结构
    return Scaffold(
      //页面的主内容区
      //可以是单独的 StatefulWidget,也可以是当前页面构建的组件,如 Text 文本组件
      body: PageView(
        //将 PageView 设置为不可滑动切换
        physics: NeverScrollableScrollPhysics(),
        //PageView 的控制器
        controller: _pageController,
        //PageView 中的 3 个子页面
        children: < Widget >[
          //第 1 个页面
          HomeItemMainPage(),
          //第 2 个页面
          HomeItemMainPage(),
          HomeItemMainPage(),
          //个人中心页面
          HomeItemMainPage(),
        ],
```

```
      ),
      //底部导航栏
      bottomNavigationBar: buildTipsBottomAppBar(),
    );
  }
}
```

一般在首页中一级页面不超过5个为最佳，常用的底部菜单导航栏的实现方式有以下4种：

(1) BottomNavigationBar和BottomNavigationBarItem的组合(2.2.4节)。

(2) BottomAppBar与TabBar的组合(2.2.4节)。

(3) BottomAppBar与自定义UI布局(本节使用)。

(4) 自定UI布局、内容区域与底部菜单栏使用Column来组合。

本节实现的自定义底部导航栏支持小红点显示，如图11-32所示。

图11-32　App首页底部导航菜单栏

在BottomAppBar中配置shape，结合Scaffold中的floatingActionButton，并且将floatingActionButtonLocation设置为FloatingActionButtonLocation.centerDocked，可实现不规则且中间凹陷的导航栏，在本节中未使用不规则排版，代码如下：

```
//代码清单 11 - 11 - 1
 BottomAppBar buildTipsBottomAppBar() {
   return BottomAppBar(
     child: Row(
       children: <Widget>[
         bottomAppBarItem(0,99),
         bottomAppBarItem(1),
         bottomAppBarItem(2),
         bottomAppBarItem(3),
       ],
       mainAxisAlignment: MainAxisAlignment.spaceAround,
     ),
     shape: CircularNotchedRectangle(),
   );
 }
```

底部菜单栏使用Row结合Expanded实现每个菜单选项按1∶1排版，每个菜单选项需要设置单击事件，实现切换图标与页面内容显示，代码如下：

```
//代码清单 11-11-2

  List<String> titles = ["首页", "分类", "阅读", "我的"];

  Expanded bottomAppBarItem(int index, [int tips = 0]) {
    //设置默认未选中的状态
    TextStyle style = TextStyle(fontSize: 12, color: Colors.grey);
    //这里使用的是图标，读者可使用自己的小图片用 Image 来加载
    IconData iconData = normalImgUrls[index];
    Color iconColor = Colors.grey;
    if (_currentIndex == index) {
      //选中
      style = TextStyle(fontSize: 13, color: Colors.blue);
      iconData = selectedImgUrls[index];
      iconColor = Colors.blue;
    }
    //小红点 UI 构建
    Widget redTips = buildRedTips(tips);
    //构造返回的 Widget
    Widget item = SizedBox(
      height: 64,
      child: GestureDetector(
        behavior: HitTestBehavior.opaque,
        child: Stack(
          alignment: Alignment.center,
          children: [
            Column(
              mainAxisSize: MainAxisSize.min,
              children: <Widget>[
                Icon(iconData, size: 25, color: iconColor),
                Text(
                  titles[index],
                  style: style,
                )
              ],
            ),
            redTips,
          ],
        ),
        onTap: () {
          if (_currentIndex != index) {
            _currentIndex = index;
            _pageController.jumpToPage(_currentIndex);
            setState(() {});
```

```
          }
        },
      ),
    );
    return Expanded(child: item);
  }
```

小红点用于未读消息的提醒，一般浮在菜单选项小图标的右上角，所以可以使用 Align 实现对齐排列，代码如下：

```
//代码清单 11-11-3 构建小红点
  Widget buildRedTips(int tips) {
    if (tips == 0) {
      return const SizedBox();
    }
    String tipsStr = tips.toString();
    if (tips >= 100) {
      tipsStr = "…";
    }
    return Align(
      alignment: const Alignment(0.3, -0.7),
      child: Row(
        mainAxisSize: MainAxisSize.min,
        children: [
          Container(
            alignment: Alignment.center,
            padding:
                const EdgeInsets.only(left: 2, right: 2, top: 1, bottom: 1),
            height: 12,
            decoration: const BoxDecoration(
                color: Colors.red,
                borderRadius: BorderRadius.all(Radius.circular(10))),
            child: Text(
              tipsStr,
              style: const TextStyle(fontSize: 6, color: Colors.white),
            ),
          )
        ],
      ),
    );
  }
```

上述小红点是通过 Stack 层叠布局实现的，在业务开发中，页面中的某一张图标或者消息类的提醒功能有时会使用小红点的功能，小红点可以使用插件 badges 来便捷地实现，添加的依赖如下：

```
dependencies:
  #小红点
  badges: ^2.0.2
```

然后在用到小红点标记的页面中导包，代码如下：

```
import 'package:badges/badges.dart';
```

对上述代码清单 11-11-2 中构建的底部导航栏的 Item 进行修改，修改后的代码如下：

```
Expanded bottomAppBarItem(int index, [int tips = 0]) {
  ...
    //构造返回的 Widget
    Widget item = SizedBox(
      height: 64,
      child: GestureDetector(
        behavior: HitTestBehavior.opaque,
        child: Column(
          mainAxisSize: MainAxisSize.min,
          children: <Widget>[

            Badge(
              //位置微调
              position: BadgePosition.topEnd(end: -1,top: -2),
              badgeContent: Text(
                "$tips",
                style: const TextStyle(color: Colors.white, fontSize: 8),
              ),
              padding: const EdgeInsets.all(2),
              showBadge: tips == 0 ? false : true,    //是否显示小红点
              badgeColor: Colors.red,                  //小红点背景色
              child: Icon(iconData, size: 25, color: iconColor),
            ),

            Text(
              titles[index],
              style: style,
            )
          ],
        ),
        onTap: () {
          controller.onBottomTap(index);
        },
      ),
    );
    return Expanded(child: item);
  }
```

11.4 小结

本章从 Flutter 的基础起步，11.1 节总体讲解打包不同平台 App 的基本配置，打包发布时 Android 需要使用签名文件，打包生成 APK 安装包，可直接安装到手机上使用，也可上传到不同的应用市场。对于 iOS，需要使用苹果开发者账号，然后发布到 App Store；11.2 节也分别讲解了一些常用工具的封装及使用，网络请求使用 Dio 框架，简易封装了网络请求工具类，对于 Dio 下载，将在 12.2.2 节应用版本检查更新中使用；11.3 节讲解了应用从 main 函数入口到启动页面、倒计时页面、引导页面、应用首页底部导航栏功能。11.3 节总体讲解了一个基础快速开发脚手架，读者可使用本章的内容做基础架构，快速构建应用。

第12章 GetX架构视频应用开发

CHAPTER 12

在本书第7章中有概述常见的状态管理框架，可以独立使用其中的一种，也可以单独使用其中的一种来构建应用程序，为本章每个页面构建的架构图如图12-1所示，实现的效果读者可观看视频讲解【12.1 App应用程序根视图配置】。

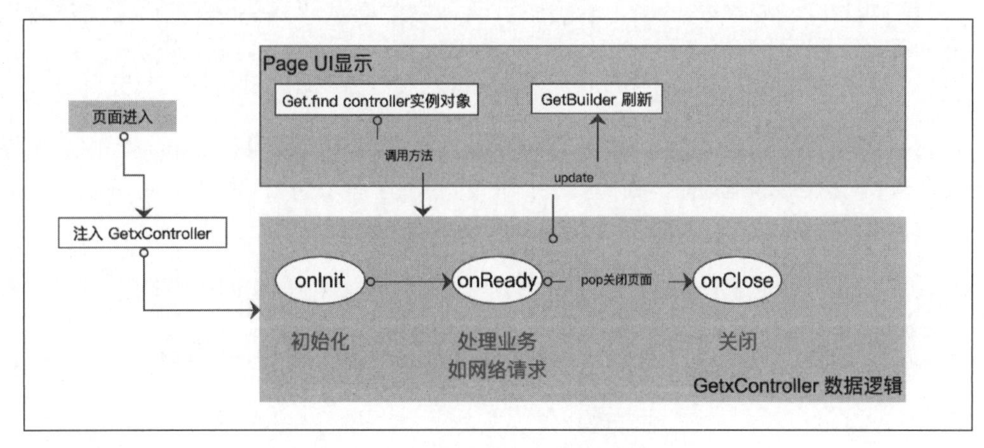

图12-1 GetX App页面架构图

使用GetX，注意需要添加的插件依赖如下：

```
dependencies:
#https://pub.flutter-io.cn/packages?q=get 查看最新版本
  get: ^4.6.1
```

在本章实战案例中，每个页面都是一个独立的小架构，在其中，GetxController用来实现数据处理（如网络请求、缓存加载），页面UI结合GetBuilder实现跨组件通信刷新页面。

GetxController（后续提到的控制器均指其对应的实现类）的update方法用于通知组件刷新，可以通过绑定id来刷新指定的GetBuilder，以便刷新内容，页面加载的时候通过Get.put方法将控制器注入当前页面，会回调控制器的onInit方法，可以在这种方法中初始化一些配置内容，State是不可用状态；在其回调方法onReady中，是State绑定后的回调，通常在这个回调方法中开始处理业务，如网络请求数据；当在页面中执行pop或者按手机物理键退出页面时，控制器会自动销毁，并回调onClose方法，可以在这种方法中处理结束动画、取消订阅等功能。

本章是在第11章封装的脚手架的基础代码上进行GetX架构改造，完整源码在本书配套项目flutter_get_kit中。

12.1 App应用程序根视图配置

在应用程序的入口main函数中执行runApp方法，加载应用的根视图，代码如下：

```
//代码清单 12-1
//程序的入口 lib/main.dart
void main() {
  //启动根目录
  runApp(const AppRootPage());
}
```

AppRootPage是自定义的StatelessWidget，用于构建GetMaterialApp，是一个无界面的根视图，如果是使用静态路由，则需要在此页面中配置静态路由，本章节使用动态路由功能，代码如下：

```
//代码清单 12-2
//lib/app/app_root_page.dart
class AppRootPage extends StatelessWidget{
  const AppRootPage({Key? key}) : super(key: key);
  @override
  Widget build(BuildContext context) {
    return GetMaterialApp(
      //应用的主题
      theme: ThemeData(
        //主背景色
        primaryColor: Colors.blue,
        //Scaffold 脚手架的背景色
        scaffoldBackgroundColor: Color(0xffeeeeee),
      ),
      //应用程序默认显示的页面
      home: const IndexPage(),
      //duBug 模式下不显示 deBug 标签
      deBugShowCheckedModeBanner: false,
      //国际化语言环境
      localizationsDelegates: const [
```

```
        //初始化默认的 Material 组件本地化
        GlobalMaterialLocalizations.delegate,
        //初始化默认的通用 Widget 组件本地化
        GlobalWidgetsLocalizations.delegate,
        //支持使用 CupertinoAlertDialog 的代理
        GlobalCupertinoLocalizations.delegate,
      ],
      //配置程序语言环境
      locale: const Locale('zh', 'CN'),
    );
  }
}
```

12.1.1 启动页面初始化配置

3min

本节实现的功能与 11.3.3 节中实现的功能一致，通过 GetX 架构，将业务处理代码与 UI 构建代码剥离，读者可观看视频讲解【12.1.1 启动页面初始化配置】，页面构建代码如下：

```
//代码清单 12 - 3 启动页面
class IndexPage extends StatelessWidget {
  const IndexPage({Key? key}) : super(key: key);

  @override
  Widget build(BuildContext context) {
    //依赖注入
    Get.put(IndexController(), permanent: true);
    return Scaffold(
      //层叠布局
      body: SizedBox(
        width: double.infinity,
        height: double.infinity,
        child: Stack(
          alignment: Alignment.center,
          children: [
            //构建背景
            Positioned(
                child: Image.asset(
                  "assets/images/welcome_bg.jpeg",
                  fit: BoxFit.fill,
                ),
                left: 0, right: 0,top: 0,bottom: 0),

            Positioned.fill(
              child: Container(
                color: const Color(0x50ffffff),
              ),
            ),
```

```
              GetBuilder(
                builder: (IndexController controller) {
                  return buildLoading(controller);
                },
              ),

            ],
          ),
        ),
      );
    }

    buildLoading(IndexController controller) {
      if (controller.loadingStatues == LoadingStatues.loading) {
        //显示一个加载小圆圈
        return const CupertinoActivityIndicator();
      }
      //加载失败时显示原因,单击重新发起加载
      if (controller.loadingStatues == LoadingStatues.faile) {
        return TextButton(
          onPressed: () {
            controller.loadingStatues = LoadingStatues.loading;
            controller.initData(update: true);
          },
          child: Text(controller.errMsg),
        );
      }
    }
  }
```

IndexController 是启动页面定义的 GetxController，在通过 Get.put 注入时，将参数 permanent 设置为 true，表示该控制器在 App 退出前不会销毁，此参数默认为 false，指控制会跟随页面销毁而销毁。

此控制器方法的执行过程如图 12-2 所示，实现了弹框申请权限、弹框提示用户协议等，代码如下：

```
//代码清单 12 - 4 启动页面对应控制器
class IndexController extends GetxController {

  @override
  void onReady() {
    super.onReady();
    checkPermissonFunction();
  }
//动态权限申请
```

```
  void checkPermissonFunction() async {
    …
    //初始化数据
    initData();
  }
//初始化数据
  void initData({bool update = false}) async {
    …
    bool isSuccess = await requestNet();
    …
    //下一步
    openUserProtocol();
  }
//网络请求配置信息
  Future<bool> requestNet() async { … }


//弹框显示用户协议
  void openUserProtocol() async {
   …
  }

  void openNext() {
    //获取配置信息
    if (_userFirst == false) {
      //第一次显示左右滑动的引导并关闭当前页面
      Get.off(() => SplashPage());
    } else {
      Get.off(() => WelcomePage());
    }
  }
}
```

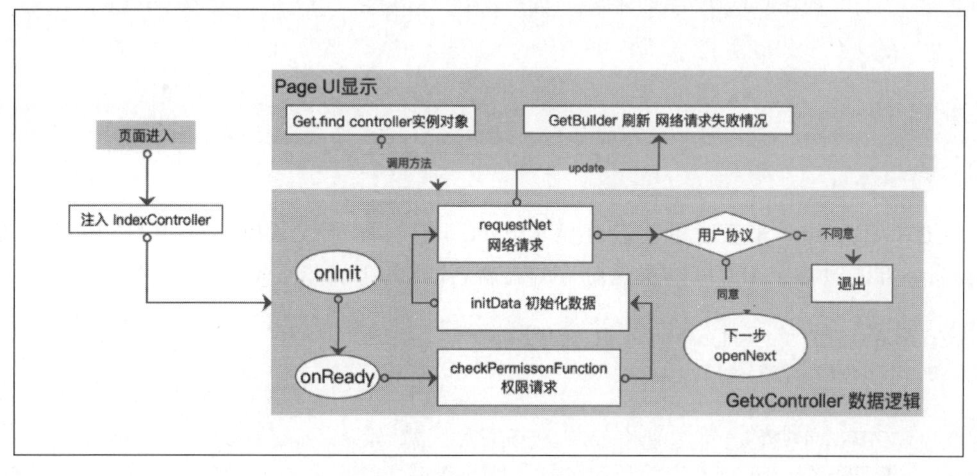

图 12-2 GetX App 页面架构图

12.1.2 倒计时页面架构剥离

3min

WelcomePage 实现的是页面左上角的倒计时，页面 UI 只负责显示，计时功能定义在对应的控制器 WelcomController 中，读者可观看视频讲解【12.1.2 倒计时页面架构剥离】，页面构建的代码如下：

```
//代码清单 12-5 广告倒计时欢迎页面
//lib/src/welcome_page.dart
class WelcomePage extends StatelessWidget {
  @override
  Widget build(BuildContext context) {
    //注入控制器
    Get.put(WelcomController());

    return Scaffold(
      body: SizedBox(
        width: double.infinity,
        height: double.infinity,
        //层叠布局
        child: Stack(
          children: [
            //背景图片
            Positioned.fill(
              child: Image.asset(
                "assets/images/welcome_bg.jpeg",
                fit: BoxFit.fill,
              ),
            ),
            //倒计时使用的进度圆圈
            buildTimerProgress(),
          ],
        ),
      ),
    );
  }}
```

通过 GetBuilder 来实时更新进度显示，GetBuilder 用于设置唯一标识 ID，在调用控制器 update 方法时指定 ID，可以实现对应的更新，代码如下：

```
//代码清单 12-5-1 可单击的倒计时进度圆圈
  Positioned buildTimerProgress() {
    return Positioned(
      //右上角对齐
      right: 20, top: 60,
```

```
        //单击事件回调
        child: InkWell(
          onTap: () {
            Get.find<WelcomController>().goHome();
          },
          child: GetBuilder(
            id: "progress",
            builder: (WelcomController controller) {
              //初始的时间
              double progress = controller.progress;
              //倒计时时间
              double totalProgress = controller.totalProgress;
              return Stack(
                alignment: Alignment.center,
                children: [
                  CircularProgressIndicator(
                    value: progress / totalProgress,
                  ),
                  Text(
                    "${progress ~/ 1000}",
                    style: const TextStyle(color: Colors.white),
                  )
                ],
              );
            },
          ),
        ),
      );
    }
```

WelcomController 实现数据计算功能，代码如下：

```
//代码清单 12-6 lib/controller/welcom_controller.dart
class WelcomController extends GetxController {
  //时间计时器
  Timer? _timer;
  //初始的时间
  double progress = 1000;
  //倒计时时间
  double totalProgress = 6000;
  @override
  void onReady() {
    super.onReady();
    startTimer();
  }
```

```
  startTimer() {
    //初始化时间计时器
    //每 100ms 执行一次
    _timer = Timer.periodic(const Duration(milliseconds: 100), (timer) {
      //进度每次累加 100,
      progress += 100;
      //更新指定的 GetBuilder
      update(["progress"]);
      //计时完成后进入首页面
      if (progress >= totalProgress) {
        //完成计时后取消计时,进入首页面
        _timer!.cancel();
        goHome();
      }
      LogUtil.e("定时器 $progress");
    });
  }

  //跳转首页面
  void goHome() {
    Get.off(() => HomeMainPage());
  }

  @override
  void onClose() {
    //取消定时
    if (_timer != null && _timer!.isActive) {
      _timer!.cancel();
    }
    super.onClose();
  }
}
```

对于引导页面，也是按上述架构方式剥离代码，在这里不再赘述，完整代码读者可查看本书配套源码。

12.2 应用首页面

本节应用首页 HomeMainPage 与 11.3.6 节实现的效果一致，不同的是本节的数据与页面通过 GetX 架构分离处理，默认显示的首页面是一个视频播放列表，本节会讲解如何在 ListView 的子 Item 中分别注入多个 GetxController 实例。

12.2.1 首页面底部菜单导栏

首页页面通过 Scaffold 的 bottomNavigationBar 配置自定义导航栏 BottomAppBar，结合 GetxController 实现刷新，读者可观看视频讲解【12.2.1 首页面底部菜单导栏】，代码如下：

```
//代码清单 12-7 主页面的根布局
//lib/home/home_main_page.dart
class HomeMainPage extends StatelessWidget {
  HomeMainPage({Key? key}) : super(key: key);

  late HomeController controller;

  @override
  Widget build(BuildContext context) {
    //依赖注入
    controller = Get.put(HomeController());

    return Scaffold(
      //页面的主内容区
      body: PageView(
        //将 PageView 设置为不可滑动切换
        physics: NeverScrollableScrollPhysics(),
        //PageView 的控制器
        controller: controller.pageController,
        //PageView 中的三个子页面
        children: <Widget>[
          //第 1 个页面
          HomeVideoPage(1),
          //第 2 个页面
          HomeItemMainPage(),
          HomeItemMainPage(),
          //个人中心页面
          HomeItemMainPage(),
        ],
      ),
      //底部导航栏
      bottomNavigationBar: GetBuilder(
        builder: (HomeController controller) {
          return buildTipsBottomAppBar();
        },
      ),
    );
  }
  …
}
```

HomeController 是首页面使用的控制器，在业务开发中，一般进入首页面会加载一些基础数据，可以在此控制器中处理加载。

12.2.2 应用版本检查更新

一般在应用首页面，可以放置应用程序检查更新功能，以及网络请求配置，有新版本时，Android 平台调用下载功能，将新版本的 APK 下载到手机目录中，然后调用安装程序进行安装；在 iOS 平台则可以跳转 App Store 对应的应用详情，读者可观看视频讲解【12.2.2 应用版本检查更新】，代码如下：

1min

```
//代码清单 12-8 首页面使用控制器
//lib/controller/home_controller.dart
class HomeController extends GetxController {

  @override
  void onReady() {
    super.onReady();
    //加载请求版本信息
    …网络请求
    //如果有新版本
    if (Platform.isAndroid) {
      downAndInstall();
    } else {
      InstallPluginCustom.gotoAppStore("应用在 App Store 的地址");
    }
  }
}
```

InstallPluginCustom 是笔者封装的一个插件，读者可在项目配置文件 pubspec.yaml 中添加依赖，依赖如下：

```
dependencies:
  install_plugin_custom: ^1.0.0
```

在安卓平台，本节使用 Dio 实现下载，读者可自行实现下载页面显示，代码如下：

```
//代码清单 12-8-1 下载 APK
void downAndInstall() async {
  //APK 网络存储链接
  String apkNetUrl = "";
  //手机中 SD 卡上 APK 下载存储路径
  String localPath = "";
```

```
    http.Dio dio = http.Dio();
    //设置连接超时时间
    dio.options.connectTimeout = 1200000;
    //设置数据接收超时时间
    dio.options.receiveTimeout = 1200000;
    http.Response response = await dio.download(apkNetUrl, localPath,
        onReceiveProgress: (int count, int total) {
      //count 当前已下载文件大小
      //total 需要下载文件的总大小
    });
    if (response.statusCode == 200) {
      print('下载请求成功');
      //"安装";
      InstallPluginCustom.installApk(
        localPath,
        '应用包名',
      ).then((result) {
        print('install apk $result');
      }).catchError((error) {
        print('install apk $error');
        //安装失败
      });
    } else {
      //"下载失败重试";
    }
  }
```

12.3 视频列表页面架构构建

在应用首页面的 PageView 中加载的第 1 个页面 HomeVideoPage 用来构建视频列表页面，单击列表页面中的视频可实现视频播放，滑动列表可停止播放视频。

12.3.1 视频列表数据与 UI 构建

视频播放列表使用 ListView 来构建，当视频列表中的 Item 在滑动视图时，需要校验一下视频是否正在播放，此时需要获取 ListView 目前显示的第 1 个 Item 的位置与屏幕上最后一个 Item 的索引，在这里使用自定义代理 CustomScrollDelegate(5.3.1 节中构建)实现，读者可观看视频讲解【12.3.1 视频列表数据与 UI 构建】，代码如下：

1min

```
//代码清单 12-9 视频列表页面
class HomeVideoPage extends StatelessWidget {
  //注入控制器
```

```
  final VideoListController videoListController =
      Get.put(VideoListController());

  HomeVideoPage({Key? key}) : super(key: key);

  @override
  Widget build(BuildContext context) {
    return Scaffold(
      backgroundColor: Colors.grey[200],
      body: ListView.custom(
        //缓存空间
        cacheExtent: 0.0,
        //自定义代理
        childrenDelegate: CustomScrollDelegate(
          (BuildContext context, int index) {
            //构建子 Item 显示布局
            return ListViewItemWidget(widgetIndex: index);
          },
          itemCount: 1000, //子 Item 个数
          //滑动回调
          scrollCallBack: (int firstIndex, int lastIndex) {
            print("firstIndex $firstIndex lastIndex $lastIndex");
            //播放的视频 Item 在滑动视图时需要关闭视频播放
            videoListController.checkBorderLimit(
              firstIndex,
              lastIndex,
            );
          },
        ),
      ),
    );
  }
}
```

本节实现视频播放所使用的插件是 video_player，读者也可以使用 fijkplayer 添加依赖，代码如下：

```
dependencies:
  #视频播放
  video_player: ^2.2.19
```

需要在 android 目录下的清单文件(/android/app/src/main/AndroidManifest.xml)中添加网络访问权限，代码如下：

```
<manifest xmlns:android = "http://schemas.android.com/apk/res/android">
<application …>

</application>

<uses-permission android:name = "android.permission.INTERNET"/>
</manifest>
```

在 ios 目录下(/ios/Runner/Info.plist)的配置文件中添加权限,代码如下:

```
<key>NSAppTransportSecurity</key>
<dict>
<key>NSAllowsArbitraryLoads</key>
<true/>
</dict>
```

视频列表控制器 VideoListController 用来加载数据,读者可在对应的 onReady 方法中实现网络获取数据的功能,在本节中网络请求部分省略,列表 VideoListController 控制器用来做 3 件事情:

(1) 记录列表中正在播放的视频信息,当单击开始播放新的视频时,需要停止上一个视频。

(2) ListView 滑动边界判断。

```
//代码清单 12-10 视频列表页面使用控制
class VideoListController extends GetxController {
  //当前播放视频的控制器
  VideoPlayerController? videoPlayerController;
  //播放视频对应的 GetxController 的 Tag
  String? videoTag;

  //停止上一个播放视频
  Future<bool> stopGlobalVideo([bool destroy = false]) async {
    if (videoPlayerController != null) {
      //校验控制是否已注册
      if (Get.isRegistered<VideoDetailController>(tag: videoTag)) {
        //通过 Tag 获取控制器
        VideoDetailController detailController =
            Get.find<VideoDetailController>(tag: videoTag);
        videoTag = null;
        //停止视频播放
        detailController.pause(destroy);
        //销毁释放
        await videoPlayerController!.dispose();
        return true;
```

```
      }
    }
    return false;
  }

  //视频开始播放时记录当前正在播放的播放控制器
  void updateVideoController(event, tag) {
    //视频组件播放使用的控制器
    videoPlayerController = event;
    //列表中正在播放视频的 Item 对应 tag
    videoTag = tag;
  }

  //列表滑动时实时回调校验
  void checkBorderLimit(int firstIndex, int lastIndex) {
    if (videoTag != null) {
      LogUtil.e("videoTag $ videoTag");
      int index = int.parse(videoTag!);
      if (index >= firstIndex && index <= lastIndex) {
        //停止播放销毁播放控制器不更新页面显示
        stopGlobalVideo(true);
      }
    }
  }
}
```

12.3.2 视频播放控制 UI 页面构建

视频播放列表 ListView 的子 Item 中 UI 页面通过 Stack 层叠两层布局，第一层为视频播放组件，第二层为展示的静态页面（控制层），一般是视频的第一帧，单击静态页面，隐藏控制层，播放视频，读者可观看视频讲解【12.3.2 视频播放功能控制】，代码如下：

2min

```
//代码清单 12-11 视频列表子 Item UI
//lib/home/video/video_details_widget.dart
class VideoDetailsWidget extends StatelessWidget {
  //当前 Item 对应的索引
  final int widgetIndex;

  const VideoDetailsWidget({Key? key, required this.widgetIndex})
      : super(key: key);

  @override
  Widget build(BuildContext context) {
```

```
    //注入依赖
    Get.put(VideoDetailController(tag: "$widgetIndex"), tag: "$widgetIndex");
    return GetBuilder(
      id: "$widgetIndex",
      tag: "$widgetIndex",
      builder: (VideoDetailController controller) {
        return SizedBox(
          width: double.infinity,
          child: Stack(
            children: [
              //第一层的视频
              buildVideoWidget(controller),
              //第二层的控制按钮
              buildControllerWidget(controller)
            ],
          ),
        );
      },
    );
  }
}
```

第二层的控制层构建一个按钮，单击按钮实现视频播放，代码如下：

```
//代码清单 12-11-1
Widget buildControllerWidget(VideoDetailController controller) {
  if (controller.isPlay) {
    //如果正在播放
    return Container();
  }
  return Positioned.fill(
    child: Container(
      //0.3 的蓝色透明度
      color: Colors.blueGrey.withOpacity(0.5),
      //图标
      child: GestureDetector(
        onTap: () {
          //播放视频
          controller.play();
        },
        child: Icon(Icons.play_circle_fill, size: 44),
      ),
    ),
  );
}
```

通过 VideoPlayer 实现视频播放功能，结合 AspectRatio 组件实现视频尺寸及比例约束，代码如下：

```
//代码清单 12-11-2
buildVideoWidget(VideoDetailController controller) {
//获取视频播放控制器
  VideoPlayerController? playerController = controller.playController;
  if (playerController == null) {
    //也可以在此实现控制层
    return Text("单击播放");
  }
  return Positioned.fill(
    child: GestureDetector(
      onTap: () {
        //视频正在播放时单击暂停
        Get.find<VideoDetailController>(tag: "$widgetIndex").pause();
      },
      child: AspectRatio(
        //视频比例
        aspectRatio: playerController.value.aspectRatio,
        child: VideoPlayer(playerController),
      ),
    ),
  );
}
```

12.3.3 视频播放功能控制

单击"播放"按钮，调用页面绑定的 VideoDetailController 中的 play 方法实现视频播放，如果反复单击同一个 Item，则直接使用已初始化的 VideoPlayerController 视频播放控制器重新播放。

如果单击的是其他未播放过的视频，则创建新的 VideoPlayerController，并校验是否有其他视频正在播放，代码如下：

```
//代码清单 12-12
class VideoDetailController extends GetxController {
  //创建视频播放控制器
  VideoPlayerController? playController;
  //当前控制器对应的 tag
  final String? tag;
  //当前播放状态
  bool isPlay = false;
```

```
  VideoDetailController({this.tag});

  //播放视频
  void play() async {
    //单击同一个视频
    if (playController != null) {
      if (playController!.value.isPlaying) {
        playController!.play();
      }
    }
    //校验是否有其他视频正在播放
    VideoListController listController = Get.find<VideoListController>();
    await listController.stopGlobalVideo();

    //创建新的视频播放控制器
playController =
    VideoPlayerController.asset('assets/video/list_item.mp4');
//更新列表中的记录
    listController.updateVideoController(playController, tag);
    //添加监听
    playController!.addListener(() {
      if (isPlay && !playController!.value.isPlaying) {
        //视频播放的当前时间进度
        Duration currentDuration = playController!.value.position;
        //视频的总时长
        Duration totalDuration = playController!.value.duration;
        if (currentDuration == totalDuration) {
          isPlay = false;
          update([tag??'']);
        }
      }
    });
    //初始化
    await playController!.initialize();
    //当前视频播放的位置
    Duration postion = playController!.value.position;
    //视频的总长度
    Duration duration = playController!.value.duration;

    if (postion == duration) {
      //播放完毕再单击"播放"按钮时,当前播放位置滑动到开始位置
      playController!.seekTo(Duration.zero);
    }
    //开始播放
    playController!.play();
    isPlay = true;
```

```
    update([tag??'']);
  }
  //暂停播放
  void pause([bool destroy = false]) {
    //视频暂停
    playController!.pause();
    isPlay = false;
    //非销毁时需要更新页面显示
    if(!destroy) {
      update([tag ?? '']);
    }
  }
}
```

12.3.4 性能优化小提示

Flutter 应用编译出的 APK、App Bundle 和 IPA 均持有应用运行时需要的所有代码和资源，是完全独立的。一个应用越大，在设备上占用的空间就越多，下载时间就越长，还可能超出 Android 即时应用等实用功能的限制。

在 build()方法中，开发者常会在其中构建大量的 UI 代码，控制 build()方法的耗时，是性能优化的一方面。

(1) 避免在 build()方法中进行重复且耗时的工作，因为当父 Widget 重建时，子 Widget 的 build()方法会被频繁地调用。

(2) 避免在一个超长的 build()方法中返回一个过于庞大的 Widget，把它们分拆成不同的 Widget，并进行封装。

从 Flutter 1.22 和 DevTools 0.9.1 版本开始，包含了一个大小分析工具，帮助开发者了解和拆分应用的发行版本，可以在构建时添加 --analyze-size 标记来调用，代码如下：

```
flutter build apk --analyze-size
flutter build appbundle --analyze-size
flutter build ios --analyze-size
flutter build Linux --analyze-size
flutter build macOS --analyze-size
flutter build Windows --analyze-size
```

Flutter 构建的应用程序在默认情况下都是高性能的，在 Flutter Performance 窗口中，勾选 Show widget rebuild information 复选框，此功能可帮助检测帧的渲染和显示时间是否超过 16ms，使用 Widget 的建议如下：

(1) 避免使用 Opacity Widget，尤其是在动画中，可以使用 AnimatedOpacity 或 FadeInImage。

(2) 使用 AnimatedBuilder 时,避免在不依赖于动画的 Widget 构造方法中构建。

(3) 避免在动画中剪裁,尽可能地在动画开始之前预先剪裁图像。

12.4 小结

本章讲解的内容是在第 11 章中的基础架构上进行 GetX 架构改造,讲解应用启动页面至倒计时页面、应用首页面、视频播放页面使用 GetX 开发业务功能,是本书 7.4 节的应用实例,在列表中,每个 Item 分别是一个单独的组件,分别注入绑定一个 GetX 控制器,这是一种开发思想,可以便捷管理。

图书推荐

书　　名	作　　者
鸿蒙应用程序开发	董昱
HarmonyOS 应用开发实战(JavaScript 版)	徐礼文
鸿蒙操作系统开发入门经典	徐礼文
鸿蒙操作系统应用开发实践	陈美汝、郑森文、武延军、吴敬征
HarmonyOS 移动应用开发	刘安战、余雨萍、李勇军 等
JavaScript 基础语法详解	张旭乾
华为方舟编译器之美——基于开源代码的架构分析与实现	史宁宁
鲲鹏架构入门与实战	张磊
华为 HCIA 路由与交换技术实战	江礼教
Android Runtime 源码解析	史宁宁
深度探索 Go 语言——对象模型与 runtime 的原理、特性及应用	封幼林
Flutter 组件精讲与实战	赵龙
Flutter 组件详解与实战	[加]王浩然(Bradley Wang)
Flutter 实战指南	李楠
Dart 语言实战——基于 Flutter 框架的程序开发(第 2 版)	亢少军
Dart 语言实战——基于 Angular 框架的 Web 开发	刘仕文
IntelliJ IDEA 软件开发与应用	乔国辉
Vue+Spring Boot 前后端分离开发实战	贾志杰
Vue.js 企业开发实战	千锋教育高教产品研发部
Python 从入门到全栈开发	钱超
Python 全栈开发——基础入门	夏正东
Python 游戏编程项目开发实战	李志远
Python 人工智能——原理、实践及应用	杨博雄 主编，于营、肖衡、潘玉霞、高华玲、梁志勇 副主编
Python 深度学习	王志立
Python 预测分析与机器学习	王沁晨
Python 异步编程实战——基于 AIO 的全栈开发技术	陈少佳
Python 数据分析实战——从 Excel 轻松入门 Pandas	曾贤志
Python 数据分析从 0 到 1	邓立文、俞心宇、牛瑶
Python Web 数据分析可视化——基于 Django 框架的开发实战	韩伟、赵盼
Python 玩转数学问题——轻松学习 NumPy、SciPy 和 matplotlib	张骞
Pandas 通关实战	黄福星
深入浅出 Power Query M 语言	黄福星
FFmpeg 入门详解——音视频原理及应用	梅会东
云原生开发实践	高尚衡
虚拟化 KVM 极速入门	陈涛
虚拟化 KVM 进阶实践	陈涛
物联网——嵌入式开发实战	连志安
人工智能算法——原理、技巧及应用	韩龙、张娜、汝洪芳